Invasive Plant
ECOLOGY

Edited by

**Shibu Jose • Harminder Pal Singh
Daizy Rani Batish • Ravinder Kumar Kohli**

CRC Press
Taylor & Francis Group
Boca Raton   London   New York

CRC Press is an imprint of the
Taylor & Francis Group, an **informa** business

Cover: Old World climbing fern (*Lygodium microphyllum*) smothering understory and overstory native plant species in a bald cypress (*Taxodium distichum*) swamp in southern Florida. Photo credit: Peggy Greb, USDA/ARS.

CRC Press
Taylor & Francis Group
6000 Broken Sound Parkway NW, Suite 300
Boca Raton, FL 33487-2742

© 2013 by Taylor & Francis Group, LLC
CRC Press is an imprint of Taylor & Francis Group, an Informa business

No claim to original U.S. Government works

Printed on acid-free paper
Version Date: 20121205

International Standard Book Number: 978-1-4398-8126-2 (Hardback)

Visit the Taylor & Francis Web site at
http://www.taylorandfrancis.com

and the CRC Press Web site at
http://www.crcpress.com

Contents

Preface

Invasion is not a novel phenomenon; however, it is one that has increased tremendously during the past few years due to rapidly expanding trade and transport among countries. It is one of the most important impacts humans have ever produced on the Earth's ecosystems. In fact, it has led to the globalization of world biota, further resulting in biotic homogenization. While some argue that invasion biologists should adopt a more objective and dispassionate stance toward invasive species, everyone agrees that it is a problem that needs attention and action.

Recognizing that a deeper understanding of biological characteristics of invasive species and the ecological principles that underpin the invasion process is critical in formulating management decisions, the scientific community has done an exemplary job in publishing original research as well as synthesis articles on the subject in recent years. Unfortunately, the science is still lagging behind, necessitating a reactive rather than a proactive response to the problem. The body of knowledge is growing so fast, including our ability to predict invasions. This will accelerate science-based decision making, whether it is for reactive or proactive management or for developing effective public policies.

The idea for this book began back in 2009 when the editors organized a regional International Union of Forestry Research Organizations (IUFRO) conference on invasive plants in Chandigarh, India. They met again in 2010 at the XXIII IUFRO World Congress in Seoul and finalized the plan for a book that would focus on invasive plant ecology. In addition to selected presentations from the IUFRO conference and the Congress, several prominent authors were invited to contribute chapters on specific topics. In the end, we accepted 18 chapters in the current volume.

Acquiring new scientific information and rapidly incorporating new knowledge and experiences into planning and actions are of the utmost importance with invasive plant management. It is critical to provide relevant and timely information to resource professionals, policy makers, and the general public so that they can make informed decisions. We recognize that one book alone cannot fill this niche. However, we hope that the current volume will serve as a reference book for students, scientists, professionals, and policy makers who are involved in the study and management of alien invasive plants in ecosystems the world over. We are grateful to a large number of individuals for assistance in accomplishing this task, particularly the authors for their commitment to the project and their original research or synthesis of the current knowledge. Also, the invaluable comments and suggestions made by the referees significantly improved the clarity and content of the chapters. We also wish to extend our sincere thanks to John Sulzycki and Jill Jurgensen of CRC Press for their timely efforts in publishing this book.

Shibu Jose
Harminder Pal Singh
Daizy R. Batish
Ravinder Kumar Kohli

Editors

Shibu Jose, PhD, an applied ecologist, is the H.E. Garrett Endowed Professor in the School of Natural Resources and director of the Center for Agroforestry at the University of Missouri. He was a professor of forest ecology in the School of Forest Resources and Conservation at the University of Florida before moving to Missouri. He received his BS (forestry) from India and MS and PhD (forest science) from Purdue University. He is the editor-in-chief of *Agroforestry Systems*, associate editor of the *International Journal of Ecology*, and associate editor of the *Journal of Forestry*. He recently served as a Fulbright Scholar (J. William Fulbright Scholarship Board and the U.S. Department of State), lecturing and conducting research in Bangladesh. Dr. Jose's research program has the overarching goal of identifying and quantifying key ecological processes and interactions that define ecological sustainability. He examines how resource availability (light, water, nutrients, and carbon) and disturbances (e.g., management interventions, fire, and exotic invasions) influence ecosystem structure and function in agroforests, natural forests, and plantation forests. He uses the ecological information in designing agroforestry systems and restoring degraded and damaged ecosystems. Over the past 20 years, Dr. Jose and his research team have conducted studies in the United States, Australia, Costa Rica, Belize, Bangladesh, Panama, and India. His awards and honors include the Aga Khan International Fellowship (Switzerland), the Nehru Memorial Award for Scholastic Excellence (India), the University of Florida Junior Faculty Award of Merit, the Award of Excellence in Research by the Southeastern Society of American Foresters (SAF), the Stephen Spurr Award by the Florida Division SAF, and the Young Leadership Award by the National SAF. His research has resulted in over 150 publications, including 6 edited books and 3 special issues.

Harminder Pal Singh, PhD, an ecophysiologist, is an assistant professor in the Department of Environmental Studies at Panjab University, Chandigarh, India. He received his PhD in botany from Panjab University. His research interests include examining the ecophysiological basis of plant interactions, evaluating the impact of exotic invasive plants on structural and functional aspects of native ecosystems, and exploring the role of natural plant products as novel pesticides. His research findings have resulted in 75 research papers in refereed journals and 6 edited books. His honors include the UNESCO-TWAS Associateship; the Young Scientist Award of the Panjab Academy of Sciences, Indian Science Congress Association, and Dalela Educational Foundation; and the Junior Environmentalist Award by the National Environment Science Academy.

Daizy R. Batish, PhD, is a professor in the Department of Botany at Panjab University, Chandigarh, India. Over the past 20 years, Dr. Batish and her research group have been involved in studying the biology and ecology of invasive weeds, ecophysiology of plant–plant interactions, and ecological weed management. Her research program aims at identifying and evaluating plant products with potential application as novel herbicides under sustainable agriculture. She teaches ecology, environmental botany, and forestry to undergraduate and postgraduate students. She has to her credit over 100 research papers in refereed journals, 6 books, and 40 research articles. Her honors include Fellow of National Environment Science Academy (India), the Rajib Goyal Young Scientist Award in Environmental Sciences, and the Research Award of the University Grants Commission, New Delhi. She is on the review committee of several high impact international journals in plant sciences.

Ravinder Kumar Kohli, PhD, a certified senior ecologist of the Ecological Society of America, is a senior professor of botany and a dean (research) at Panjab University, Chandigarh, India. He was an adjunct professor at the Chinese Academy of Sciences for two years and held a SAARC (South Asian Association for Regional Cooperation) chair in Bangladesh. He is a fellow of all four national official academies of science (Indian National Science Academy [INSA], New Delhi; Indian Academy of Science [IAS], Bangalore; National Academy of Sciences, India [NASI], Allahabad; National Academy of Agricultural Science [NAAS], New Delhi). He is the coordinator of IUFRO unit 8.02.04-Ecology of Alien Invasives and 4.02.02-Multipurpose Inventories. His research career of 32 years remains focused on the ecological implications of introduced alien trees and invasive alien plants in India. He is on the editorial boards of several journals devoted to crop production, ecology, and environment and a reviewer of several international journals. His honors and awards include the B.P. Pal National Environment Fellowship Award 2001 (by the Ministry of Environment and Forests for work on biodiversity); the Best Scientist Award 2010 (by the Government of Haryana state); Punjab Ratan, a state honor from the government of Chandigarh for teaching and research on environment; and the Nanda Memorial National Young Scientist award for work on eucalypts. He has published about 260 research papers and 12 books.

Contributors

E. G. Achigan-Dako
Plant Resources of Tropical
 Africa
and
Department of Plant Production
Faculty of Agronomic Sciences
University of Abomey-Calavi
Benin Republic, Africa

J. D. Ackerman
Department of Biology
Faculty of Natural Sciences
University of Puerto Rico
San Juan, Puerto Rico

Damian C. Adams
School of Forest Resources and
 Conservation
University of Florida
Gainesville, Florida

Edward B. Barbier
Department of Economics and
 Finance
University of Wyoming
Laramie, Wyoming

Sougata Bardhan
Center for Agroforestry
School of Natural Resources
University of Missouri
Columbia, Missouri

Daizy R. Batish
Department of Botany
Panjab University
Chandigarh, India

Alexandra Robin Collins
Department of Plant Biology
University of Vermont
Burlington, Vermont

Pervaiz A. Dar
Department of Botany
University of Kashmir
Jammu and Kashmir, India

Erich K. Dodson
Department of Forest Management
College of Forestry and Conservation
University of Montana
Missoula, Montana

Lindsay M. Dreiss
Department of Natural Resources and
 the Environment
University of Connecticut
Storrs, Connecticut

S. J. Farrington
Missouri Department of Conservation
West Plains, Missouri

Carl E. Fiedler
Department of Forest Management
College of Forestry and Conservation
University of Montana
Missoula, Montana

Coert J. Geldenhuys
Department of Forest and Wood Science
Stellenbosch University
Pretoria, South Africa

Qinfeng Guo
USDA Forest Service
Eastern Forest Environmental Threat
    Assessment Center
Asheville, North Carolina

Johnson Gwatipedza
Department of Economics and Finance
University of Wyoming
Laramie, Wyoming

Donald L. Hagan
School of Agriculture, Forest, and
    Environmental Sciences
Clemson University
Clemson, South Carolina

Allan D. Hollander
Information Center for the Environment
Department of Environmental
    Science and Policy
University of California
Davis, California

Roland A. Y. Holou
Monsanto Company
St. Louis, Missouri

Eric J. Holzmueller
Department of Forestry
Southern Illinois University
Carbondale, Illinois

Shibu Jose
Center for Agroforestry
School of Natural Resources
University of Missouri
Columbia, Missouri

Shalinder Kaur
Department of Environment Studies
Panjab University
Chandigarh, India

Duncan Knowler
School of Resource and Environmental
    Management
Simon Fraser University
Burnaby, British Columbia, Canada

Ravinder Kumar Kohli
Department of Botany
Panjab University
Chandigarh, India

Donna J. Lee
DJL Economic Consulting
Honolulu, Hawaii

Kerry L. Metlen
Division of Biological Sciences
University of Montana
Missoula, Montana

Jane Molofsky
Department of Plant Biology
University of Vermont
Burlington, Vermont

M. C. Muñoz
Grupo de Investigación en Biodiversidad
    Neotropical (INCIVA)
Museo Departamental de Ciencias
    Naturales
Cali, Colombia
and
Biodiversity and Climate Research Center
    (BiK-F)
Frankfurt, Germany

R. M. Muzika
Department of Forestry
University of Missouri
Columbia, Missouri

Steve Norman
USDA Forest Service
Eastern Forest Environmental Threat
    Assessment Center
Asheville, North Carolina

Jason R. Parent
Department of Natural Resources and
    the Environment
University of Connecticut
Storrs, Connecticut

Priyanka
Department of Botany
Panjab University
Chandigarh, India

James F. Quinn
Information Center for the Environment
Department of Environmental
    Science and Policy
University of California
Davis, California

A. S. Raghubanshi
Department of Botany
Banaras Hindu University
Varanasi, India

Nazima Rasool
Department of Botany
University of Kashmir
Jammu and Kashmir, India

Waheeda Rehman
Department of Botany
University of Kashmir
Jammu and Kashmir, India

Sarah H. Reichard
College of Forest Resources
University of Washington
Seattle, Washington

Zafar A. Reshi
Department of Botany
University of Kashmir
Jammu and Kashmir, India

Manzoor A. Shah
Department of Botany
University of Kashmir
Jammu and Kashmir, India

Gyan P. Sharma
Department of Environmental Studies
University of Delhi
Delhi, India

Harminder Pal Singh
Department of Environment Studies
Panjab University
Chandigarh, India

Brice Sinsin
Laboratoire d'Ecologie Appliquée
Faculté des Sciences Agronomiques
Université d'Abomey-Calavi
Cotonou, Bénin

Priyanka Srivastava
Department of Botany
Banaras Hindu University
Varanasi, India

R. S. Tripathi
National Botanical Research Institute
Rana Pratap Marg
Lucknow, India

Emma C. Underwood
Information Center for the
    Environment
Department of Environmental
    Science and Policy
University of California
Davis, California

John C. Volin
Department of Natural Resources and
    the Environment
University of Connecticut
Storrs, Connecticut

chapter one

Invasive plant ecology
The horse behind the cart?

Shibu Jose, Harminder Pal Singh, Daizy R. Batish,
Ravinder Kumar Kohli, and Sougata Bardhan

Contents

1.1   Introduction

Invasive alien species present a severe human dilemma due to their collective threat of replacing and damaging ecosystems (Mack et al. 2000, Pimentel et al. 2005). Human migration and goods trade around the world have provided pathways for introduction of invasive alien species in regions other than their place of origin (Osyczka et al. 2012). The "tens rule" (Williamson 1996), which states that 10% of imported species spread, 10% of these establish, and 10% of the established species cause problems, is applicable to alien plants. Not all alien plants or even invasive plants carry the same risks. Consequently, the challenge is to identify the 0.1% of species that can be harmful among the plant species introduced into a country or a region and prioritize control efforts based on the specific threats they pose.

Osyczka et al. (2012) reported that timber supply meant for a Polish Antarctic expedition team was contaminated with several species of fungi and insects. Although the survival and establishment of these organisms in the harsh Antarctic conditions are unlikely, it is a glaring example of how human activities can unintentionally result in the introduction of alien species that can lead to problems associated with biological invasion. Under optimum conditions, such introduced species develop traits that enable them to outcompete and dominate in the recipient environment, thus becoming invasive species. Once they become invasive, they alter the ecosystem, threaten the existence of native species, reduce biodiversity, cause economic losses, and degrade the environment. This phenomenon is ubiquitous (Allan et al. 2010) and has been widely described as biological invasion; it is blamed for species extinctions, loss of habitat (Kimbro et al. 2009), and loss of biodiversity (Pimentel et al 2005) and may result in human health hazards (Allan et al. 2010). Invasive species include plants, insects, arthropods, mollusks, fishes, reptiles and amphibians, birds and mammals, and pathogens.

Biological invasions have negatively influenced human civilization in the areas of agriculture, aquaculture, wildlife, forestry, and human health. The combined effects of such invasions have altered many ecological landscapes and eliminated native species around the globe.

1.2   Ecological and economic impacts

When the introduction of alien species becomes a problem, either through intentional or unintentional release and establishment, the impact can be overwhelming for native species, human beings, and the ecosystem as a whole. Invasive plants have wreaked havoc in ecosystems the world over with far-reaching consequences. They are considered to be one of the leading threats to the ecological integrity of native flora and fauna. In fact, invasive alien species are the second leading cause of global biodiversity loss, after habitat fragmentation (Drake et al. 1989). Many invasive plants can outcompete native flora and establish monospecific stands. Changing species composition can also change the functional processes of an ecosystem. There are well-documented examples of invasive alien plant species altering the structure and function of ecosystems (Vitousek et al. 1996). They may alter soil chemistry, nutrient cycling, hydrology, and disturbance regimes of the infested ecosystem. This can have serious direct and indirect implications on resource availability and seedlings recruitment of indigenous species. In addition to lowering the resource availability of an ecosystem, many invasive alien plants can act as a physical barrier to native seedling establishment. Seedlings of indigenous species including trees have to penetrate through the thick mat of root and rhizomes belowground or dense canopy aboveground to establish a foothold. Even if they successfully establish themselves in a dense stand of invasive alien plants, intense competition for resources can result in reduced vigor and growth rate. This poses serious challenges to both natural and artificial regeneration in natural ecosystems infested with invasive plants.

Of the 20,000 nonnative free-living plant species now in the United States, about 4,500 have invasive tendencies, and thousands more reside in our gardens, increasingly in the expanding urban fringe, with unknown consequences to adjoining lands (Pimental 2002, Miller and Schelhas 2009). The impact of the *Acacia* spp. invasion in the African continent has jeopardized the livelihood of millions of people. *Acacia* spp. originating from Australia have established themselves by outcompeting local vegetation for water, nitrogen, and other resources. *Acacia* trees are not only harmful to native vegetation but also notoriously adept in utilizing vast amounts of water resources, which is especially important in a water-scarce continent like Africa (Dye and Jarmin 2004). Other examples of invasive plant species creating major economic and environmental impact are *Ageratum conyzoides* (Kohli et al. 2006), *Eichhornia crassipes* (Perez et al. 2011), *Eupatorium odoratum* (Tripathi et al. 2006), *Lantana camara* (Lin 2007, Kohli et al. 2009), *Ailanthus altissima* (Burch and Zedaker 2003), and *Imperata cylindrica* (Daneshgar and Jose 2009).

Estimates of annual costs to the U.S. economy due to invasive alien species (including weeds, invertebrates, vertebrates, and pathogens) range between $1.1 billion (Office of Technology Assessment 1993) and $120 billion (Pimentel et al. 2005). The large discrepancy between these two estimates was primarily due to the greater number of species used by Pimentel et al. (more than 10 times as many). Pimentel et al. (2005) also reported higher numbers for some of the same species used in the Office of Technology Assessment (OTA) estimate. Pimentel et al. (2005) present annual costs due to invasive alien species to some of the important sectors:

- Weeds (crops): $27 billion annually

   Based on an annual crop production loss of $33 billion due to weeds and the estimate that about 73% of agricultural weeds are alien, Pimentel et al. (2005) conclude that the annual loss from exotic weeds is $24 billion. In addition to these losses, approximately $4 billion in herbicides are applied to U.S. crops annually, of which $3 billion are used for controlling exotic weeds.

- Weeds (pastures): $6 billion annually

   An estimated $2 billion in forage losses are caused by weeds in pastures every year. Of these weeds, 45% are exotic, so nearly $1 billion in forage losses can be attributed to exotic weeds. An additional $5 billion per year is spent on controlling exotic weeds in pastures and rangelands.

- Vertebrate pests: $1.6 billion annually

   Introduced vertebrates are another source of significant crop losses in the United States. Most notable are European starlings, which cause an estimated $800 million per year in damage to grain and fruit crops, and feral pigs, which are also believed to cause around $800 million in yearly damages to grain, peanut, soybean, cotton, hay, and vegetable crops.

- Insect pests: $13.5 billion annually

   Insects destroy approximately 13% of potential crop yields annually at a cost of $33 billion. Since 40% of insects are exotic, $13 billion per year in crop losses can be attributed to exotic insects. Additionally, of the $1.2 billion spent on pesticides every year, $500 million can be attributed to exotic insect control.

- Plant pathogens: $21.5 billion annually

   Plant pathogens cause crop losses of $33 billion per year. Of this, an estimated $21 billion per year is attributable to alien plant pathogens. Additionally, $500 million per year is spent on controlling alien plant pathogens.

- Livestock diseases: $5 billion annually

   Microbes and parasites infecting livestock cause approximately $5 billion in damages and control costs annually.

- Forests: $4.2 billion annually

   Insects and pathogens together cause $14 billion in losses to the forest products industry. Since 30% of these organisms are exotic, damages of approximately $4.2 billion can be attributed to exotic invasive organisms.

- Aquatic systems: $7.5 billion annually

   An estimated total of $145 million is invested annually in the control of invasive exotic aquatic and wetland plants. The estimated loss due to negative impacts of exotic fish on native fish and native ecosystems is $5.4 billion annually. The annual damage and control cost of zebra and quagga mussels is about $1 billion. The Asian clam also costs an estimated $1 billion in damages to waterways and native species.

1.3   Management strategies: Is there enough science-based information?

Although the estimated losses caused by invasive organisms are in billions of dollars worldwide, there is a lack of serious, coordinated approaches to stem the problem. Deficiencies in policy, consistent research, and management funding, lack of social organization to counter these invasions, and persistent gaps in scientific knowledge have all been identified as root causes of our current invasive dilemma (Simberloff et al. 2005, Miller and

Schelhas 2009). Natural resource managers and the general public are often unaware of the autecology of alien plants and their ecological impacts. Although control strategies are in place for some species, they are often inadequate to eradicate the alien invasive species. Eradicating invasive species is an attainable goal if new introductions are detected early. However, eradication may not be feasible when populations grow beyond a threshold. When limited resources or the degree of infestation preclude eradication, a more realistic management goal is to control the alien invasive species by reducing its population to a level that does not compromise the integrity of the ecosystem.

Novel control strategies are critical to gain an upper hand against invasive alien species and to achieve significant results. It is even more critical today since invasive species have evolved as global warming and human-induced climate change have contributed to variable biotic and abiotic conditions. Therefore, it is essential to comprehend the underlying mechanisms of invasive species and their ecology to successfully combat their burgeoning expansion. Although control programs should employ an integrated pest control approach with all the available tools such as manual, mechanical, chemical, biological, and cultural control techniques, they should be based on the life cycle characteristics of particular invasive alien species and the best available science to determine which control method or combination of methods will be most effective and economical for a species. Unfortunately, the "science" is still developing in the rapidly evolving field of invasive plant ecology. Natural resource managers often do not have all the required scientific information on specific species to make appropriate management decisions. In most cases, the horse is behind the cart, but "doing nothing" is not an option. Waiting until all the scientific information is generated may exacerbate the situation. Furthermore, in many cases invasive plants are recognized as a problem not during the lag phase following their introduction but during their exponential population growth phase.

The lack of public awareness and continued widespread sale and planting of invasive plants in many parts of the world suggest that invasive plant control requires an integrated process involving adaptive management cycles carried out through collaborative networks across landscapes for the containment of invasive alien plants and the restoration of impacted stands and ecosystems (Miller and Schelhas 2009). This process is called "adaptive collaborative restoration" (sensu Miller and Schelhas 2009). It is adaptive because we must learn as we go; collaborative because it requires coordinated individual efforts across ownership boundaries and among landowners, managers, and scientists; and restoration because our aim is to restore both sustainable food and fiber production systems and the wildlife habitats and cultural values associated with them. Adaptive techniques targeted at specific invaders have to be developed based on the physiology and ecology of invader species. Acquiring new scientific information and rapidly incorporating new knowledge and experiences into planning and actions is of utmost importance with invasive plant management due to the number of new species arriving on the scene, evolving perspectives and laws, and the current lack of developed strategies. Instilling adaptive management cycles into an integrated approach can turn reactive management of invasive plants into a proactive mode (Foxcroft 2004, Miller and Schelhas 2009).

1.4   *Why another book?*

It is reasonable to ask the following question: Why another book on invasive plants? We intend to complement existing information by presenting the latest body of knowledge using original research as well as synthesis articles related to invasive plant ecology and ecological economics. We acknowledge that one book alone cannot fill all the knowledge gaps in this discipline. As we advance the ecological science behind alien plant invasions, one study at a time, we will move closer toward making science-based decisions for combating the issue and formulating effective policies.

This book is a collection of original research or synthesis articles from around the globe that goes beyond a simple discussion of the status of some of the invasive plants and their impacts on different ecosystems. There are 18 chapters in this volume. They collectively provide a foundation in invasion ecology through an examination of ecological theories and case studies that explain plant invasions, their impacts, management strategies, and the ecological economics. Contributors discuss about ecological attributes, mutualistic associations, microbial communities, and disturbance regimes that influence the establishment and spread of alien invasive plants. Case studies are included to provide readers with an appreciation for the magnitude of the issue in various parts of the world. The book also covers spatial analysis and predictive modeling of invasive plants. It concludes by providing some principles and guidelines for ecological management and restoration of invaded areas along with overviews of ecological economics associated with alien plants.

1.5   Conclusion

We have come a long way since Charles Elton, a pioneer in population ecology, wrote in 1958 about how ecological explosions were threatening the world. Half a century later, his early warning has become one of the most important environmental crises of our time. We recognize invasive alien plants as an issue that needs to be addressed not only at the local but also at the global scale. There are also views expressed by several scientists that invasion biologists should adopt a more objective and dispassionate stance toward invasive species. Biological invasion is not only an ecological issue but also a complex social issue. It is imperative that scientists collaborate with nonscientists, including landowners, to help solve the problem. This will enhance the value of science-based information that governments, policymakers, resource professionals, and the general public need to make informed decisions about managing invasive species so that serious economic and ecological threats can be alleviated.

References

Allan BF, Dutra HP, Goessling LS, Barnett K, Chase JM, Marquis RJ, Pang G, Storch G, Thach RE, Orrock JL. (2010). Invasive honeysuckle eradication reduces tick-borne disease risk by altering host dynamics. *PNAS* 107(43): 18523–18527.

Burch P, Zedaker SM. (2003). Removing the invasive tree *Ailanthus altissima* and restoring natural cover. *Journal of Arboriculture* 29(1): 203.

Daneshgar P, Jose S. (2009). *Imperata cylindrica*, an alien invasive grass, maintains control over N availability in an establishing pine forest. *Plant and Soil* 320: 209–218.

Drake JA, Mooney HA, di Castri F, Groves RH, Kruger FJ, Williamson M. (1989). *Biological Invasions: A Global Perspective*. Wiley, New York.

Dye P, Jarman C. (2004). Water use by black wattle (*Acacia mearnsii*): Implications for the link between removal of invading tree and catchment streamflow response. *South African Journal of Science* 100: 1–5.

Foxcroft LC. (2004). An adaptive management framework for linking science and management of invasive alien plants. *Weed Technology* 18: 1275.

Kimbro DL, Grosholz ED, Baukus A, Nesbitt NJ, Travis NM, Attoe S, Coleman-Hulbert C. (2009). Invasive species cause large-scale loss of native California oyster habitat by disrupting trophic cascades. *Oecologia* 160(3): 563–575.

Kohli RK, Batish DR, Singh HP, Dogra KS. (2006). Status, invasiveness and environmental threats of three tropical American invasive weeds (*Parthenium hysterophorus* L., *Ageratum conyzoides* L., *Lantana camara* L.) in India. *Biological Invasions* 8: 1501–1510.

Kohli RK, Batish DR, Singh HP, Dogra KS. (2009). Ecological status of some invasive plants of Shiwalik Himalayas in Northwestern India. In Kohli, R., Jose, S., Singh, H.P., Batish, D.R. (Eds.). *Invasive Plants and Forest Ecosystems*. CRC Press, Boca Raton, Florida, 437p.

Lin S. (2007). The distribution and role of an invasive plant species, *Lantana camara*, in disturbed roadside habitats in Moorea, French Polynesia. http://escholarship.org/uc/item/2cv2z8zr, accessed April 4, 2012.

Mack RN, Simberloff D, Lonsdale WM, Evans H, Clout M, Bazzaz FA. (2000). Biotic invasions: Causes, epidemiology, global consequences and control. *Ecological Applications* 10: 689–710.

Miller JH, Schelhas J. (2009). Adaptive collaborative restoration: A key concept in invasive plant management. In Kohli, R., Jose, S., Singh, H.P., Batish, D.R. (Eds.). *Invasive Plants and Forest Ecosystems*. CRC Press, Boca Raton, Florida, 437p.

Office of Technology Assessment, US Congress. (1993). Harmful Non-indigenous Species in the United States, OTA-F-565, U.S. Government Printing Office, Washington, DC.

Osyczka P, Mleczko P, Karasiński D, Chlebicki A. (2012). Timber transported to Antarctica: A potential and undesirable carrier for alien fungi and insects. *Biological Invasions* 14(1): 15–20.

Perez AE, Tellez RT, Guzman SJM. (2011). Influence of physico-chemical parameters of the aquatic medium on germination of *Eichhornia crassipes* seeds. *Plant Biology* 13(4): 643–648.

Pimentel D, Zuniga R, Morrison D. (2005). Update on the environmental and economic costs associated with alien invasive species in the United States. *Ecological Economics* 52: 273–288.

Simberloff D, Parker IM, Windle PN. (2005). Introduced species policy, management, and future research needs. *Frontiers in Ecology and the Environment* 3: 12.

Tripathi RS, Kushwaha SPS, Yadav AS. (2006). Ecology of three invasive species of *Eupatorium*: A review. *International Journal of Ecology and Environmental Sciences* 32(4): 301–326.

Vitousek PM, D'Antonio CM, Loope LL, Westbrooks R. (1996). Biological invasions as global environmental change. *American Scientist* 84: 468.

Williamson M. (1996). *Biological Invasions*. Chapman & Hall, London.

chapter two

What makes alien plants so successful?
Exploration of the ecological basis*

Eric J. Holzmueller and Shibu Jose

Contents

2.1   Introduction

Thousands of exotic plant species have been introduced, both intentionally and unintentionally, to native ecosystems across the globe. However, not all exotic plant species that are introduced become invasive. Knowing which exotic plant species are most likely to become invasive is critical for the control of invasive plants in native ecosystems; however, often this can be difficult to determine. This chapter attempts to use ecological hypotheses to explain invasive plant mechanisms that are associated with successful plant invasions. Understanding these mechanisms will assist land managers with controlling invasive species by allowing them to rapidly respond before the species becomes widespread in native ecosystems.

2.2   Efficient resource uptake and use

In order for an invasive plant species to become successfully established, the species must be able to efficiently capture water, light, and nutrient resources. Generally, species-rich communities are thought to utilize all available resources, making it difficult for invasive species to become established in such communities, as compared to less diverse communities (diversity–invasibility hypothesis) (Elton 1958). If a community does not utilize all available resources (water, nutrients, and light), there is an "empty niche" that leaves it susceptible to invasion (Elton 1958, MacArthur 1970). For example, in the western United States, *Centaurea solstitialis* L. (yellow star thistle) is an invasive species that displaces native vegetation by utilizing water resources below the rooting zone of native species (Holmes

---

* This chapter has been modified from an article previously published in the *Journal of Tropical Agriculture*, 2009, 47:18–29.

and Rice 1996). However, when the species was grown in a microcosm experiment with *Hemizonia congesta* DC. (hayfield tarweed), its growth was reduced compared to when it was grown with other native plants (Dukes 2002). The authors hypothesized that this was because *H. congesta* and *C. solstitialis* have similar morphology and growth habits, which led to a competition between the two for similar resources. *C. solstitialis* became invasive when it was grown with native plants that did not fully utilize all resources, leaving an empty niche for this species to become established.

Similar to the empty niche hypothesis, the "fluctuating resource" hypothesis states that as resources become available within a given area, that area will become more susceptible to invasion (Davis et al. 2000). Excessive nutrients can become available in two primary ways (Davis et al. 2000): (1) The first is when native plants decrease their uptake because of decreased populations following a disturbance or predatory outbreak. (2) The second occurs when more nutrients become available from external or internal sources, such as increased precipitation or accelerated mineralization. Because invasive plants are often more successful in capturing excessive nutrients compared to native species, the community becomes vulnerable to invasion. Using *Lantana camara* L. (lantana) as an example, Duggin and Gentle (1998) illustrated this concept in a series of experiments conducted in New South Wales, Australia. Native to Central and South America, *L. camara* is an invasive shrub that has been introduced in over 60 countries around the world. It can grow in a variety of soil types and habitats, but it generally does not occur in undisturbed forests (Sharma et al. 2005). Duggin and Gentle (1998) tested the impacts of fertilization, biomass removal, and fire on *L. camara* germination, survival, and growth. These authors reported that whereas fertilization alone had little effect on plant development, fertilization combined with biomass removal or burning significantly increased *L. camara* germination, survival, and growth (Figure 2.1). This increase in invasion success was correlated with increased light, water, and nutrient availability, and it explains why the species performs better in disturbed areas. Similar results have been observed for *Imperata cylindrica* (L.) Beauv. (cogongrass), an invasive C_4 perennial grass species native to Asia. The species often exists only in sparse patches in undisturbed forests (MacDonald 2004), but it becomes dominant following fire or overstory removal (Holzmueller and Jose 2011).

2.3   Rapid growth and reproduction

Invasive plants often form dense monocultures with high productivity, sometimes at rates much higher than what they are capable of in their native range (Hierro et al. 2005). In addition to exhibiting higher rates of productivity, some invasive plants produce larger individuals in invaded populations. The "evolution of increased competitive ability" (EICA) hypothesis (Blossey and Nötzold 1995) states that reduced herbivory in the introduced range causes an evolutionary shift in resource allocation from herbivore defense to growth. As a result, according to EICA, introduced genotypes are expected to grow more vigorously than conspecific native genotypes. The authors hypothesized that in the absence of specialized enemies invasive plant genotypes allocate more resources to biomass production and reduce resources to defense mechanisms, thereby increasing the abundance of high-biomass-producing individuals. This hypothesis has been supported by several recent studies. In a common garden experiment in the southern United States, Siemann and Rogers (2001) used *Triadica sebifera* (L.) Small (Chinese tallow tree) to test the EICA hypothesis. *T. sebifera* is a fast-growing tree species native to Asia that has been introduced in the United States (Bruce et al. 1997). Results from the 14-year-old experiment

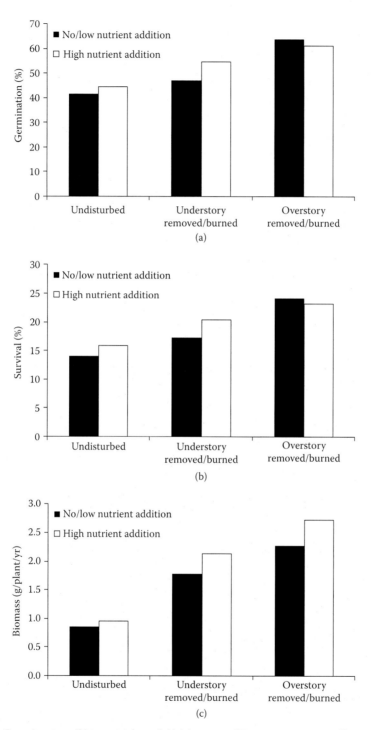

Figure 2.1 (a) Germination, (b) survival, and (c) biomass of *Lantana camara* seedlings in disturbed and undisturbed forests with no/low fertilizer and high fertilizer additions: Values were greater for all three criteria in disturbed areas. (Adapted from Duggin, J.A., and C.B. Gentle. *Forest Ecology Management* 109, 279–292, 1998.)

indicated that *T. sebifera* trees had greater biomass and fewer leaf defense chemicals in invasive genotypes compared to native genotypes (Siemann and Rogers 2001). Jakobs et al. (2004) compared the populations of *Solidago gigantea* Aiton (giant goldenrod), a rhizomatous perennial herb, in its native range, North America, to its invaded range, Europe. The authors reported that average population density and total plant biomass were greater in the invaded range (Figure 2.2). Evolution, however, is not the only way in which plants increase resource allocation in favor of defense. Cheplick (2005) reported an increase in resource allocation to growth in *Microstegium vimineum* that was attributed to phenotypic plasticity in resource allocation rather than evolution.

Many invasive plants are ruderal (r-strategists) species, such as *Lythrum salicaria* L. (purple loosestrife), *I. cylindrica*, and *Schinus terebinthifolius* Raddi (Brazilian pepper), that can quickly occupy a large area in a short period of time (Ewel 1979, Mullin 1998, Jose et al. 2002). The ability to spread rapidly and grow quickly helps to ensure that invasive species dominate these disturbed areas. However, not all invasive species require disturbance to become productive. *Pueraria montana* var. *lobata* (kudzu) is a vine species native to Asia. It was first introduced into the United States as an ornamental plant in 1876, but it became more widespread when it was encouraged to be used as a forage crop and soil stabilizer by the U.S. government in the 1930s and 1940s (Forseth and Innis 2004). The species is capable of occupying undisturbed areas by allocating resources to growth instead of structure. The species can grow up to 20 cm in a single day, easily overtopping any tree species; produces up to 1900 g of biomass annually; and has a leaf area index of up to 7.8 (Forseth and Innis 2004).

The ability to produce an abundance of long-lasting viable seeds also increases species invasiveness (propagule pressure hypothesis) (Williamson 1996). *L. salicaria* is a wetland perennial herb native to Eurasia, but it has invaded most of North America. In its native range the species occupies less than 5% of vegetative cover, but in invaded areas it can form dense, monospecific stands (Mullin 1998). Although any wetland is susceptible to *L. salicaria* invasion, disturbed areas with bare soil are the most vulnerable (Mullin 1998). Once established, the species produces abundant quantities of seeds with each stem producing over 100,000 seeds, 60% of which remains viable for up to 20 years (Mullin 1998). This high amount of viable seeds makes long-term control efforts difficult, and this

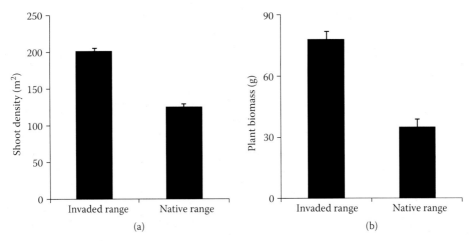

Figure 2.2 (a) Population shoot density and (b) individual plant biomass of *Solidago gigantea* in its invaded and native ranges. Values for both criteria were greater in the invaded range. (Adapted from Jakobs et al. *Diversity and Distributions*, 10, 11–19, 2004.)

characteristic is not unique to *L. salicaria*. *Acacia mearnsii* De Wild. (black wattle), *Alliaria petiolata* (M. Bieb.) Cavara & Grande (garlic mustard), and *M. vimineum* (Trin.) Camus (Japanese stilt grass) are three examples of other plant species that are difficult to control because of high seed production and viability (Anderson et al. 1996, Gibson et al. 2002, de Neergaard et al. 2005).

2.4   *Environmental modification*

Invasive plants can increase their competitive ability by modifying the invaded environment. One of the ways in which this is done is through the production of allelochemicals by invasive plants that inhibit the growth of native plants; this is also referred to as the "novel weapons hypothesis" (Callaway and Ascheoug 2000, Bias et al. 2003). For example, field observations indicate seedling recruitment to be minimal underneath *L. camara*, which has been attributed to allelochemicals produced by the species (Sharma et al. 2005). Maiti et al. (2008) used extracts from *L. camara* leaves to suppress *Mimosa pudica* L. (sleeping grass) seed germination and growth in a laboratory experiment that supports these field observations. Not all aspects of allelochemical production, however, are negative. Kong et al. (2006) suggested allelochemicals produced by *L. camara* could be used to improve the control of invasive plants in aquatic ecosystems.

In addition to the production of allelochemicals that inhibit native plant growth, some invasive plants can also lower soil pH and alter nutrient cycling within the native community (Callaway and Ascheoug 2000, Drenovsky et al. 2007). Decreased pH can lower nutrient availability and lead to decreased native plant growth, particularly in nutrient-poor sites. Collins and Jose (2008) reported that *I. cylindrica* decreased soil pH in recently invaded areas in the southeastern U.S. *Pinus* forests. Although *Mikania micrantha* H.B.K. (mile-a-minute vine), a fast-growing perennial vine native to Central America and South America, decreased soil pH in a subtropical forest in China. Chen et al. (2009) reported increased NH_4^+ and net soil nitrification. Allelochemicals released by *M. micrantha* actually increased soil fertility, which probably enabled the species to become invasive. It is obvious that allelochemicals produced by invasive plants exert both direct and indirect influence on native species. Daneshgar and Jose (2008) proposed a new hypothesis, the "rhizochemical dominance" hypothesis, which integrates several of these mechanisms in explaining the success of invasive plants. This hypothesis attributes invasive success to allelopathy (novel weapons) and alteration of soil chemical properties by the rhizosphere exudates of the invader, which in turn favors its own growth while inhibiting the growth of competing vegetation. These chemical alterations may include changes in soil pH and nutrient levels and availability.

Modifications to the invaded environment can also occur aboveground. This is frequently accomplished by invasive plants through the alteration of the native community fire regime (Brooks et al. 2004). In many cases, fire can promote invasive species and invasive species can promote fire, resulting in a positive feedback cycle that decreases native species abundance as the invasive species become more dominant (Figure 2.3) (Brooks et al. 2004). *Tamarix ramosissima* Ledeb. (tamarisk) is an invasive perennial shrub native to Eurasia. Litter produced from *T. ramossisima* burns readily and increases fire frequency in riparian ecosystems, resulting in an increased dominance of *T. ramossisima* (Busch and Smith 1993). Invasive species can also decrease fire frequency in native communities. Stevens and Beckage (2009) observed a lower density of *S. terebinthifolius* in areas that were burned frequently in *Pinus*–savanna ecosystems in southern United States. In fire-suppressed areas, *S. terebinthifolius* became more abundant and inhibited fires within

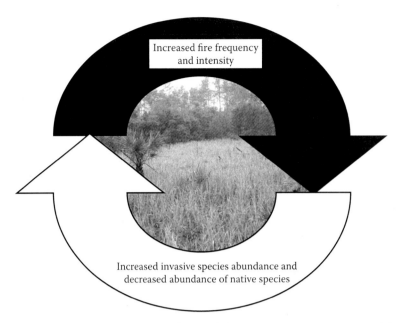

Figure 2.3 Conceptual diagram of positive feedback between invasive species and fire behavior: The background photograph shows a patch of *Imperata cylcindrica* that has invaded a *Pinus palustris* ecosystem. There is a paucity of native species in the invaded area.

the community resulting in further dominance of the invasive species. It is also possible for invasive species to affect ecosystem processes in such a way as to facilitate invasion by other alien invasive species (invasional meltdown hypothesis) (Simberloff and Von Holle 1999). Among the examples the authors use to support this hypothesis are invasive plants that modify the environment, for example, increased nitrogen fixation or alteration of the fire regime, which then facilitates invasion by other invasive plant species and by invasive animals that encourage the spread of invasive plant species through seed dispersal and selective browsing of native plants.

2.5 Genetic variability and evolutionary genetics

The ability of a potentially invasive plant to adapt to the invaded environment is critical to its success as an invasive species (Lee 2002, Ren and Zhang 2009). Some invasive plants have the phenotypic plasticity to tolerate a broad range of environmental conditions, which increases the potential number of sites they can invade (Richards et al. 2006). Annapurna and Singh (2003) illustrated this concept with *Parthenium hysterophorus* L. (congress grass), an invasive plant in India. The authors reported significant phenotypic plasticity in response to a gradient of soil conditions. If a species does not exhibit phenotypic plasticity to environmental conditions and only a few individuals of a species are introduced, a population bottleneck can occur. This can result in lower genetic diversity of the invasive plant compared to that of the same plant in its native range (Sakai et al. 2001). During this time, the invasive plant population may undergo a lag period until genetic diversity increases by evolution, additional introductions occur, or hybridization occurs with a native species (Sakai et al. 2001). However, although hybridization may increase the genetic diversity of an invasive species, it may result in a decreased abundance of the native

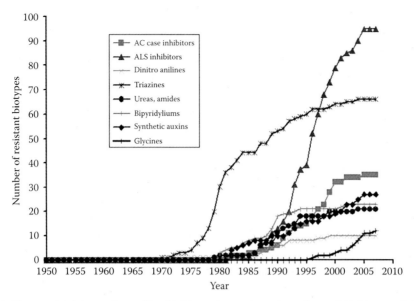

Figure 2.4 Change in the number of herbicide-resistant invasive plants (332 biotypes) from 1950 to 2007: The biotypes are grouped by herbicide mode of action. The increase in herbicide-resistant invasive plants has been dramatic over the past 30 years and correlates with the increase in herbicide use throughout the world. (From Heap, I.M., *International Survey of Herbicide Resistant Weeds*, http://www.weedscience.org, 2005.)

species, which is particularly damaging to native species with small populations (Levin et al. 1996). For example, *Cercocarpus traskaie* Eastw., a rare shrub species located only off an island in southern California, is hybridizing with the invasive plant *C. betuloides* Torrey & A. Gray (birch-leaf mountain mahogany) and is threatened with assimilation.

Plants may also be able to persist as invaders despite control efforts that attempt to eliminate them. Herbicides are a common management tool used to control invasive species throughout the world. However, nearly 200 invasive plants (332 biotypes) are considered resistant to herbicides and resistance has increased dramatically over the past 30 years with the increase in herbicide use (Heap 2005) (Figure 2.4). Nine invasive plants are resistant to glyphosate, which limits the effectiveness of glyphosate-resistant crops. Crop mimicry is another way in which invasive plants are able to persist despite control efforts. *Echinochloa crus-galli* (L.) P. Beauv. (barnyard grass) is a C_4 wetland grass native to Eurasia. When grown with *Oryza sativa* L. (rice), it is difficult to identify and remove *E. crus-galli* during manual weedings because of its close resemblance with the crop (Barrett 1983).

2.6   Enemy resistance

Almost all plant species have enemies that reduce their populations; these enemies can include fungi that cause plant tissue necrosis or herbivores that remove the plants entirely. The enemies can be either specialists, attacking only a single species, or generalists, attacking multiple species. Within their native range, invasive plants are subject to specialized enemies that evolve with the plant and limit plant population growth. Outside the native range, invasive plants are often not suppressed by these specialized

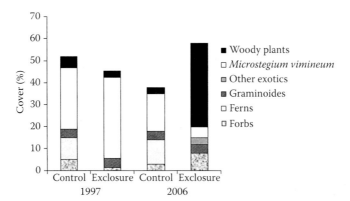

Figure 2.5 Species group abundance between 1997 and 2006: Prior to the establishment of deer exclosures in 1997, the species composition was similar among all plots. In 2006, exclosure plots had a significant increase in native woody vegetation and decrease in the invasive species, that is, *Microstegium vimineum*. (Adapted from Webster et al., *Oecol*, 157, 497–508, 2008.)

enemies and are only preyed on by generalist enemies. Invasive plants are believed to gain an advantage over native species in the invaded range because of their apparent lack of specialized enemies. The increased performance of invasive plants because of the absence of specialized enemies is referred to as the "enemy release" hypothesis (Elton 1958, Keane and Crawley 2002).

Several examples of the enemy release hypothesis exist for multiple species. Wolfe (2002) tested this hypothesis on *Silene latifolia* Poir. (white campion), a small (<1 m tall) short-lived perennial plant, using populations from native European and invaded U.S. ranges. The author recorded damage by phloem-feeding insects, floral herbivory, fungal disease, and fruit and seed predation and observed greater damage on European plants for each group. In addition, Wolfe observed much damage caused by specialist enemies on European plants, but little specialist enemy damage on invasive U.S. populations. Mitchell and Power (2003) sampled viruses and fungi (rust, smut, and powdery mildew) on 473 plants that were invasive to the United States but native to Europe. They found that invasive plants had 84% fewer fungi and 24% fewer viruses compared with the same plants in their native range.

While a lack of specialized enemies contributes to increased invasiveness, resistance to generalist enemies can increase invasiveness as well. For example, in the eastern United States, *M. vimineum* is an invasive, shade-tolerant, annual C_4 grass species native to Asia that frequently outcompetes native species, reducing tree growth and biodiversity (Gibson et al. 2002). This problem is exacerbated with the presence of *Odocoileus virginianus* Zimm. (white-tailed deer). *O. virginianus*, a generalist herbivore whose populations have risen exponentially over the past century, is known to browse over 100 plant species native to the eastern United States (Rooney 2001). Although large populations of *O. virginianus* regularly decrease native herbaceous and woody species understory vegetation, they are reluctant to browse on *M. vimineum*. Webster et al. (2008) reported that preferential foraging by *O. virginianus* reduced native vegetation cover and density, resulting in increased abundance of *M. vimineum* because of reduced competition for light in their study conducted at Great Smoky Mountains National Park, Tennessee, in the United States (Figure 2.5). Furthermore, chronic browsing by *O. virginianus* inhibited the ability of native species to recover from *M. vimineum* invasion during times of drought, when the species was susceptible to becoming overtopped. This resulted in a successional state that resisted transition, which in turn contributed to dominance by *M. vimineum* (Webster et al. 2008).

2.7 Summary

This chapter illustrates multiple mechanisms used by invasive species to successfully invade native ecosystems. In general, we found that characteristics of successful invaders often include broad ecological requirements and tolerances, sometimes reflected in large geographical ranges (Rejmanek 1996, Sax and Brown 2000); r-selected life histories (Tominaga 2003); associations with disturbed or anthropogenic habitats; and origins from large continents with diverse biota (Elton 1958). In analyzing invasive plant mechanisms, we can also identify several common characteristics of invaded environments, including geographical and historical isolation; underutilized resources; high levels of natural disturbance or human activities; and absence of coadapted enemies, including competitors, predators, herbivores, parasites, and diseases (Davis et al. 2000). By understanding invasive plant mechanisms, land managers can develop strategies to prevent the establishment or stop the spread of invasive species. For example, they can monitor disturbed areas that may have an abundance of resources that are not being utilized, avoid repeated herbicide treatments with chemicals that have similar modes of action to prevent the evolution of herbicide resistance, and avoid significant changes to the historic disturbance regime in native ecosystems. Species may exhibit multiple invasive mechanisms, such as cogongrass, and such species can be particularly problematic to control (Holzmueller and Jose 2011). In addition, it is important to remember that invasive species do not recognize political boundaries and, hence, efforts must be made to coordinate control among various landowners and agencies (Miller and Schelhas 2008).

References

Anderson, R.C., S.S. Dhillion, and T.M. Kelley. 1996. Aspects of the ecology of an invasive plant, garlic mustard (*Alliaria petiolata*), in central Illinois. *Restoration Ecology* 4:181–191.

Annapurna, C., and J.S. Singh. 2003. Phenotypic plasticity and plant invasiveness: Case study of congress grass. *Current Science* 85:197–201.

Barrett, S.C.H. 1983. Crop mimicry in weeds. *Economic Botany* 37:255–282.

Bias, H.P., R. Vepachedu, S. Gilroy, R.M. Callaway, and J.M. Vivanco. 2003. Allelopathy and exotic plant invasion: From molecules and genes to species interactions. *Science* 301:1377–1380.

Blossey, B., and R. Nötzold. 1995. Evolution of increased competitive ability in invasive nonindigenous plants: A hypothesis. *Journal of Ecology* 83:887–889.

Brooks, M.L., C.M. D'Antonio, D.M. Richardson, J.B. Grace, J.E. Keeley, J.M. DiTomaso, R.J. Hobbs, M. Pellant, and D. Pyke. 2004. Effects of invasive alien plants on fire regimes. *BioScience* 54:677–688.

Bruce, K.A., G.N. Cameron, P.A. Harcombe, and G. Jubinsky. 1997. Introduction, impact on native habitats, and management of a woody invader, the Chinese Tallow Tree, *Sapium sebiferum* (L.) Roxb. *Natural Areas Journal* 17:255–260.

Busch, D.E., and S.D. Smith. 1993. Effects of fire on water and salinity relations of riparian woody taxa. *Oecologia* 94:186–194.

Callaway, R.M., and E.T. Aschehoug. 2000. Invasive plants versus their new and old neighbors: A mechanism for exotic invasion. *Science* 290:521–523.

Chen, B.M., S.L. Peng, and G.Y. Ni. 2009. Effects of the invasive plant *Mikania micrantha* HBK on soil nitrogen availability through allelopathy in South China. *Biological Invasions* 11:1291–1299.

Cheplick, G.P. 2005. Biomass partitioning and reproductive allocation in the invasive, cleistogamous grass *Microstegium vimineum*: Influence of the light environment. *Journal of the Torrey Botanical Society* 132:214–224.

Collins, A.R., and S. Jose. 2008. Cogongrass invasion alters soil chemical properties of natural and planted forestlands. In *Invasive Plants and Forest Ecosystems*, eds. R.K. Kohli, S. Jose, H.P. Singh, and D.R. Batish, pp. 295–323. Boca Raton, FL: CRC Press.

Daneshgar, P., and S. Jose. 2008. Mechanisms of plant invasion: A review. In *Invasive Plants and Forest Ecosystems*, eds. R.K. Kohli, S. Jose, H.P. Singh, and D.R. Batish, pp. 11–28. Boca Raton, FL: CRC Press.

Davis, M.A., J.P. Grime, and K. Thompson. 2000. Fluctuating resources in plant communities: A general theory of invisibility. *Journal of Ecology* 88:528–534.

de Neergaard, A., C. Saarnak, T. Hill, M. Khanyile, A.M. Berzosa, and T. Birch-Thomsen. 2005. Australian wattle species in the Drakensberg region of South Africa – An invasive alien or a natural resource? *Agricultural Systems* 85:216–233.

Drenovsky, R.E., and K.M. Batten. 2007. Invasion by *Aegilops triuncialis* (Barb Goatgrass) slows carbon and nutrient cycling in a serpentine grassland. *Biological Invasions* 9:107–116.

Duggin, J.A., and C.B. Gentle. 1998. Experimental evidence on the importance of disturbance intensity for invasion of *Lantana camara* L. in dry rainforest-open forest ecotones in north-eastern NSW, Australia. *Forest Ecology and Management* 109:279–292.

Dukes, J.S. 2002. Species composition and diversity affect grassland susceptibility and response to invasion. *Ecological Applications* 12:602–617.

Elton, C.S. 1958. *The Ecology of Invasion by Animals and Plants*. London: T Metheun and Co.

Ewel, J.J. 1979. Ecology of *schinus*. In *Schinus - Technical Proceedings of Techniques for Control of Schinus in South Florida: A Workshop for Natural Area Managers*, ed. R. Workman, pp. 7–21. Sanibel, FL: The Sanibel-Captiva Conservation Foundation, Inc.

Forseth, Jr. I.N., and A.F. Innis. 2004. Kudzu (*Pueraria montana*): History, physiology, and ecology combine to make a major ecosystem threat. *Critical Reviews in Plant Sciences* 23:401–413.

Gibson, D.J., G. Spyeareas, and J. Benedict. 2002. Life history of *Microstegium vimineum* (Poaceae), an invasive grass in southern Illinois. *Journal of the Torrey Botanical Society* 129:207–219.

Heap, I.M. 2005. *International Survey of Herbicide Resistant Weeds*. Available online at URL: http://www.weedscience.org (accessed September 2009).

Hierro, J.L., J.L. Maron, and R.M. Callaway. 2005. A biogeographical approach to plant invasions: The importance of studying exotics in their introduced and native range. *Journal of Ecology* 93:5–15.

Holmes, T.H., and K.J. Rice. 1996. Patterns of growth and soil-water utilization in some exotic annuals and native perennial bunchgrasses of California. *Annals of Botany* 78:233–243.

Holzmueller, E.J., and S. Jose. 2011. Invasion success of cogongrass, an alien C_4 perennial grass, in the southeastern United States: Exploration of the ecological basis. *Biological Invasions* 13:435–442.

Jakobs, G., E. Weber, and P.J. Edwards. 2004. Introduced plants of the invasive *Solidago gigantea* (Asteraceae) are larger and grow denser than conspecifics in the native range. *Diversity and Distributions* 10:11–19.

Jose, S., J. Cox, D.L. Miller, D.G. Shilling, and S. Merritt. 2002. Alien plant invasions: The story of cogongrass in southeastern Florida. *Journal of Forestry* 100:41–44.

Keane, R.M., and M.J. Crawley. 2002. Exotic plant invasions and the enemy release hypothesis. *Trends in Ecology and Evolution* 17:164–170.

Kong, C.H., P. Wang, C.X. Zhang, M.X. Zhang, and F. Hu. 2006. Herbicidal potential of allelochemicals from *Lantana camara* against *Eichhornia crassipes* and the alga *Microcystis aeruginosa*. *Weed Research* 46:290–295.

Lee, C.E. 2002. Evolutionary genetics of invasive species. *Trends in Ecology and Evolution* 17:386–391.

Levin, D.A., J. Francisco-Ortega, and R.K. Jansen. 1996. Hybridization and the extinction of rare plant species. *Conservation Biology* 10:10–16.

MacArthur, R.H. 1970. Species packing and competitive equilibruium for many species. *Theoretical Population Biology* 1:1–11.

MacDonald, G.E. 2004. Cogongrass (*Imperata cylindrica*) – Biology, ecology and management. *Critical Reviews in Plant Sciences* 23:367–380.

Maiti, P.P., R.K. Bhakat, and A. Bhattacharjee. 2008. Allelopathic effects of *Lantana camara* on physio-biochemical parameters of *Mimosa pudica* seeds. *Allelopathy Journal* 22:59–67.

Miller, J.H., and J.W. Schelhas. 2008. Adaptive collaborative restoration: A key concept for invasive plant management. In *Invasive Plants and Forest Ecosystems*, eds. R.K. Kohli, S. Jose, H.P. Singh, and D.R. Batish, pp. 251–265. Boca Raton, FL: CRC Press.

Mitchell, C.E., and A.G. Power. 2003. Release of invasive plants from fungal and viral pathogens. *Nature* 421:625–627.

Mullin, B.H. 1998. The biology and management of purple loosestrife (*Lythrum salicaria*). *Weed Technology* 12:397–401.

Rejmanek, M. 1996. A theory of seed plant invasiveness: The first sketch. *Biological Conservation* 78:171–181.

Ren, M.X., and Q.G. Zhang. 2009. The relative generality of plant invasion mechanisms and predicting future invasive plants. *Weed Research* 49:449–460.

Richards, C.L., O. Bossdorf, N.Z. Muth, J. Gurevitch, and M. Pigliucci. 2006. Jack of all trades, master of some? On the role of phenotypic plasticity in plant invasions. *Ecological Letters* 9:981–993.

Rooney, T.P. 2001. Deer impacts on forest ecosystems: A North American perspective. *Forestry* 74:201–208.

Sakai, A.K., F.W. Allendorf, J.S. Holt, D.M. Lodge, J. Molofsky, K.A. With, S. Baughman, et al. 2001. The population biology of invasive species. *Annual Review of Ecology and Systematics* 32:305–332.

Sax, D.F., and J.H. Brown. 2000. The paradox of invasion. *Global Ecology and Biogeography* 9:363–371.

Sharma, G.P., A.S. Raghubanshi, and J.S. Singh. 2005. *Lantana* invasion: An overview. *Weed Biology and Management* 5:157–165.

Siemann, E., and W.E. Rogers. 2001. Genetic differences in growth of an invasive tree species. *Ecology Letters* 4:514–518.

Simberloff, D., and B. Von Holle. 1999. Positive interactions of nonindigenous species: Invasional meltdown? *Biological Invasions* 1:21–32.

Stevens, J.T., and B. Beckage. 2009. Fire feedbacks facilitate invasion of pine savannas by Brazilian pepper (*Schinus terebinthifolius*) *New Phytologist* 184:365–375.

Tominaga, T. 2003. Growth of seedlings and plants from rhizome pieces of cogongrass (*Imperata cylindrica* (L.) Beauv. *Weed Biology Management* 3:193–195.

Webster, C.R., J.H. Rock, R.E. Froese, and M.A. Jenkins. 2008. Drought-herbivory interaction disrupts competitive displacement of native plants by *Microstegium vimineum*, 10-year results. *Oecologia* 157:497–508.

Williamson, M. 1996. *Biological Invasions*. London: Chapman & Hall.

Wolfe, L.M. 2002. Why alien invaders succeed: Support for the escape-from-enemy hypothesis. *American Naturalist* 160:705–711.

chapter three

Novel weapon hypothesis for the successful establishment of invasive plants in alien environments
A critical appraisal

Daizy R. Batish, Harminder Pal Singh, Shalinder Kaur, Priyanka, and Ravinder Kumar Kohli

Contents

3.1   Introduction

It is well established that invasive exotic plants cause much harm to native ecosystems (Mack et al. 2000). Such invasive plants alter ecosystem functions, harm local biodiversity, pose a risk of extinction to threatened native species, and lead to economic loss (Wilcove et al. 1998, Pimentel et al. 2005, Vilà et al. 2010). Therefore, worldwide efforts are being made to manage them. However, in order to make management effective, it is important to understand the factors that provide invasion success to certain species on introduction to or entry into alien environments. Such factors could be related to the species themselves (e.g., superior traits) or the habitats they invade. Species-related traits include fast growth rate, high reproductive potential, alternative means of reproduction, phenotypic plasticity, seed size, number and dispersal, and high competitive ability (Baker 1965, Sharma et al. 2005). Habitat-related factors, on the other hand, include high disturbance, availability of empty niches, and availability of a high resource environment (Davis et al. 2000, Shea and Chesson 2002, Blumenthal 2005). Based on these factors, several hypotheses/mechanisms have been proposed for the invasiveness of exotic species in alien environments (Alpert 2006, MacDougall et al. 2009). However, invasiveness is not linked to just one mechanism; several mechanisms may be involved in this process, apart from many traits of the species themselves (Lamarque et al. 2011). Some important hypotheses/mechanisms proposed in this regard include the "biodiversity resistance" hypothesis (Elton 1958), "enemy release" hypothesis (Keane and Crawley 2002), "evolution of increased competitive ability" (EICA) hypothesis (Blossey and Nötzold 1995), and "novel weapon hypothesis" (hereafter referred to as NWH) (Hierro and Callaway 2003). Among these hypotheses, NWH has caught the attention of ecologists and thus has been a subject of debate in the recent past. Briefly, this hypothesis is based on "allelopathy" (an ecological phenomenon in which a plant releases chemicals that often deleteriously affect the growth of other plants), which plays a significant role in providing selective advantage to an invading species so that it dominates in the new environment. However, in the original native range, such chemicals fail to exhibit toxic effects on other species because of the tolerance that evolves as a consequence of long association. Since this hypothesis has generated much interest over the past few years, this chapter critically reviews NWH as a mechanism for successful invasion.

3.2   Inception of novel weapon hypothesis

According to NWH, invasive species possess novel weapons in the form of chemicals that suppress the growth of neighboring plants in an alien environment, allowing them to spread and form their own monocultures. In the native range, on the other hand, such species grow normally in association with other plants. Seemingly, the native plants' tolerance evolves toward chemicals or the so-called novel weapons on account of their long association (Figure 3.1). The hypothesis was proposed by Callaway and Aschehoug (2000) while they were working with *Centaurea diffusa* (commonly known as diffuse knapweed), an invasive Eurasian forb in North America. These investigators found that *C. diffusa* forms monocultures in invaded areas of North America and outcompetes several grassy neighbors that otherwise coexist with it in its native areas in Eurasia. Based on exhaustive experiments, the investigators concluded that *C. diffusa* releases some novel chemicals through root exudates, which act as weapons against native species in invaded regions, and outcompetes native species. On the other hand, in its native Eurasian habitats, the same species growing in the vicinity of *C. diffusa* remain unaffected owing to evolved tolerance for its root exudates. Thus, this hypothesis is primarily based on allelopathy. In fact, NWH

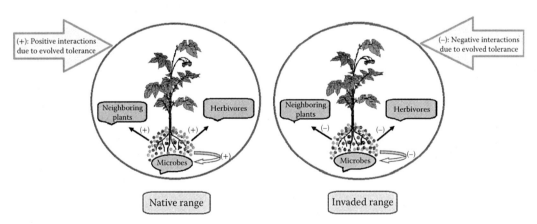

Figure 3.1 **(See color insert.)** Pictorial representation of the novel weapon hypothesis: In the native range, chemicals released by the plants exhibit positive interactions toward neighboring plants, herbivores, and microbes possibly due to tolerance acquired over the years. In the invaded range, on the other hand, chemicals released by the exotic plants serve as novel weapons to the other naive neighboring plants, herbivores (generalists), and microbes that were not earlier exposed to such chemicals. Thus, there is no evolved tolerance in them toward novel chemicals; this results in negative interactions that provide selective advantage to the exotic plant (donor of novel chemicals).

provides support in favor of allelopathy, which also faced criticism from a section of ecologists regarding its field demonstration (Fitter 2003). Two factors are thus very important in the inception of NWH: (1) Novel chemicals or allelochemicals and (2) long-term associations among species that evolve mutual tolerance. Allelopathic plants thus have an ability to dominate nonallelopathic plants in the alien environment through their chemicals that serve as novel weapons, although in the native range no such dominance is evident. In fact, NWH gets support from earlier assertions made by Rabotnov (1977, 1982) that allelopathic plants strongly inhibit other plants in association in exotic range but not in native range where the species develop tolerance to allelochemicals. *Eucalyptus* and *Juglans* species were cited as examples as they exhibit strong allelopathic influence on associated species in the introduced range in comparison with little or no effect in the native range (Rabotnov 1982). Later, Mallik and Pellissier (2000) supported this view based on their study on the allelopathic effect of the ericaceous understory plant *Vaccinium myrtillus* on the regeneration of conifers (*Picea mariana* [subalpine spruce] and *P. abies* [black spruce]). Both coniferous species, *P. abies* and *P. mariana*, failed to regenerate in the presence of *V. myrtillus* under laboratory as well as field conditions. It is noted that *P. mariana*, an exotic tree in North America, was affected more than *P. abies*, a native conifer; this supports Rabotnov's assertion.

Novel chemicals are effective against not only native plants cooccurring with invasive plants but also other organisms such as generalist herbivores, pathogens, and microbes (Callaway and Ridenour 2004, Cappuccino and Arnason 2006). This further strengthens NWH. Owing to the increasing number of reports favoring the release of novel chemicals by plant invaders, proponents of the hypothesis elaborated it further to be called as "allelopathic advantage against resident species" (AARS) hypothesis. It states that plant invaders with novel chemicals are better defended than native species and, thus, selection pressure favors them and results in the evolution of better genotypes that are more allelopathic and thus better defended (Callaway and Ridenour 2004). In fact, the AARS hypothesis goes parallel with the EICA hypothesis.

3.3   *Evidence in favor of novel weapon hypothesis*

Many studies support NWH as a mechanism for successful invasion by some species (Table 3.1). Vivanco et al. (2004) identified a chemical, 8-hydroxyquinoline, from the root exudates of *C. diffusa*. This chemical inhibited the growth of plants from the invasive range to a greater extent than from the native range in experimental plots. This study depicted an evolved resistance in the native species toward a novel chemical, 8-hydroxyquinoline. Other evidence for NWH comes from an allied congeneric species, *C. maculosa*, also known to be invasive in North America. A novel chemical, (–)-catechin, isolated from the roots of *C. maculosa* was phytotoxic to native species of North America (invaded zone) compared to European species that were tolerant to its phytotoxic effects (Bais et al. 2003, Weir et al. 2003, 2006, Callaway et al. 2005, Perry et al. 2005a, b). Despite the doubts raised by Blair et al. (2006) regarding the effective concentrations of catechin in the soil, several other studies provide support to NWH in *C. maculosa* (Ridenour et al. 2008, Thorpe et al. 2009, He et al. 2009). Ridenour et al. (2008) in a way also supported the hypothesis and pointed out that new neighbors (plants or herbivores) encountered by *C. maculosa* in the invaded range exert strong directional selection on the defensive and allelopathic traits of the plant. Thorpe et al. (2009) provided field evidence for NWH by conducting experiments in native and invaded ranges. He et al. (2009) pointed out that biochemicals in root exudates, leaves, or plant litter may be ineffective against their natural plant, herbivore, or microbial neighbors, but they are inhibitory toward new such neighbors in invaded

Table 3.1 List of Invasive Plants Exhibiting Novel Biochemistry in Invaded Ranges

Invasive plant	Native range	Invaded range	Possible novel chemicals	References
Acroptilon repens (Russian knapweed)	Eurasia	North America	Sesquiterpene lactones, polyacetylenes, thiophene derivatives, α-naphthoflavone	Stevens (1982), Stermitz et al. (2003), Ni et al. (2010), Callaway et al. (2011)
Alliaria petiolata (garlic mustard)	Europe, Africa, temperate and tropical Asia	North America	Cyanide, glucosinolates (allyl isothiocyanate) and flavonoids	Prati and Bossdorf (2004), Barto et al. (2010), Cantor et al. (2011)
Centaurea diffusa (diffuse knapweed)	Temperate Asia and Europe	North America	8-Hydroxyquinoline	Vivanco et al. (2004)
Centaurea maculosa (= *C. stoebe*) (spotted knapweed)	Eurasia	North America	(±)-Catechin	Thorpe et al. (2009), He et al. (2009), Thorpe and Callaway (2011)
Solidago canadensis (Canada goldenrod)	North America	Europe	Four compounds, i.e., polyactylene derivative and 3 diterpene lactone derivatives, total saponins, flavones, and phenolics	Abhilasha et al. (2008), Zhang et al. (2009), Zhang et al. (2011)
Typha angustifolia (narrow-leaf cattail)	Europe	North America	Water-soluble phenolics	Jarchow and Cook (2009)

areas. These investigators demonstrated the role of catechin released from the exudates of *C. maculosa* in providing invasive success to the donor plant. They pointed out that compositions of plant communities being invaded also play an important role in determining the invasive potential of species (He et al. 2009). Thorpe and Callaway (2011) tested NWH on *C. stroebe* (a synonym of *C. maculosa*) that releases the chemical compound (±)-catechin through root exudates. The investigators demonstrated that (±)-catechin affects soil microbial communities that facilitate ammonification and nitrification in both nonnative and native ranges of this species. In Montana (nonnative range), soil nitrate (NO_3^-) concentrations were lower in invaded grasslands. The reason for this was attributed to the effect of (±)-catechin on nitrifying bacteria. On the other hand, in native Romanian soils, a lesser effect on nitrogen-related processes was seen. The study concluded that biogeographic isolation resulted in greater effect of allelochemicals on nitrogen cycling that indirectly affects native vegetation.

Another evidence for NWH was provided by Prati and Bossdorf (2004) who took the example of *Alliaria petiolata* (commonly known as garlic mustard). *A. petiolata* is an aggressive invader of forests in North America and is native to Europe. North American populations of *A. petiolata* strongly inhibited the germination of *Geum laciniatum*, a cooccurring species. However, it had a little effect on another allied species, *G. urbanum*, cooccurring in its native region. The European population of *A. petiolata*, on the other hand, reduced the germination of both species of *Geum* by the same proportion. This partially supported NWH. NWH was further extended to plant-microbial interactions wherein the success of invasive plants in alien environments was attributed to plant-microbial interactions that at least partially depended on soil rhizosphere chemistry (Callaway and Ridenour 2004). Thus, novel chemicals produced by invasive plants were not only phytotoxic but also antimicrobial to the extent of providing an advantage to donor plants. In *A. petiolata*, the plant extracts prepared from invasive populations were found to be inhibitory to mutualistic mycorrhizae of invaded soils but not to mycorrhizae of native soils. Some flavonoid glycosides were probably associated with these inhibitory effects. The study concluded that probably the sensitive mutualistic mycorrhizae were beneficial to plants cooccurring in the invaded area (Callaway et al. 2008). Barto et al. (2010) provided the phytochemical profile of *A. petiolata*, and it was found that some chemicals such as cyanide, glucosinolates, and flavonoids are specific to the plant and distinguish it from other brassicaceous plants native to North America, which provides support for NWH.

Acroptilon repens (Russian knapweed), a native of Eurasia and invasive in North America, is another example where novel chemicals favor its invasiveness outside the native range. Its invasive success in the biogeographically isolated North America could be attributed to the presence of novel chemicals that inhibit North American plants more than the plants of Uzbekistan, that is, the native range of the plant (Ni et al. 2010). *A. repens* possesses novel biochemistry such as sesquiterpene lactones (Stevens 1982), polyacetylenes and thiophene derivatives (Stevens 1986), and 7,8-benzoflavone (α-naphthoflavone) in its roots (Stermitz et al. 2003). Alford et al. (2007) reported the presence of 7,8-benzoflavone in soil invaded by *A. repens*, indicating its possible role as a novel phytotoxic chemical.

Solidago canadensis is a native of North America, and it has invaded different parts of the world including Europe, Asia, and Australia. Abhilasha et al. (2008) reported that novel compounds released by this plant as root exudates inhibit the growth of native European flora and provide support for NWH. Four different types of chemicals were identified from the root extracts of *S. canadensis*. Although their identification was not confirmed, these chemicals could be polyacetylene derivative or diterpene lactone derivatives (Abhilasha et al. 2008). Later, the effect of *S. canadensis* was also studied on soil-borne pathogens in China. Extracts prepared from the root and rhizome of *S. canadensis* (possibly rich in allelochemicals) inhibited soilborne pathogens such as *Pythium ultimum* and *Rhizoctonia solani*, which may provide selective advantage to *S. canadensis* for successful spread in the invaded range (Zhang et al. 2009). Furthermore, specific studies were conducted to find out the nature of metabolites released by *S. canadensis* that affect soil-borne pathogens. The plant released phenolics, flavones, and saponins into the soil; they accumulated and persisted in the soil and affected soilborne pathogens (Zhang et al. 2011). The concentration of these metabolites increased with increasing density of the plant.

Evidence in favor of novel weapons were provided not only by the studies conducted on different invasive plants suppressing the growth of plants from the native range as well as the invasive range but also by comparing novel phytochemicals in invasive and noninvasive plants. Cappuccino and Arnason (2006) compared phytochemicals from various exotic species, both invasive and noninvasive, of North America. They revealed that invasive exotics possess novel chemicals compared to noninvasive exotics that had phytochemicals widespread among other plants. Furthermore, the unique chemicals reported from invasive exotics also possessed various biological activities such as antiherbivory, antimicrobial, antifungal, and allelopathic properties. Kim and Lee (2011) compared the phenolic content of nine native and nine invasive species of East Asia and determined their allelopathic and antimicrobial potential. The mean content of total phenolics in all nine invasive species was 2.6 times higher than the mean concentration of phenolics in native species. Invasive plants were found to be inhibitory to test plants compared to native plants and, likewise, they exhibited greater antimicrobial effect. The effect of total suits of phenolics produced by invasive plants differed from those produced by native plants. Scopoletin was the novel compound found to be present in all invasive plants. The study demonstrated that invasive plants produce more biochemicals that exhibit strong effects on other plants and microbes. Therefore, the study was consistent with NWH.

3.4   Criticism of the hypothesis

NWH has been a subject of criticism for many ecologists, particularly regarding the role of (±)-catechin as a novel chemical produced by *C. stoebe* (= *C. maculosa*). Blair et al. (2005) pointed out that the amount of (±)-catechin was much lower in soil than what was previously reported (Bais et al. 2003, Perry et al. 2005a, b, Thelen et al. 2005). Besides, it failed to exhibit phytotoxicity on the plant species reported to be sensitive to (±)-catechin even at much higher concentrations. The study also revealed that (±)-catechin was unstable at neutral and basic pH, irrespective of whether in water or soil, and also at acidic pH under moist conditions. Later, it was found that (±)-catechin degrades fast in wet soils compared to dry soils (Blair et al. 2006). These studies ruled out the role of (±)-catechin as a novel chemical providing invasion success to *C. stoebe*.

The involvement of (–)-catechin in the invasiveness of *C. stoebe* was also rejected outright (Duke et al. 2009a, b). Duke et al. (2009a), based on their studies, pointed out that (–)-catechin is rather an antioxidant and doubted its role as a novel chemical. These investigators further suggested that other chemicals/factors must be explored to understand the invasiveness of *C. stoebe* in North America. Duke et al. (2009b) further argued that (–)-catechin is a weakly phytotoxic, unstable compound and, thus, unlikely to contribute toward the invasiveness of donors. Chobot et al. (2009) demonstrated the role of (±)-catechin as an antioxidant rather than an allelochemical or a novel chemical to provide an advantage to the invasive species *C. stoebe*.

Lind and Parker (2010) tested NWH on 19 exotic plants and 21 cooccurring native plants using herbivore feeding assays and pointed out that the invasive plant metabolites were no more deterrent to herbivores than the cooccurring native species. Further analysis of chemical deterrence revealed that both native and invasive species had similar trade-offs among traits. The investigators concluded that although some invasive plants may possess novel chemical weapons that may provide invasive success, it is not a general pattern (Lind and Parker 2010).

Zhu et al. (2011) also rejected the involvement of NWH in the case of the highly invasive weed *Eupatorium adenophorum*. According to their study, the allelochemicals (9-oxo-10,11-dehydroageraphorone and 9b-ageraphorone) released by the plant in the soil are degraded by soil microbes, leading to little or no effect on native plants. Thus, novel chemicals proved ineffective toward native species, which rejects their involvement in suppressing neighboring plants. The investigators indicated the significant role of soil biota in determining the inhibitory potential of allelochemicals (Zhu et al. 2011).

3.5 Conclusions

NWH provides strong evidence that chemicals released by invasive plants in a new alien environment prove toxic to the native species and other organisms such as microbes and herbivores and help the donor plants expand their territories. In the native range, however, such species occur normally in the diverse communities since the associated species in the vicinity remain unaffected by their chemicals owing to tolerance developed as a result of their co-evolution. In other words, novel biochemistry of a plant is likely to affect geographic coevolutionary trajectories and thus determine coexistence of species (Callaway et al., 2004). Since allelopathy is the basis of this hypothesis, there are quite a few plants for which this hypothesis has been found to hold good, although it is yet to be tested in other allelopathic invasive plants. *Parthenium hysterophorus*, *Lantana camara*, and *Ageratum conyzoides*, fast-spreading invaders from tropical America in the Indian subcontinent and elsewhere (Kohli et al. 2006), provide a strong foothold for this hypothesis. Furthermore, it is not the only hypothesis that solves the mystery of plant invasiveness. Cafford et al. (2009) pointed out that no single hypothesis/mechanism could be held responsible for the invasion success of species; rather, an integrated approach to test many factors/hypotheses should be considered. Considering that invasion is a complicated and multifaceted process, many hypotheses should be tested to understand the invasion success of introduced species rather than relying on a single hypothesis.

In conclusion, NWH can provide strong evidence for allelopathic nature as well as the invasiveness of plants; thus, it needs to be tested in other plant species in order to have a strong basis. A few suggestions could be helpful in this regard:

- More allelopathic plants should be studied for their novel chemicals in native and invaded areas.
- There should be more collaborative efforts among scientists to facilitate studies across different biogeographic areas.
- Choice of test species should be based on the species actually occurring in native or nonnative areas.
- Studies must be carried out in the laboratory as well as the field to ensure the involvement of chemicals present in the rhizosphere.

References

Abhilasha, D., N. Quintana, and J. Vivanco. 2008. Do allelopathic compounds in invasive *Solidago canadensis* SL restrain the native European flora? *Journal of Ecology* 96:993–1001.

Alford, É.R., L.G. Perry, B. Qin, J.M. Vivanco, and M.W. Paschke. 2007. A putative allelopathic agent of Russian knapweed occurs in invaded soils. *Soil Biology and Biochemistry* 39:1812–1815.

Alpert, P. 2006. The advantages and disadvantages of being introduced. *Biological Invasions* 8:1523–1534.

Bais, H.P., R. Vepachedu, S. Gilroy, R.M. Callaway, and J.M. Vivanco. 2003. Allelopathy and exotic plant invasion: From molecules and genes to species interactions. *Science* 301:1377–1380.

Baker, H.G., and G.L. Stebbins. 1965. *The genetics of colonizing species*. New York: Academic Press.

Barto, E.K., J.R. Powell, and D. Cipollini. 2010. How novel are the chemical weapons of garlic mustard in North America forest understories? *Biological Invasions* 12:3465–3471.

Blair, A.C., B.D. Hanson, G.R. Brunk, R.A. Marrs, P. Westra, S.J. Nissen, and R.A. Hufbauer. 2005. New techniques and findings in the study of a candidate allelochemical implicated in invasion success. *Ecology Letters* 8:1039–1047.

Blair, A.C., S.J. Nissen, G.R. Brunk, and R.A. Hufbauer. 2006. A lack of evidence for an ecological role of the putative allelochemical (±)-catechin in spotted knapweed invasion success. *Journal of Chemical Ecology* 32:2327–2331.

Blossey, B., and R. Nötzold. 1995. Evolution of increased competitive ability in invasive non-indigenous plants: A hypothesis. *Journal of Ecology* 83:887–889.

Blumenthal, D.M. 2006. Interactions between resource availability and enemy release in plant invasion. *Ecology Letters* 9:887–895.

Cafford, J.A., R. Jansson, and C. Nilsson. 2009. Reducing redundancy in invasion ecology by integrating hypotheses into a single theoretical framework. *Diversity and Distributions* 15:22–40.

Callaway, R.M., and E.T. Aschehoug. 2000. Invasive plants versus their new and old neighbours: A mechanism for exotic invasion. *Science* 290:521–523.

Callaway, R.M., and W.M. Ridenour. 2004. Novel weapons: Invasive success and the evolution of increased competitive ability. *Frontiers in Ecology and the Environment* 2:436–443.

Callaway, R.M., D. Cipollini, K. Barto, G.C. Thelen, S.G. Hallett, D. Prati, K. Stinson, and J. Klironomos. 2008. Novel weapons: Invasive plant suppresses fungal mutualists in America but not in its native Europe. *Ecology* 89:1043–1055.

Callaway, R.M., G.C. Thelen, A. Rodriguez, and W.E. Holben. 2004. Soil biota and exotic plant invasion. *Nature* 427:731–733.

Callaway, R.M., L.P. Waller, A. Diaconu, R. Pal, A.R. Collins, H. Mueller-Schaerer, and J.L. Maron. 2011. Escape from competition: Neighbors reduce *Centaurea stoebe* performance at home but not away. *Ecology* 92:2208–2213.

Callaway, R.M., W.M. Ridenour, T. Laboski, T. Weir, and J.M. Vivanco. 2005. Natural selection for resistance to the allelopathic effects of invasive plants. *Journal of Ecology* 93:576–583.

Cantor, A., A. Hale, J. Aaron, M.B. Traw, and S. Kalisz. 2011. Low allelochemical concentrations detected in garlic mustard-invaded forest soils inhibit fungal growth and AMF spore germination. *Biological Invasions* 13:3015–3025.

Cappuccino, N., and J.T. Arnason. 2006. Novel chemistry of invasive exotic plants. *Biology Letters* 2:189–193.

Chobot, V., C. Huber, G. Trettnhahn, and F. Hadacek. 2009. (±)-Catechin: Chemical weapon or stress regulator? *Journal of Chemical Ecology* 35:980–996.

Davis, M.A., J.P. Grime, and K. Thompson. 2000. Fluctuating resources in plant communities: A general theory of invasibility. *Journal of Ecology* 88:528–534.

Duke, S.O., A.C. Blair, F.E. Dayan, R.D. Johnson, K.M. Meepangala, D. Cook, and J. Bajsa. 2009a. Is (−)-catechin a novel weapon of spotted knapweed (*Centaurea stoebe*)? *Journal of Chemical Ecology* 35:141–153.

Duke, S.O., F.E. Dayan, J. Bajsa, K.M. Meepangala, R.A. Hufbauer, and A.C. Blair. 2009b. The case against (−)-catechin involvement in allelopathy of *Centaurea stoebe* (spotted knapweed). *Plant Signaling and Behavior* 4:422–424.

Elton, C.S. 1958. *The ecology of invasions by animals and plants.* London: Methuen.

Fitter, A. 2003. Making allelopathy respectable. *Science* 301:1337–1338.

He, W.-M., Y. Feng, W. M. Ridenour, G.C. Thelen, J.L. Pollock, A. Diaconu, and R.M. Callaway. 2009. Novel weapons and invasion: Biogeographic differences in the competitive effects of *Centaurea maculosa* and its root exudate (±)-catechin. *Oecologia* 159:803–815.

Hierro, J.L., and R.M. Callaway. 2003. Allelopathy and exotic plant invasion. *Plant and Soil* 256:25–39.

Jarchow, E.J., and B.J. Cook. 2009. Allelopathy as a mechanism for the invasion of *Typha angustifolia*. *Plant Ecology* 204:113–124.

Keane, R.M., and M.J. Crawley. 2002. Exotic plant invasions and the enemy release hypothesis. *Trends in Ecology and Evolution* 17:164–170.

Kim, Y.O., and E.J. Lee. 2011. Comparison of phenolic compounds and the effects of invasive and native species in East Asia: Support for the novel weapons hypothesis. *Ecological Research* 26:87–94.

Kohli, R.K., D.R. Batish, H.P. Singh, and K.S. Dogra. 2006. Status, invasiveness and environmental threats of three tropical American invasive weeds (*Parthenium hysterophorus* L., *Ageratum conyzoides* L., *Lantana camara* L.) in India. *Biological Invasions* 8:1501–1510.

Lamarque, L.J., S. Delzon, and C.J. Lortie. 2011. Tree invasions: A comparative test of the dominant hypotheses and functional traits. *Biological Invasions* 13:1969–1989.

Lind, E.M., and J.D. Parker. 2010. Novel weapons testing: Are invasive plants more chemically defended than native plants? *PLoS ONE* 5:e10429.

MacDougall, A.S., B. Gilbert, and J.M. Levine. 2009. Plant invasions and the niche. *Journal of Ecology* 97:609–615.

Mack, R.N., D. Simberloff, W.M. Lonsdale, H. Evans, M. Clout, and F.A. Bazzaz. 2000. Biotic invasions: Causes, epidemiology, global consequences and control. *Ecological Applications* 10:689–710.

Mallik, A.U., and F. Pellissier. 2000. Effects of *Vaccinium myrtillus* on spruce regeneration: Testing the notion of coevolutionary significance of allelopathy. *Journal of Chemical Ecology* 26:2197–2209.

Ni, G.-Y., U. Schaffner, S.-L. Peng, and R.M. Callaway. 2010. *Acroptilon repens*, an Asian invader, has stronger competitive effects on species from America than species from its native range. *Biological Invasions* 12:3653–3663.

Perry, L.G., C. Johnson, E.R. Alford, J.M. Vivanco, and M.W. Paschke. 2005a. Screening of grassland plants for restoration after spotted knap-weed invasion. *Restoration Ecology* 13:725–735.

Perry, L.G., G.C. Thelen, W.M. Ridenour, T.L. Weir, R.M. Callaway, M.W. Paschke, and J.M. Vivanco. 2005b. Dual role for an allelochemical: (±)-catechin from *Centaurea maculosa* root exudates regulates conspecific seedling establishment. *Journal of Chemical Ecology* 93:1126–1135.

Pimentel, D., R. Zuniga, and D. Morrison. 2005. Update on the environmental and economic costs associated with alien invasive species in the United States. *Ecological Economics* 52:273–288.

Prati, D., and O. Bossdorf. 2004. Allelopathic inhibition of germination by *Alliaria petiolata* (Brassicaceae). *American Journal of Botany* 91:285–288.

Rabotnov, T.A. 1977. The significance of the coevolution of organisms for the formation of phytocoenoses. *Bulletin of Moscow Naturalist's Society, Department of Biology* 82:91–102. (In Russian)

Rabotnov, T.A. 1982. Importance of the evolutionary approach to the study of allelopathy. *Ékologia* 3:5–8 (translated from Russian).

Ridenour, W.M., J.M. Vivanco, Y. Feng, J.-I. Horiuchi, and R.M. Callaway. 2008. No evidence for tradeoffs: *Centaurea* plants from America are better competitors and defenders. *Ecological Monographs* 78:369–386.

Sharma, G.P., J.S. Singh, and A.S. Raghubanshi. 2005. Plant Invasions: Emerging trends and future implications. *Current Science* 88:726–734.

Shea, K., and P. Chesson. 2002. Community ecology theory as a framework for biological invasions. *Trends in Ecology and Evolution* 17:170–176.

Stermitz, F.R., H.P. Bais, T.A. Foderaro, and J.M. Vivanco. 2003. 7,8-Benzoflavne: A phytotoxin from root exudates of invasive Russian knapweed. *Phytochemistry* 64:493–497.

Stevens, K.L. 1982. Sesquiterpene lactones from *Centaurea repens*. *Phytochemistry* 21:1093–1098.

Stevens, K.L. 1986. Allelopathic polyacetylenes from *Centaurea repens* (Russian knapweed). *Journal of Chemical Ecology* 12:1205–1211.

Thelen, C.C., J.M. Vivanco, B. Newingham, W. Good, H.P. Bais, P. Landres, A. Caesar, and R.M. Callaway. 2005. Insect herbivory stimulates allelopathic exudation by an invasive plant and the suppression of natives. *Ecology Letters* 8:209–217.

Thorpe, A.S., and R.M. Callaway. 2011. Biogeographic differences in the effects of *Centaurea stoebe* on the soil nitrogen cycle: Novel weapons and soil microbes. *Biological Invasions* 13:1435–1445.

Thorpe, A.S., G.C. Thelen, A. Diaconu, and R.M. Callaway. 2009. Root exudate is allelopathic in invaded community but not in native community: Field evidence for the novel weapons hypothesis. *Journal of Ecology* 97:641–645.

Vilà, M., C. Basnou, P. Pyšek, M. Josefsson, P. Genovesi, S. Gollasch, W. Nentwig, et al. 2010. How well do we understand the impacts of alien species on ecosystem services? A Pan-European, cross-taxa assessment. *Frontiers in Ecology and the Environment* 8:135–144.

Vivanco, J.M., H.P. Bais, F. Stermitz, G.C. Thelen, and R.M. Callaway. 2004. Bio-geographical variation in community response to root allelochemistry: Novel weapons and exotic invasion. *Ecology Letters* 7:285–292.

Weir, T.L, H.P. Bais, and J.M. Vivanco. 2003. Intraspecific and interspecific interactions mediated by a phytotoxin, (±)-catechin, secreted by the roots of *Centaurea maculosa* (spotted knapweed). *Journal of Chemical Ecology* 29:2379–2393.

Weir, T.L., Bais, H.P., Stull, V.J., Callaway, R.M., Thelen, G.C., Ridenour, W.M., Bhamidi, S., Stermitz, F.R., and Vivanco, J.M. 2006. Oxalate contributes to the resistance of *Gaillardia grandiflora* and *Lupinus sericeus* to a phytotoxin produced by *Centaurea maculosa*. *Planta* 223:785–795

Wilcove, D.S., D. Rothstein, J. Dubow, A. Phillips, and E. Losos. 1998. Quantifying threats to imperiled species in the United States: Assessing the relative importance of habitat destruction, alien species, pollution, overexploitation and disease. *Bioscience* 48:607–615.

Zhang, S. W. Zhu, B. Wang, J. Tang, and X. Chen. 2011. Secondary metabolites from the invasive *Solidago canadensis* L. accumulation in soil and contribution to inhibition of soil pathogen *Pythium ultimum*. *Applied Soil Ecology* 48:280–286.

Zhang, S., Y. Jin, J. Tang, and X. Chen. 2009. The invasive plant *Solidago canadensis* L. suppresses local soil pathogens through allelopathy. *Applied Soil Ecology* 41:215–222.

Zhu, X., J. Zhang, and K. Ma. 2011. Soil biota reduce allelopathic effects of the invasive *Eupatorium adenophorum*. *PLoS ONE* 6(9):e25393.

chapter four

Functional basis for geographical variation in growth among invasive plants

John C. Volin, Jason R. Parent, and Lindsay M. Dreiss

Contents

4.1   Introduction

The ability for aggressive invasion is relatively rare among nonnative plants introduced to a new environment. The "tens rule" proposed by Williamson and Brown (1986) predicts that 10% of nonnative plants become casual, 10% of these become naturalized, and only 10% of naturalized plants become invasive. Although this rule is a very general approximation, it illustrates the point that nonnative plants rarely become invasive (Rejmánek et al. 2005); however, it illustrates the point that nonnative plants rarely become invasive. Despite this fact, plants that do become invasive can cause serious harm to both ecosystems and economies (Vitousek et al. 1997, Pimentel et al. 2000). Consequently, a great deal of research has been directed toward understanding why certain nonnative plant species become invasive outside their native range and what physiological characteristics provide these species with an advantage over native species occupying the same niche (Daehler 2003, Callaway and Maron 2006, Richardson and Pyšek 2006, Van Kleunen et al. 2010). The goal of such research is to gain sufficient knowledge to facilitate the ability of scientists to identify potential invasive plants as well as invasible ecosystems before invasions occur. Early identification of threats will vastly improve our ability to mitigate invasion episodes by allowing options for prevention or early treatment. This chapter discusses some of the key hypotheses believed to explain how invasive traits may arise in the introduced ranges of nonnative plants and explores some of the physiological traits that are common among invasive species. In our discussion on invasive traits, we present our research on the highly invasive plant *Lygodium microphyllum* as a case study. *L. microphyllum* is a climbing fern that is an aggressive invader in the New World and exhibits many of the traits believed to be characteristic of invasive plants.

4.2   *Hypotheses explaining why plants become invasive*

Many hypotheses have been proposed to explain why certain plants become invasive when introduced to a new environment. One of the earliest and most commonly cited hypotheses is the "enemy release" hypothesis (e.g., Elton 1958). Its premise is that reduced pressure within a species' introduced range, whether from decreased herbivory or disease, may allow an introduced species to achieve greater growth and reproductive success, which may eventually lead to the species becoming invasive. Release from enemies is believed to be a primary or contributing factor in the success of many invasive species (Elton 1958, Keane and Crawley 2002, Mitchell and Power 2003, Callaway et al. 2004, DeWalt et al. 2004). Our research with *L. microphyllum* clearly shows that release from natural soilborne enemies is a major factor in the invasiveness of this plant in its introduced range (Volin et al. 2010).

Many cases of invasion by nonnative plants arise in areas where the native ecology has been disturbed by natural or anthropogenic means (Gill and Williams 1996, Lake and Leishman 2004, Alston et al. 2006). Disturbances typically cause fluctuations in the available resources, especially light and nutrients. Davis et al. (2000) proposed the "fluctuating resource theory of invasibility" in which invading species are more successful where and when they do not encounter intense competition from native species for the available resources. After a disturbance, reduced competition and increased resource availability may result in a disequilibrium of resource supply and uptake that leads to higher susceptibility of the community to invasion. In other words, greater net availability of resources results in greater susceptibility to invasion (Davis and Pelsor 2001). Several studies support this theory by showing that invasion is facilitated by increases in the availability of water and nitrogen (Seabloom et al. 2003), as well as light (Matlack 1994, Panrendes and Jones 2000).

Closely related to the hypothesis by Davis et al. (2000) is the "empty niche" hypothesis, which posits that an open niche is equivalent to having a net surplus of resources that may be utilized with little or no competition (Elton 1958, MacArthur 1970, Levine and D'Antonio 1999, Mack et al. 2000). A community that has an open niche should, in theory, have a greater susceptibility to invasion. Intuitively, one may expect the probability of an unused niche in an ecosystem to decline with increasing diversity. From this idea, Elton (1958) proposed the "diversity resistance" hypothesis, which suggests that invasion susceptibility is negatively related to the diversity of native species. Studies have found evidence both for and against this hypothesis with experiment-based studies tending to provide support (e.g., Rejmánek 1989, Tilman 1997, 1999, Knops et al. 1999, Levine 2000, Naeem et al. 2000, Dukes 2002, Kennedy et al. 2002) and many large-scale observation studies tending to counter the hypothesis and suggest that native diversity is positively correlated with invasibility (Timmins and Williams 1991, Planty-Tabacchi et al. 1996, Lonsdale 1999, Stohlgren et al. 1999, Stadler et al. 2000, McKinney 2001, Pyšek et al. 2002). The contradictory evidence found in the observational studies may be explained by the occurrence of external factors that covary with species richness (Shea and Chesson 2002). Factors such as habitat diversity, nutrient availability, climate, and propagule pressure may promote high diversity of both native and nonnative plants (Siemann and Rogers 2003, Leishman and Thomson 2005). To date, observation-based studies tend to not control for external factors (Levine and D'Antonio 1999, Shea and Chesson 2002), which makes it difficult to discern the relationship between native diversity and invasibility. On the other hand, small-scale experimental studies tend to control external factors by omitting them altogether. Thus, due to the complexities of accounting for covarying factors in both small- and large-scale studies, our knowledge of the relationship between native plant diversity and invasibility remains open to debate (Hierro et al. 2005).

The invasiveness of a nonnative species may depend on its ability to tolerate a wide range of environmental conditions. Such tolerance may result from a high degree of phenotypic plasticity, which allows individuals to acclimate themselves to conditions by altering their physiological processes. Studies have shown that invasive species tend to have greater phenotypic plasticity than cooccurring native species (Daehler 2003). Nonnative plants that do not have sufficient plasticity to acclimate may still become invasive through evolution (Thompson 1998). Evolution may occur rapidly in founder populations by genetic drift, inbreeding, hybridization, and selective pressures imposed by the novel environment (Bossdorf et al. 2005). The "evolution of increased competitive ability" (EICA) hypothesis (Blossey and Nötzold 1995), a nuance of the enemy release hypothesis, suggests that a lack of pressure from predation in an introduced range will favor the reallocation of energy from defense mechanisms to increased growth and reproduction in nonnative plants. Several studies have supported the EICA hypothesis (e.g., Siemann and Rogers 2001, Blair and Wolfe 2004, Meyer et al. 2005), although several others have found no support or even contradictory evidence for it (e.g., Willis et al. 2000, Richardson and Pyšek 2006, Franks et al. 2008, Volin et al. 2010). However, few of these studies included a full test of the EICA hypothesis by investigating growth and defense in the same species (Bossdorf et al. 2005). Our research of *L. microphyllum* populations in both native and introduced environments, which also included returning the introduced population back to its native range in the study, provided a rigorous test of the EICA hypothesis, but it still did not find supporting evidence for EICA (Volin et al. 2010). Further research comparing invasive plants in their native and introduced ranges is needed to definitively assess the degree to which the EICA hypothesis, as well as other hypotheses, plays a role in invasion ecology.

4.3 *Physiological characteristics of invasive plants*

The hypotheses proposed to explain the invasiveness of species in their introduced ranges are often based on studies comparing the physiological attributes of invasive and noninvasive plants. These studies provide a mechanistic understanding of the competitive advantages of invasive plants over cooccurring native plants. Herbert Baker (1965) attempted to classify the "ideal weed" nearly 50 years ago by identifying a discrete set of characteristics that account for the weed's success. Baker's (1965) research is analogous to current attempts to classify the "ideal invader." Over the decades, many comparative multispecies studies, congeneric/confamilial studies, and meta-analyses spanning various temporal and spatial scales have continued to assess the associations between physiology and invasiveness (Daehler 2003, Bossdorf et al. 2005, Lortie and Callaway 2006, Leishman et al. 2007, Van Kleunen et al. 2010).

The superior physiological performance of many invasive plants may be due to one or both of the following: (1) the invader may exhibit a high degree of phenotypic plasticity or (2) the invading species may have the capability of rapidly evolving advantageous traits (Bossdorf et al. 2005, Richardson and Pyšek 2006). The ability to acclimate or adapt is extremely advantageous for a species introduced to a new habitat. Assessing plasticity and rapid evolution in invasive plants is most effectively accomplished in studies considering a biogeographical approach. Many studies have addressed physiological adaptations as a link to invasive plant success (Harrington et al. 1989, Bossdorf et al. 2005, Finnoff and Tschirhart 2005, van Kleunen et al. 2010). However, relatively few studies have investigated how plants become invasive through comparisons in native and introduced environments

(Sheppard et al. 2000, Grigulis et al. 2001, Volin et al. 2010). We argue that concomitantly testing the allocational, morphological, and physiological mechanisms of species invasiveness in both native and introduced ranges of a species will fill in an important knowledge gap on the origin of invasive characteristics. Our study of *L. microphyllum* uses cross-continental comparisons to explain how this relatively benign constituent in its native environment becomes a highly invasive species in its introduced environment.

4.4 Case study: The growth and physiology of Lygodium microphyllum

Lygodium microphyllum, a vine-like climbing fern native to the pantropics of the Old World, was likely introduced into North America as an ornamental landscape plant in the southern peninsula of Florida in the mid-twentieth century (Figure 4.1). By the 1960s, in a relatively quick manner, it was already noted as having escaped cultivation (Beckner 1968). However, it was not until the mid-1990s that the dramatic spread and dominance of *L. microphyllum* in many habitats across southern Florida drew the serious attention of scientists and land managers. Since then, research has been ongoing to understand its ecology and management including methods of biological, chemical, and mechanical control.

In *L. microphyllum*'s introduced habitat in Florida, the climbing fern can develop a near monoculture, smothering the native understory beneath a mat of vegetation that can be greater than 1 m thick (Pemberton and Ferriter 1998) (Figure 4.2). The stems are indeterminate single compound fronds that climb high into the native overstory, eventually shading the canopy trees and resulting in their collapse. In some cases, the fronds may act as a "fire ladder,"

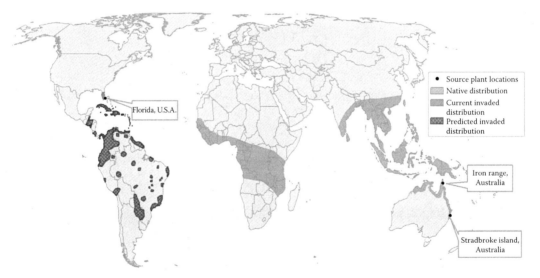

Figure 4.1 **(See color insert.)** Native, introduced, and predicted distributions of *Lygodium microphyllum*: Spore–source populations of plants used in the two studies include the introduced Florida population, the native Iron Range population (reputed source of the plants originally introduced to Florida), and the native Stradbroke Island population. The study conducted in Florida used only plants originating from the Florida source population, whereas the study conducted in southeast Queensland included plants from all the three source populations. (Adapted from Pemberton, R., and A. Ferriter, *Am. Fern J.*, 88(4), 165–175, 1998; J. A. Goolsby, *Nat. Areas J.*, 24, 351–353, 2004; Volin et al. *Diversity and Distributions*, 10, 439–446, 2004; Volin et al. *Plant Ecol.*, 208(2), 223–234, 2010.)

Figure 4.2 **(See color insert.)** Old World climbing fern (*Lygodium microphyllum*) smothering understory and overstory native plant species in a bald cypress (*Taxodium distichum*) swamp in southern Florida. (Courtesy of Peggy Greb, United States Department of Agriculture/Agricultural Research Service.)

allowing fire to climb into the canopy and kill the host trees. Until recently, the reasons for the success of *L. microphyllum* in its introduced environment have been elusive. However, after more than a decade of intensive study, we have now begun to develop a clearer understanding of the qualities that have made this plant one of Florida's most aggressive invaders.

To truly understand why an invasive plant is successful in its introduced environment, one is often required to examine the species across a range of scales in both its native and introduced habitats. We have used this strategy to explore the invasive success of *L. microphyllum*. Initially, we focused on studying its reproduction, including its mode of reproduction, seasonality of sporulation, and spore production. Our subsequent studies examined the fern's community ecology and its growth, both in situ and under controlled conditions. Various environmental treatments, including the influence of different light and hydrological conditions, were tested for effects on growth. Cross-continental controlled and field common garden studies were carried out to compare the growth of the fern in its native and introduced environments. Finally, several of these field and controlled studies were linked with landscape-scale studies to develop a spatial model of the fern's potential spread across its introduced environment.

Invasive plants tend to have highly effective means of dispersing propagules. For instance, in comparison with native species, invasive species have been shown to have lighter seeds facilitating greater wind dispersal (*Banksia ericaifolia*) (Honig et al. 1992), more appetizing fruits for animal dispersal (*Carpobrotus edulis*) (Vila and D'Antonio 1998), or greater reproductive output (*Spartina alterniflora*) (Callaway and Josselyn 1992). One striking aspect of *L. microphyllum* is its ability to establish itself and grow far from the nearest known source population. Therefore, our first studies were focused on the fern's reproductive biology. *L. microphyllum* is homosporous, which means that it produces bisexual gametophytes and, consequently, can potentially reproduce by three different reproductive strategies: (1) Intragametophytic selfing, (2) intergametophytic selfing, and (3) outcrossing. Intragametophytic selfing allows long-distance dispersal because it only requires one gametophyte to establish new colonies. The second and third strategies require two spores to land close together and germinate into gametophytes so that a sperm can swim from one gametophyte to the other. In the case of intergametophytic selfing, the two gametophytes are from the same parent plant; with outcrossing, the two gametophytes are from different parent plants. Plasticity of reproductive techniques is not unique to *L. microphyllum* and has often been thought to play a role in the invasiveness of many species. For example, *Fallopia japonica*'s breeding system may involve male, female, or hermaphroditic individuals (Schweitzer and Larson 1999). It can reproduce by seed or by clonal spreading via rhizomes, which results in dense monocultures. In addition to plasticity in individuals, studies have found that microevolution can lead to favorable changes in the reproductive strategy used by some invasive plants. Evolution from outcrossing to selfing in *Eichhornia paniculata* (Barrett et al. 2008) and evolutionary increases in reproductive capacity of *Butomus umbellatus* (Brown and Eckert 2004) are clear evidence of adaptations of invasive species to new environments resulting in the evolution of advantageous morphological traits.

In the case of ferns, most species that have been studied to date appear to reproduce via intergametophytic outcrossing (Hedrick 1987, Soltis and Soltis 1992, Hooper and Haufler 1997), whereas a few fern species seem to possess a mixed mating system (Klekowski 1982, Soltis and Soltis 1987). Consequently, intragametophytic selfing and outcrossing may represent stable mating systems in homosporous ferns (Crist and Farrar 1983, Peck et al. 1990, Korpelainen 1996). However, *L. microphyllum* is atypical in that it is able to reproduce by all three mating strategies with an interesting twist that further contributes to its invasiveness (Lott et al. 2003). When a gametophyte germinates from a spore, it is almost always female. These female gametophytes produce a pheromone that makes all younger nearby gametophytes become male and thereby increases the likelihood of outcrossing. On the other hand, if after a few weeks no spores or gametophytes land nearby, the female becomes bisexual by producing male organs and thus ensures fertilization (Lott et al. 2003). This antheridiogen system is thought to facilitate intergametophytic crossing whenever possible and assures self-fertilization when such crossing is not possible (Schneller et al. 1990, Yamauchi et al. 1996, Wynne et al. 1998, Kurumatani et al. 2001). Consequently, the reproductive plasticity of *L. microphyllum* greatly contributes to its long-distance dispersal capability.

Propagule pressure and reproductive output have also been found to contribute greatly to the success of invasive plants (Kolar and Lodge 2001). Our investigations of *L. microphyllum*'s seasonal spore patterns, spore production, and community ecology revealed its exceptional capacity for spore production in terms of both quantity and duration (Volin et al. 2004). In its introduced environment, dense populations of *L. microphyllum* produce spores throughout the year, with peak production in the wet season. *L. microphyllum* produces an estimated 15,000 spores per square centimeter of fertile leaf area, which, for a typical invaded site in Florida, yields approximately 3 billion spores at any given time. High reproductive capacity was also found for *Echium plantagineum* for which seed banks are two to five times greater for plants in the introduced range (Grigulis et al. 2001).

In both its native and introduced environment, *Lygodium microphyllum* is often found in wetlands, but may tolerate a wide range of hydrological conditions in both native and introduced environments. Hydrology has a strong influence on the composition and productivity of wetland plant communities. A greater plasticity for hydrological conditions favors the success of *L. microphyllum* as it has done for other wetland invasives, such as *Phalaris arundinacea* and *Typha latifolia* in North America (Kercher and Zedler 2004). These species are able to outgrow native plants as they are able to produce more shoot, root, and whole-plant biomass under a wide range of hydrological conditions. We therefore conducted a glasshouse study to assess *L. microphyllum*'s growth and its allocational, morphological, and physiological response to flood, drought, and field capacity treatments (Gandiaga et al. 2009). We found that while flooding significantly reduced relative growth rate by 55% compared with the other treatments, *L. microphyllum* still showed a positive growth rate after more than 2 months in inundated soils. Flooding significantly decreased specific leaf area and area-based leaf photosynthesis, which resulted in an overall lower mass-based photosynthesis. In other words, the fern's growth response to differences in hydrological conditions was largely explained by changes in its morphological and physiological determinants of growth. Interestingly, we found no significant differences in growth rates among plants subjected to field capacity or drought conditions (Gandiaga et al. 2009). The differing response of *L. microphyllum* to the hydrological treatments makes clear the difficulties encountered in assessing the traits of invasive plants through field observations: The performance of the plant is dependent on the environmental conditions occurring during the study period. These difficulties are also evident in a study comparing growth rates of native willows (Salix spp.) and the invasive salt cedar (*Tamarix ramosissima*). Under conditions of reduced salinity the native willows had higher growth rates, whereas under more saline conditions the invasive salt cedar had higher growth rates (Glenn et al. 1998).

In our community-level studies, we found that the presence of *L. microphyllum* coincided with wet but not permanently inundated environments, and the coverage was greatest in a low-light understory environment where it appears to have a competitive advantage. However, *L. microphyllum* can be found throughout the range of light conditions that exist in Florida's native ecosystems, from low to high irradiances. Evidence from controlled studies showed that small plants of *L. microphyllum* have the same relative growth rate after 3 months whether it grows in 20%, 50%, or 70% full sunlight (Figure 4.3) (Volin 2010). The ability of the fern to maintain high growth under different light environments appears to be related to its ability to allocate carbon to stems in a highly plastic and favorable manner (Figure 4.4). For instance, we found that in the highest light environment, the plants grow

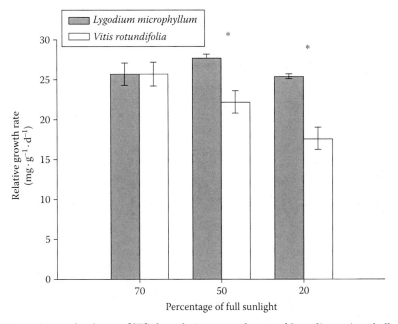

Figure 4.3 Means (±standard error [SE]) for relative growth rate of *Lygodium microphyllum* and *Vitis rotundifolia* plants under three different light treatments (70%, 50%, and 20% of full sunlight): In the figure, "*" indicates significant ($p < .05$) differences within treatments.

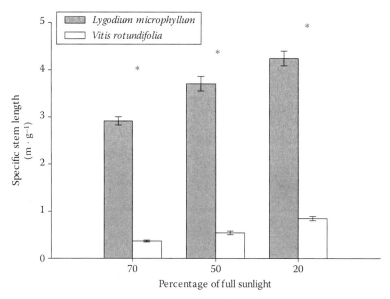

Figure 4.4 Means (±SE) for specific stem length of *Lygodium microphyllum* and *Vitis rotundifolia* plants under three different light treatments (70%, 50%, and 20% of full sunlight): In the figure, "*" indicates significant ($p < .05$) differences within treatments.

on average 2.9 m for every gram of carbon invested; in the lowest light environment, the plants grew 4.2 m per gram of carbon invested (Volin 2010). These findings were consistent with 2 years of field observations where under high light conditions the height of the actual climbing mat (i.e., not individual stems) increased by 1.43 m/yr on host canopy trees (Volin et al. 2004). Thus, *L. microphyllum*'s ability to optimize its morphological and physiological characteristics allows it to maximize photosynthetic area while minimizing carbon costs in tissue construction. Similar optimization was found in studies of *Oxalis corymbosa* and *Peperomia pellucida* in which lower leaf expense favored increased biomass, nitrogen, and phosphorus allocation to aboveground structures. Such an allocation yielded higher rates of light-saturated photosynthesis, allowed greater carbon assimilation, and facilitated avoidance of belowground predators, which allowed the species to achieve higher relative growth rates (Baruch and Goldstein 1999, Daehler 2003, Funk and Vitousek 2007, Zou et al. 2007, Feng et al. 2008,). Some invasive plants have been found to exhibit superior growth rates when compared with native cogeners by virtue of their higher specific leaf areas and lower leaf construction costs (Baruch and Goldstein 1999, Feng et al. 2008, Leishman et al. 2007), although this is not always the case (Nagel and Griffin 2001, Bastlová and Květ 2002).

In controlled light experiments, we found that *L. microphyllum* allocated, on average, 48% of its carbon to belowground tissues, whereas a cooccurring native vine, *Vitis rotundifolia*, averaged only 27% (Figure 4.5). The latter is a typical investment in belowground biomass for native vines, which characteristically maximize their allocation to aboveground tissues in order to achieve greater heights and thus optimize light interception (Condon et al. 1992, Toledo-Aceves and Swaine 2008). Lower root-to-shoot ratios are also more typical of introduced populations given that carbon and nutrients are often allocated to increase aboveground productivity (Erfmeier and Bruelheide 2004, Zou et al. 2007, van Kleunen et al. 2010). Although plants benefit from a greater photosynthetic area, this allocation may also allude to greater efficiency in soil nutrient uptake or adaptation to the lack of underground herbivory threats (Zou et al. 2007). On the other hand, *L. microphyllum* is

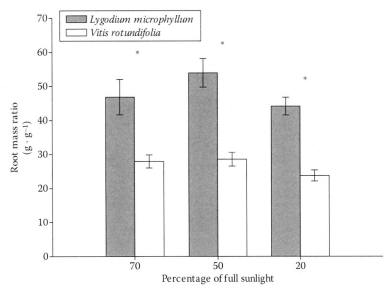

Figure 4.5 Means (±SE) for root mass ratio of *Lygodium microphyllum* and *Vitis rotundifolia* plants under three different light conditions (70%, 50%, and 20% of full sunlight): In the figure, "*" indicates significant (*p* < .05) differences within treatments.

able to reach great heights with relatively little investment in aboveground carbon. This observation led us to hypothesize that *L. microphyllum* is a successful invasive species because it has been released from its natural belowground enemies. This release allows the plant to allocate more resources to belowground tissues, which may confer a competitive advantage in the low-nutrient environment of southern Florida. *Ageratina adenophora* is another invasive plant that has been shown to change resource allocation in its introduced range, where it is not under pressure from its native enemies, by increasing nitrogen allocated to photosynthesis and reducing carbon allocated to cell walls (Feng et al. 2009).

We tested the hypothesis that *L. microphyllum* has been released from its underground enemies in its introduced range through a series of coordinated common garden studies in Florida (introduced range) and in Australia (native range). We examined the fern's growth across soil treatments, including soil sterilization to eliminate belowground natural enemies and nutrient amendments to examine the possible interaction with soil nutrient availability. A unique aspect of this experiment was that in the native range study we used three different fern source populations: (1) One from Florida; (2) one from the study location in southeastern Queensland, Australia; and (3) the third from northeastern Queensland. The third source is the reputed original location of the plants introduced to Florida (Volin et al. 2010). All populations tended to have relatively poor growth in unaltered soil, but growth for all was stimulated by nutrient amendment and soil sterilization. The overall effect of sterilization, however, was muted under high-nutrient conditions except for the fern population that originated from the same region in Australia as the soil used in the experiment. Regardless of the nutrient treatment, the southeastern Queensland population grew significantly faster in its native soil when it had been sterilized. Overall, these results supported our hypothesis that the invasiveness of *L. microphyllum* in Florida is in part mediated by its release from soilborne enemies. Furthermore, given the different response of the southeastern Queensland population in Australia as compared to the other populations, it appears that a region-specific response may be occurring as well.

The future spread of *L. microphyllum* across its introduced landscape is of great interest as the invasive potential of the species has come to light. We developed a cellular automaton model, based on the known attributes and locations of infestations, to predict the spread of the fern (Volin et al. 2004). Known locations of *L. microphyllum* invasions were taken from a survey of individual naturalized populations published by Nauman and Austin (1978), as well as from aerial monitoring surveys conducted by the South Florida Water Management District at 2-year intervals beginning in 1993 (Pemberton and Ferriter 1998). The cellular automaton landscape model was calibrated based on 1978 and 1993 survey data. The model was then validated independently using 1993 and 1999 aerial survey data. Finally, by initializing our model with 1999 aerial survey data, we predicted the spread of the fern out to 2014. The model predicted that 37% of the 40,000-km^2 grid cells covering southern Florida would have *L. microphyllum* present by 2014 (Volin et al. 2004). In other words, in the absence of aggressive control measures, the coverage of *L. microphyllum* could exceed the coverage of the top five invasive species in Florida by 2014. Fortunately, aggressive biological, chemical, and mechanical control efforts for this species have been ongoing (Goolsby et al. 2003, 2006, Boughton and Pemberton 2009, Boughton et al. 2011).

From numerous studies conducted at various scales under different environmental conditions, as well as from comparisons in plant performance between introduced and native habitats, we eventually developed an understanding of why *L. microphyllum* is such a successful invader in its introduced range. In essence, *L. microphyllum* possesses several life history characteristics, including plastic reproductive strategies and prolific and continuous spore production, that enhance its competitive ability in Florida. The fern also appears to optimize its morphological and physiological characteristics to maximize photosynthetic area and minimize carbon costs in tissue construction, thereby gaining the ability to grow rapidly across gradients in both light and hydrology (Gandiaga et al. 2009, Volin 2010). In its native range, this ability to grow across environmental gradients and reproductive output may be constrained by natural enemies belowground (Volin et al. 2010). Our cross-continental comparisons support our hypothesis that release from natural enemies belowground may also help in explaining the success of this fern in its introduced environment and, provocatively, given the different source population responses, it appears that a regional-scale version of the enemy release hypothesis might play a role in *L. microphyllum*'s invasive success.

Our findings demonstrate the importance of a biogeographical approach in studying invasive species, which contributes not only to community ecology theory in general but also to the understanding of the specific role played by physiological adaptations in plant invasiveness. Although our knowledge of the physiology of invasive plants has increased rapidly in recent decades, a greater understanding of the functional basis for geographical variation in growth among invasive species will greatly enhance our knowledge and ability to address invasive species in the future.

References

Alston, K.P., and D.M. Richardson. 2006. The roles of habitat features, disturbance, and distance from putative source populations in structuring alien plant invasions at the urban/wildland interface on the Cape Peninsula, South Africa. *Biological Conservation* 132:183–198.

Baker, H.G. 1965.Characteristics and modes of origin of weeds. In *The genetics of colonizing species*, eds. H.G. Baker, and G.L.Stebbins, pp. 147–72. New York: Academic Press.

Barrett, S.C.H., R.I. Colauti, and C.G. Eckert. 2008. Plant reproductive systems and evolution during biological invasion. *Molecular Ecology* 17:373–383.

Baruch, Z., and G. Goldstein. 1999. Leaf construction cost, nutrient concentration, and net CO_2 assimilation of native and invasive species in Hawaii. *Oecologia* 121:183–192.

Bastlová, D., and J. Květ. 2002. Differences in dry weight partitioning and flowering in phenology between native and non-native plants of purple loosestrife (*Lythrum salicaria* L.). *Flora* 197(5):332–340.

Beckner, J. 1968. *Lygodium microphyllum*, another fern escaped in Florida. *American Fern Journal* 58(2):93.

Blair, A.C., and L.M. Wolfe. 2004. The evolution of an invasive plant: An experimental study with *Silene latifolia*. *Ecology* 85:3035–3042.

Blossey, B., and R. Nötzold. 1995. Evolution of increased competitive ability in invasive nonindigenous plants: A hypothesis. *Journal of Ecology* 83:887–889.

Bossdorf, O., H. Auge, L. Lafuma, W.E. Rogers, E. Siemann, and D. Prati. 2005. Phenotypic and genetic differentiation between native and introduced plant populations. *Oecologia* 144:1–11.

Boughton, A.J., G.R. Buckingham, C.A. Bennett, R. Zonneveld, J.A. Goolsby, R.W. Permberton, and T.D. Center. 2011. Laboratory host range of *Austromusotima camptozonale* (Lepidoptera: Crambidae), a potential biological control agent of Old World climbing fern, *Lygodium microphyllum* (Lygodiaceae). *Biocontrol Science and Technology* 21(6):643–676.

Boughton, A.J., and R.W. Pemberton. 2009. Establishment of an imported natural enemy, *Neomusotima conspurcatalis* (Lepidoptera: Crambidae) against an invasive weed, Old World climbing fern, *Lygodium microphyllum*, in Florida. *Biocontrol Science and Technology* 19(7):769–772.

Brown, J.S., and C.G. Eckert. 2004. Evolutionary increase in sexual and clonal reproductive capacity during biological invasion in an aquatic plant *Butomus umbellatus* (Butomaceae). *American Journal of Botany* 92:495–502.

Callaway, R.M., and M.N. Josselyn. 1992. The introduction and spread of smooth cordgrass (*Spartina alterniflora*) in South San Francisco Bay. *Estuaries* 15:218–226.

Callaway, R.M., and J.L. Maron. 2006. What have exotic plant invasions taught us over the past 20 years? *Trends in Ecology and Evolution* 21(7):369–374.

Callaway, R.M., G.C. Thelan, A. Rodriguez, and W.E. Holben. 2004. Soil biota and exotic plant invasion. *Nature* 427:731–733.

Condon, M.A., T.W. Sasek, and B.R. Strain. 1992. Allocation patterns in two tropical vines in response to increased atmospheric CO_2. *Functional Ecology* 6(6):680–685.

Crist, K., and D. Farrar. 1983. Genetic load and long distance dispersal in *Asplenium platyneoron*. *Canadian Journal of Botany* 61:1809–1814.

Daehler, C.C. 2003. Performance's comparisons of co-occurring native and alien invasive plants: Implications for conservation and restoration. *Annual Review of Ecology and Systematics* 34:183–211.

Davis, M.A., J.P. Grime, and K. Thompson. 2000. Fluctuating resources in plant communities: A general theory of invasibility. *Journal of Ecology* 88:528–534.

Davis, M.A., and M. Pelsor. 2001. Experimental support for a resource-based mechanistic model of invasibility. *Ecology Letters* 4:421–428.

DeWalt, S.J., J.S. Denslow, and K. Ickes. 2004. Natural-enemy release facilitates habitat expansion of the invasive tropical shrub *Clidemia hirta*. *Ecology* 85:471–483.

Dukes, J.S. 2002. Species composition and diversity affect grassland susceptibility and response to invasion. *Ecological Applications* 12:602–617.

Elton, C.S. 1958. The ecology of invasions by animals and plants. *Progress in Physical Geography* 31(6):659–666.

Erfmeier, A., and H. Bruelheide. 2004. Comparison of native and invasive *Rhododendron ponticum* populations: Growth, reproduction and morphology under field conditions. *Flora* 199:120–133.

Feng, Y.L., G.L. Fu, and Y.L. Zheng. 2008. Specific leaf area related to the differences in leaf construction cost, photosynthesis, nitrogen allocation, and use efficiencies between invasive and noninvasive alien congeners. *Planta* 228:383–390.

Feng, Y.L, Y.B. Lei, R.F. Wang, R.M. Callaway, A. Valiente-Banuet, Inderjit, Y.P. Li, and Y.L. Zheng. 2009. Evolutionary tradeoffs for nitrogen allocation to photosynthesis versus cell walls in an invasive plant. *Proceedings of the National Academy of Sciences, USA* 106(6):1853–1856.

Finnoff, D., and J. Tschirhart. 2005. Identifying, preventing and controlling invasive plant species using their physiological traits. *Ecological Economics* 52(3):397–416.

Franks, S.J., P.D. Pratt, F.A. Dray, and E.L. Simms. 2008. No evolution of increased competitive ability or decreased allocation to defense in *Melaleuca quinquenervia* since release from natural enemies. *Biological Invasions* 10:455–466.

Funk, J.L., and P.M. Vitousek. 2007. Resource-use efficiency and plant invasion in low-resource systems. *Nature* 446:1079–1081.

Gandiaga, S., J.C. Volin, E.L. Krueger, and K. Kitajima. 2009. Effects of hydrology on the growth and physiology of an invasive exotic, *Lygodium microphyllum* (Old World climbing fern). *Weed Research* 49:283–290.

Gill, A.M., and J.E. Williams. 1996. Fire regimes and biodiversity: The effects of fragmentation on southeastern Australian eucalypt forests by urbanization, agriculture and pine plantations. *Forest Ecology and Management* 85:261–278.

Glenn, E., R. Tanner, S. Mendez, T. Kehret, D. Moore, J. Garcia, and C. Valdes. 1998. Growth rates, salt tolerance and water use characteristics of native and invasive riparian plants from the delta of the Colorado River, Mexico. *Journal of Arid Environments* 40:281–294.

Goolsby, J.A. 2004. Potential distribution of the invasive Old World climbing fern, *Lygodium microphyllum* in North and South America. *Natural Areas Journal* 24:351–353.

Goolsby, J.A., P.J. De Barro, J.R. Makinson, R.W. Pemberton, D.M. Hartley, and D.R. Frohlich. 2006. Matching the origin of an invasive weed for selection of a herbivore haplotype for a biological control program. *Molecular Ecology* 15:287–297.

Goolsby, J.A., A.D. Wright, and R.W. Pemberton. 2003. *Exploratory surveys in Australia and Asia for natural enemies of Old World climbing fern, Lygodium microphyllum: Lygodiaceae.* Paper 347. Publications from USDA-ARS/UNL Faculty.

Grigulis, K., A.W. Sheppard, J.E. Ash, and R.H. Groves. 2001. The comparative demography of the pasture weed *Echium plantagineum* between its native and invaded ranges. *Journal of Applied Ecology* 38:281–290.

Harrington, R.A., B.J. Brown, P.B. Reich, and J.H. Fownes. 1989. Ecophysiology of exotic and native shrubs in Southern Wisconsin. *Oecologia* 80:368–373.

Hedrick, P.W. 1987. Genetic load and the mating system in homosporous ferns. *Evolution* 41(6):1282–1289.

Hierro, J.L., J.L. Maron, and R.M. Callaway. 2005. A biogeographical approach to plant invasions: The importance of studying exotics in their introduced and native range. *Journal of Ecology* 93:5–15.

Honig, M.A., R.M. Cowling, and D.M. Richardson. 1992. The invasive potential of Australian Banksias in South African fynbos: A comparison of the reproductive potential of *Banksia ericifolia* and *Leucadendron laureolum*. *Australian Journal of Ecology* 17:305–324.

Hooper, E.A., and A.H. Haufler. 1997. Genetic diversity and breeding system in a group of neotropical epiphytic ferns (*Pleopeltis*; Polypodiaceae). *American Journal of Botany* 84(12):1664–1674.

Keane, R.M., and M.J. Crawley. 2002. Exotic plant invasions and the enemy release hypothesis. *Trends in Ecology and Evolution* 17(4):164–170.

Kennedy, T.A., S. Naeem, K.M. Howe, J.M.H. Knops, D. Tilman, and P.B. Reich. 2002. Biodiversity as a barrier to ecological invasion. *Nature* 417:636–38.

Kercher, S.M., and J.B. Zedler. 2004. Flood tolerance in wetland angiosperms: A comparison of invasive and noninvasive species. *Aquatic Botany* 80:89–102.

Klekowski, Jr., E.J. 1982. Genetic load and soft selection in ferns. *Heredity* 92:191–197.

Knops, J.M.H., D. Tilman, N.M. Haddad, S. Naeem, C.E.J. Mitchell, M. Haarstad, E. Ritchie, K.M. Howe, P.B. Reich, E. Siemann, and J. Groth. 1999. Effects of plant species richness on invasion dynamics, disease outbreaks, insect abundances and diversity. *Ecology Letters* 2:286–293.

Kolar, C.S., and D.M. Lodge. 2001. Progress in invasion biology: Predicting invaders. *Trends in Ecology and Evolution* 16(4):199–204.

Korpelainen, H. 1996. Intragametophytic selfing does not reduce reproduction in *Dryopteris filix-max*. *Sexual Plant Reproduction* 9:117–122.

Kurumatani M., K. Yagi, T. Murata, M. Tezuka, L. Mander, M. Nishiyama, and H. Yamane. 2001. Isolation and identification of antheridiogens in the ferns, *Lygodium microphyllum* and *Lygodium reticulatum*. *Bioscience, Biotechnology and Biochemistry* 65(10):2311–2314.

Lake, J.C., and M.R. Leishman. 2004. Invasion success of exotic plants in natural ecosystems: The role of disturbance, plant attributes and freedom from herbivores. *Biological Conservation* 117(2):215–226.

Leishman, M.R., T. Haselhurst, A. Ares, and Z. Baruch. 2007. Leaf trait relationships of native and invasive plants: Community- and global-scale comparisons. *New Phytologist* 176:635–643.

Leishman, M.R., and V.P. Thomson. 2005. Experimental evidence for the effects of additional water, nutrients and physical disturbance on invasive plants in low fertility Hawkesbury Sandstone soils, Sydney, Australia. *Journal of Ecology* 93:38–49.

Levine, J., and C.M. D'Antonio. 1999. Elton revisited: A review of evidence linking diversity and invasibility. *Oikos* 87:15–26.

Levine, J.M. 2000. Species diversity and biological invasions: Relating local process to community pattern. *Science* 288:852–854.

Lonsdale, W.M. 1999. Global patterns of plant invasions and the concept of invasibility. *Ecology* 80:1522–1536.

Lortie, C.J., and R.M. Callaway. 2006. Re-analysis of meta-analysis: Support for the stress-gradient hypothesis. *Journal of Ecology* 94:7–16.

Lott, M.S., J.C. Volin, R.W. Pemberton, and D.F. Austin. 2003. The reproductive biology of *Lygodium microphyllum* and *L. japonicum* (Schizaeaceae) and its implications for invasive potential. *American Journal of Botany* 90:1144–1152.

MacArthur, R.H. 1970. Species packing and competitive equilibrium for many species. *Theoretical Population Biology* 1:1–11.

Mack, R.N., D. Simberloff, W.M. Lonsdale, H. Evans, M. Clout, and F.A. Bazzaz. 2000. Biotic invasions: Causes, epidemiology, global consequences, and control. *Ecological Applications* 10:689–710.

Matlack, G.R. 1994. Vegetation dynamics of the forest edge—trends in space and successional time. *Journal of Ecology* 82:113–123.

McKinney, M.L. 2001. Effects of human population, area, and time on non-native plant and fish diversity in the United States. *Biological Conservation* 100:243–252.

Meyer, G., R. Clare, and E. Weber. 2005. An experimental test of the evolution of increased competitive ability hypothesis in goldenrod, *Solidago gigantea*. *Oecologia* 144:299–307.

Mitchell, C.E., and A.G. Power, 2003. Release of invasive plants from fungal and viral pathogens. *Nature* 421:625–627.

Naeem, S., J.M.H. Knops, D. Tilman, K.M. Howe, T. Kennedy, and S. Gale. 2000. Plant diversity increases resistance to invasion in the absence of covarying extrinsic factors. *Oikos* 91:97–108.

Nagel, J.M., and K.L. Griffin. 2001. Construction cost and invasive potential: Comparing *Lythium salicaria* (Lythraceae) with co-occurring native species along pond banks. *American Journal of Botany* 88:2252–2258.

Nauman C.E., and D.F. Austin. 1978. Spread of the exotic fern *Lygodium microphyllum* in Florida. *American Fern Journal* 68(3):65–66.

Panrendes, L.A., and J.A. Jones. 2000. Role of light availability and dispersal in exotic plant invasion along roads and streams in the H. J. Andrews Experimental forest, Oregon. *Conservation Biology* 14:64–75.

Peck, J., C. Peck, and D. Farrar. 1990. Influences of life history attributes on formation of local and distant fern populations. *American Fern Journal* 80(4):126–142.

Pemberton, R., and A. Ferriter. 1998. Old World climbing fern (*Lygodium microphyllum*), a dangerous invasive weed in Florida. *American Fern Journal* 88(4):165–175.

Pimentel, D., L. Lach, R. Zuniga, and D. Morrison. 2000. Environmental and economic costs of non-indigenous species in the United States. *Bioscience* 50(1):53–65.

Planty-Tabacchi, A., E. Tabacchi, R.J. Naiman, C. Deferrari, and H. Decamps. 1996. Invasibility of species-rich communities in riparian zones. *Conservation Biology* 10:598–607.

Pyšek, P., V. Jarosík, and T. Kucera. 2002. Patterns of invasion in temperate nature reserves. *Biological Conservation* 104:13–24.

Rejmánek, M. 1989. Invasibility of plant communities. In *Biological invasions. A global perspective*, eds. J.A. Drake, H. Mooney, F. di Castri, R. Groves, F. Kruger, M. Rejmánek, and M. Williamson, pp. 369–388. Chichester: Wiley.

Rejmánek, M., D.M. Richardson, and P. Pyšek. 2005. Plant invasions and invasibility of plant communities. In *Vegetation ecology*, ed. E. van der Maarel, pp. 332–355. Oxford: Blackwell.

Richardson, D.M., and P. Pyšek. 2006. Plant invasions: Merging the concepts of species invasiveness and community invisibility. *Progress in Physical Geography* 30(3):409–431.

Seabloom, E.W., E.T. Borer, V.L. Boucher, R.S. Burton, K.L. Cottingham, L. Goldwasser, W.K. Gram, B.E. Kendall, and F. Micheli. 2003. Competition, seed limitation, disturbance, and reestablishment of California native annual forbs. *Ecological Applications* 13:575–592.

Shea, K., and P. Chesson. 2002. Community ecology theory as a framework for biological invasions. *Trends in Ecology and Evolution* 17:170–176.

Sheppard, A.W., P. Hodge, and Q. Paynter. 2000. Factors affecting broom regeneration in Australia and their management implications. *Plant Protection Quarterly* 15:156–161.

Siemann, E., and W.E. Rogers. 2001. Genetic differences in growth of an invasive tree species. *Ecology Letters* 4:514–518.

Siemann, E., and W.E. Rogers. 2003. Increased competitive ability of an invasive tree may be limited by an invasive beetle. *Ecological Applications* 13:1503–1507.

Schneller, J., C. Haufler, and T. Ranker. 1990. Antheridiogen and natural gametophyte populations. *American Fern Journal* 80(4):143–152.

Schweitzer, J.A., and K.C. Larson. 1999. Greater morphological plasticity of exotic honeysuckle species may make them better invaders than native species. *Journal of the Torrey Botanical Society* 126:15–23.

Soltis, D., and P. Soltis. 1987. Breeding system of the fern *Dryopteris expansa*: Evidence for mixed mating. *American Journal of Botany* 74(4):504–509.

Soltis, D., and P. Soltis. 1992. The distribution of selfing rates in homosporous ferns. *American Journal of Botany* 79(1):97–100.

Stadler, J., A. Trefflich, S. Klotz, and R. Brandl. 2000. Exotic plant species invade diversity hot spots: The alien flora of northwestern Kenya. *Ecography* 23:169–176.

Stohlgren, T.J., D. Binkley, G.W. Chong, M.A. Kalkhan, L.D. Schell, K.A. Bull, Y. Otsuki, G. Newman, M. Bashkin, and Y. Son. 1999. Exotic plant species invade hot spots of native plant diversity. *Ecological Monographs* 69:25–46.

Thompson, J.N. 1998. Rapid evolution as an ecological process. *Trends in Ecology and Evolution* 13:329–332.

Tilman, D. 1997. Community invasibility, recruitment limitation, and grassland biodiversity. *Ecology* 78:81–92.

Tilman, D. 1999. The ecological consequences of changes in biodiversity: A search for general principles. *Ecology* 80:1455–1474.

Timmins, S.M., and P.A. Williams. 1991. Weed numbers in New Zealand's forest and scrub reserves. *New Zealand Journal of Ecology* 15:153–162.

Toledo-Aceves, T., and M.D. Swaine. 2008. Biomass allocation and photosynthetic responses of liana and pioneer tree seedlings to light. *Acta Oecologia* 34(1):38–49.

Van Kleunen, M., E. Weber, and M. Fischer. 2010. A meta-analysis of trait differences between invasive and non-invasive plant species. *Ecology Letters* 13:235–245.

Vila, M., and C.M. D'Antonia. 1998. Fitness of invasive *Carpobrotus* (Aizoaceae) hybrids in coastal California. *Ecoscience* 5:191–199.

Vitousek, P.M., H.A. Mooney, J. Lubchenco, and J.M. Melillo. 1997. Human Domination of Earth's Ecosystems. *Science* 277(5325):494–499.

Volin, J.C. 2010. *Gestalt* of an invader. *Current Conservation* 4:29–32.

Volin, J.C., E.L. Kruger, V.C. Volin, M.F. Tobin, and K. Kitajima. 2010. Does release from natural belowground enemies help explain the invasiveness of *Lygodium microphyllum*? A cross-continental comparison. *Plant Ecology* 208(2):223–234.

Volin, J.C., M.S. Lott, J.D. Muss, and D. Owen. 2004. Predicting rapid invasion of the Florida Everglades by Old World Climbing Fern (*Lygodium microphyllum*). *Diversity and Distributions* 10:439–446.

Williamson, M., and K.C. Brown. 1986. The analysis and modeling of British invasions. *Philosophical Transactions of the Royal Society London B* 314:505–521.

Willis, A.J., J. Memmott, and R.I. Forrester. 2000. Is there evidence for the post-invasion evolution of increased size among invasive plant species? *Ecology Letters* 3:275–283.

Wynne, G., L. Mander, N. Oyama, N. Murofushi, and H. Yamane. 1998. An antheridiogen, 13-Hydroxy-GA$_{73}$ Methyl Ester (GA$_{109}$) from the fern *Lygodium circinnatum*. *Phytochemistry* 47(2):1177–1182.

Yamauchi T., N. Oyami, H. Yamane, N. Murofushi, H. Schraudolf, M. Pour, M. Furber, and L. Mander. 1996. Identification of antheridiogens in *Lygodium circinnatum* and *Lygodium flexuosum*. *Plant Physiology* 111:741–745.

Zou, J., W.E. Rodgers, and E. Siemann. 2007. Differences in morphological and physiological traits between native and invasive populations of *Sapium sebiferum*. *Functional Ecology* 21:721–730.

chapter five

Aboveground–belowground interactions
Implication for invasiveness

Priyanka Srivastava, Gyan P. Sharma, and A.S. Raghubanshi

Contents

5.1   Introduction

All terrestrial ecosystems comprise aboveground (AG) and belowground (BG) subsystems. The BG subsystem of a terrestrial ecosystem is considered to be one of the most complex systems, and little information exists about its organization and functioning; which are on the same lines like questions why various communities or ecosystems are different in terms of their composition and functioning (e.g., Hutchinson 1961, MacArthur and Wilson 1967, Grime 1973, Tilman 1982). Many ecosystem-related questions remain unanswered because traditionally AG and BG components of ecosystems have been considered to be separate entities. Most ecologists study AG and BG communities and processes separately and do not focus on the interactions between them (Wardle et al. 1999, Bonkowski et al. 2001, Masters et al. 2001, van der Putten et al. 2001, Brown and Gange 2002, Wardle 2002). However, studies that try to relate these two components in the ecological literature are mostly from temperate ecosystems (Gange and Brown 1989, Masters et al. 1993, 2001, Masters 1995, Masters and Brown 1997, Bezemer et al. 2002). Findings indicate that a feedback relation exists between the AG and BG subsystems that influence each other significantly and is responsible for controlling overall ecosystem processes and properties (Bardgett and Wardle 2010, Grime 2001, van der Putten et al. 2001, Wardle 2002).

The AG–BG-coupled influence acts through a variety of interactions, which could be positive, negative, or neutral (Zak et al. 1994, Wardle 2002, Bardgett and Wardle 2003). Plants comprise a major component of AG and are the source of nutrients to soil biota, which is considered to be the major component of BG. Plants provide organic carbon to surface and subsurface soil decomposers and resources to root herbivores, pathogens, and symbiotic mutualists. Soil is a complex mix of organic and inorganic matter and has abundant microorganisms for nutrient mineralization and cycling, which regulates the availability of nutrients for plants. Decomposers break down dead plant materials, whereas root-associated organisms influence the nutrient quality, and flow of energy between plant and decomposer subsystems (Scheu and Setala 2002, Wardle 2002, Porazinska et al. 2003). The complex interaction between these systems maintains the dynamic ecosystem processes and thereby ecosystem services. Wardle (2002) suggests that exploring linkages between AG and BG communities will enlighten and broader ecological implications of species invasions the second important threat to biodiversity. Research focusing on AG–BG interactions with plant invasion perspective could provide information about shifts in community composition between plant and soil communities (De Deyn and Putten 2005, Ball et al. 2009). Most of the interactive components of AG–BG subsystems still remain unexplored, and studies are needed to fill this knowledge gap.

In this chapter, we provide a synoptic overview of how the AG subsystem influences the BG subsystem and vice versa, and how the changing global scenario affects AG–BG interactions. The review also focuses on an important global change component, "biological invasion," and its effects on AG–BG interactions and discusses future research needs in the area of AG–BG interactions.

5.2   Aboveground subsystem influences on belowground subsystems

Plants provide organic matter for the functioning of the decomposer subsystem (Swift et al. 1979) and also resources for obligate root-associated organisms like root herbivores, pathogens, and symbiotic mutualists (Ruiter De et al. 1995, Wardle 2002). Different plant species exert varied effect on soil biota and their process as they differ in both the quantity and quality of resources that they return to the soil. For example, plant species of various grassland ecosystems differ in the composition and abundance of soil microbial communities around their roots (Griffiths et al. 1992, Bardgett and Shine 1999), whereas differences in the quality of litter produced by existing tree species in forest ecosystems show patchy distribution of soil organisms (Saetre and Baath 2000). Studies involving manipulation of grassland plant community composition have shown varied effects on the community composition of soil organisms, which could be positive (Bever et al. 1997) or neutral (Wardle et al. 1999). AG trophic interactions also indirectly influence soil organisms by affecting the quantity and quality of resources produced by plants. For example, AG herbivory promotes release of exudates into the rhizosphere, which positively affects microbial communities by increasing nitrogen avalability in soil (Hamilton and Frank 2001), whereas a decrease in plant productivity, release of secondary metabolites, and delivery of poor quality litter are some negative effects (Wardle et al. 2001). Recently, Dubey et al. (2011) studied how AG plant traits affect herbivory, which in turn alters the BG ecosystem functioning. However, generally the effect of trophic interaction through herbivory may be positive or negative (Pastor et al. 1993). Figure 5.1 illustrates how the AG subsystem interacts with BG subsystem functioning and vice versa, through varied interactive pathways.

5.3   Belowground subsystem influences on aboveground subsystems

Studies have shown that the AG subsystem plays an important role in influencing the BG subsystem and, in response, the BG subsystem also influences AG community structure and functioning (Klironomos 2002, Bever 2003) (Figure 5.1). Primary productivity of

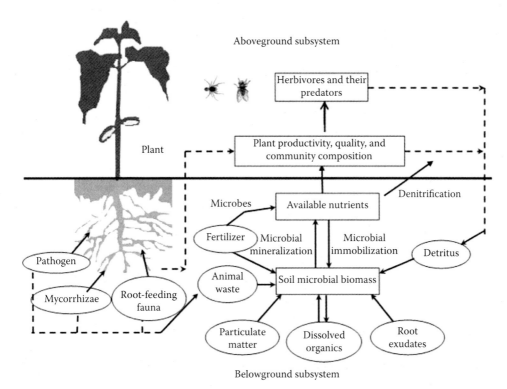

Figure 5.1 Interactive influence of aboveground and belowground subsystems on each other (solid lines and broken lines represent directly affected and indirectly affected components, respectively).

plant community is regulated by BG interactions (Wardle 2002), this regulation may be positive due to increase in nutrient availability by the detrital food web (Moore et al. 2003) and through feedback of mutualistic symbionts (Smith and Read 1997) or negative due to feedback of soil parasites, pathogens, root herbivores (Masters et al. 1993, Bever et al. 1997, van der Putten 2003), and microbial immobilization (Singh 1993). Biotic interactions in soil could induce changes in the structure and functioning of the AG community through successional replacement of plant species (van der Putten et al. 1993). Invasion of plant species in a community may also result due to the interaction of soil biota (i.e., soil pathogen) (Klironomos 2002) as rare plants tend to accumulate pathogens faster, which limit their growth, whereas abundant and invasive species accumulate pathogens more slowly (Klironomos 2002, Reinhart et al. 2003). Mutualistic symbionts found in soil also influence plant community structure through different degrees of feedback (Grime et al. 1987, van der Heijden et al. 1998), which has been reflected in studies with mycorrhizal manipulation that changed biomass, nutrient status, and abundance of plants (van der Heijden et al. 1998). These could be a few of the potential ways in which the BG subsystem influences the AG subsystem. Recently, Inderjit and van der Putten (2010) suggested three potential pathways through which soil communities influence exotic plant invasions. Pathway I suggests that the result of plant–soil feedback interactions in an invaded area could be neutral to positive and native flora is greatly affected by the negative feedback of soil. Pathway II recommends that invasive plants can interfere with the composition of local soil biota by increasing pathogen levels or affecting communities of root symbionts, which harms the native plants. Pathway III reveals that invasive plants produce toxic allelochemicals that are harmful to native plants and that these chemicals cannot be detoxified by the existing local soil communities, which results in an altered soil microbial community. These pathways provide some insight into how alien plants affect soil communities that in turn affect native plant species.

5.4   Global change and aboveground–belowground interactions

Theories of ecology recognize a variety of abiotic and biotic factors that shape the structure and dynamics of terrestrial ecosystems. Various global change phenomena, such as invasions by alien species of new communities and alteration of global cycle by anthropogenic activities that result in CO_2 enrichment, nitrogen deposition, and change in land use (e.g., afforestation), have altered the abiotic and biotic components of ecosystems and resulted in the transformation of Earth's land surface (Vitousek et al. 1997).

Studies have reported that various alterations (i.e., environmental heterogeneity, climate, disturbance, and biotic interactions) influence plant community composition and diversity of ecosystems (Tilman and Pacala 1993, Rosenzweig 1995). Many of these phenomena directly or indirectly affect soil biota and their processes also (Frey et al. 2004, Bradley et al. 2006), which ultimately influence AG biota by changing plant community composition, nutrient quantity and quality, and productivity (Virginia et al. 1992, Jackson et al. 2002). For example, the invasion of plants in an ecosystem results in the modification of soil biota, nutrient pool, and processes, which in turn affect the AG subsystem (Vitousek and Walker 1989, Hendrix and Bohlen 2002, Batten et al. 2006, Addison 2009, Sharma and Raghubanshi 2009). Similarly, CO_2 and nitrogen enrichment in the atmosphere indirectly affects soil biota through change in quantity and quality of plant litter, root exudates, and soil nutrients, which in feedback affects the AG biota (Korner 2000, Bradley et al. 2006, Sharma et al. 2010). However, the significance of these phenomena in ecosystem performance is well known (Eviner and Chapin 2003), whereas much less attention has been paid to evaluating the ecological responses to these changes (Lee 1998) and also to the mechanism behind the responses (Wardle et al. 2004). Recently, Verma and Sagar (2011) evaluated the effect of nitrogen enrichment on a few of the dry deciduous herbaceous species and concluded that nitrogen enrichment increases the herbaceous diversity initially, but as the concentration of nitrogen increases the herbaceous diversity decreases. Therefore, an intermediate level of enrichment might have selective positive response of herbaceous species that could have an impact on BG communities.

5.5   Biological invasion and aboveground–
belowground interactions

The invasion of exotic species in terrestrial ecosystems has become a global problem as they strongly affect the invaded ecosystems in several different ways (Mooney and Drake 1986, Vitousek et al. 1996, Ehrenfeld 2003, Sharma et al. 2005, Peltzer et al. 2010) (Table 5.1). Over the past few decades, biological invasion of ecosystems has become a topic of major concern in the field of ecological studies for the conservation of natural and managed areas (Mooney and Drake 1986, Domenech et al. 2006). Invasion affects plant species diversity and composition (Vivrette and Muller 1977, Christian and Wilson 1999) and is a leading cause of species extinctions. Other ecosystem consequences of invasion include changes in soil pools and processes by interfering with mineralization, nutrient cycling processes, nutrient

Table 5.1 Ecological Effects of Some Invasive Species on Ecosystem Processes and
Their Impact on Ecological Properties

Sl. no.	Invasive species	Affected ecological parameter	References
		Floral invasion	
1	*Bromus inermis, Agropyron cristatum*	Modifies soil condition, which facilitates its expansion and that of other invasive plants	Jordan et al. (2008)
2	*Aegilops triuncialis*	Slows carbon and nitrogen cycling in soil through changes in plant tissue chemistry and productivity	Drenovsky and Batten (2007)
3	*Bromus tectorm*	Alters nitrogen dynamics in soil by increasing the nitrate concentration	Sperry et al. (2006)
4	*Parthenium hysterophorus, Lantana camara, Hyptis suaveolens*	Increases available nitrogen content of soil	Raizada et al. (2008), Sharma and Raghubanshi (2009)
5	*Imperata cylindrica*	Causes shift in nitrogen pool from aboveground to belowground	Daneshgar and Jose (2009)
6	*Chromolaena odorata*	Accumulates soilborne fungi *Fusarium semitectum* in their rhizosphere, which affects native flora	Mangla et al. (2008)
		Faunal invasion	
7	*Lymantria dispar* (gypsy moth)	Causes defoliation and decline of forest trees	Gandhi and Herms (2010)
8	*Solenopsis invicta*	Alters the phylogenetic structure of ant communities	Lessard et al. (2009)
9	*Bipalium adventitium*	Predates on earthworm and affects soil structure	Fiore et al. (2004)
10	*Phytophthora kernoviae*	Causes bleeding stem lesions on forest trees and foliar necrosis of ornamental shrubs	Brasier et al. (2005)

loss (Hooper et al. 2000, Ehrenfeld et al. 2001, Sharma and Raghubanshi 2009), soil moisture, energy flux, and soil temperature (Blank and Young 2002, Prater and DeLucia 2006, Mason et al. 2007).

Much attention has been paid to the effect of invasion of AG plant and animal communities, taking into consideration losses in biodiversity (Brussaard et al. 1997). However, BG invasions may have equally far-reaching effects and be a matter of serious consideration. Interactive study linking AG and BG communities clarifies the broader ecological implications of species invasions (Wardle 2002). Therefore, to understand ecosystem conseqences, we need to focus on both the AG producer community and BG decomposer community equally. In Sections 5.5.1 and 5.5.2, we provide a conceptual outline of related effects of biological invasions on ecological systems, taking into account both AG and BG components and the links between the two subsystems.

5.5.1 Aboveground biological invasion: Ecological effects

Invasive species affect ecosystems in two lines of attack: (1) they interfere with the community structure and composition and (2) they alter the ecological properties and functioning of the invaded ecosystem. A complex interaction of the potential pathways of AG–BG interactions is shown in Figure 5.2. Invasive species influence shifts in the plant community, change the diversity and composition (Vivrette and Muller 1977, Christian and Wilson 1999) of plant species, and are a leading cause of species extinction. Invading species with different physiological traits from native flora affect the normal feedback mechanisms between AG and BG biota and interfere with the functioning of the ecosystem. Initially, interaction takes place between invasive species and root-associated biota (Klironomos 2002, Reinhart et al. 2003) and later the decomposers affected by the quantity and quality of resource input of the invading species get involved in the interaction (Ehrenfeld 2003). Shifts in plant community lead to changes in quantity and quality of litter provided to the decomposers, which results in the alteration of nutrient cycling, nutrient accumulation, and nutrient release in the ecosystem (Stark and Firestone 1996, Cavigelli and Robertson 2000, Strickland et al. 2010). Invasive species may also alter plant–microbe interactions in

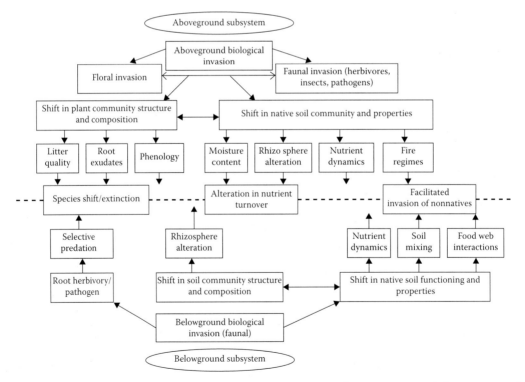

Figure 5.2 Figure illustrating various effects of invaders on the structure and functioning of aboveground and belowground subsystems and their feedback effects on each other. (Modified from Bohlen, P.J., *Appl. Soil Ecol.*, 32, 1–5, 2006.)

the soil (Allen et al. 1989, Vitousek and Walker 1989, Bever 2002) by changing microbial community structure and processes (Klironomos 2002, Kourtev et al. 2002, Batten et al. 2005). The impact of invasion on soil communities and nutrient cycling processes is considered to be an important mechanism of the invasion success of species (Ehrenfeld and Scott 2001, Kourtev et al. 2002, Chen et al. 2005, Raizada et al. 2008) and it further facilitates the invasive success of the same species (Kourtev et al. 2002, Mitchell et al. 2002, DeWalt et al. 2004). Invasive plants influence ecosystem processes, especially the nitrogen dynamics of a BG subsystem (Mack et al. 2000, Hawkes et al. 2005), and particularly when it is a nitrogen-fixing species (Mack and D'Antonio 2003). Studies on various invasive species show that they have significant effects on ecosystem nitrogen dynamics in soil, for example, Myrica faya exhibited significant alteration in nitrogen dynamics in areas invaded by it by harboring symbiotic nitrogen fixers within their root system (Vitousek and Walker 1989). Other invasive species such as Lantana camara (Singh et al. 2005, Sharma and Raghubanshi 2009) and Phragmites australis (Windham and Ehrenfeld 2003) had shown higher nitrogen-extraction efficiency, Lepidium latifolium had shown higher available nitrogen and greater net nitrogen potential in soil (Blank and Young 2002), similarly Schizachyrium condensatum, Meterosideros polymorpha invasion had shown increased ammonification and nitrification rate in invaded soils (Mack and D'Antonio 2003). Daneshgar and Jose (2009) also reported shift in nitrogen pool in pine forest of United States due to invasion of *Imperata cylidrica*. In some nitrogen limited ecosystems such as Hawaii, invasion resulted in manifold increase in soil nitrogen input and thereby altered the plant productivity of the invaded system (Vitousek and Walker 1989). On the other hand, invasive plants may inhibit nitrogen fixation, which could further decrease nitrogen inputs to the soil (Wardle et al. 1994). Invasive plants can manipulate the moisture content in the soil of invaded ecosystem because of its shading regime and differential nutrient requirement, which could hamper the nutrient turnover in the ecosystem (Enloe et al. 2004). Figure 5.2 illustrates various effects of invaders on the structure and functioning of AG and BG subsystems and their feedback effects on each other. Table 5.1 summariz on ecosystem processes and their impact on ecological properties. Invasive plants can also modify interactions in the rhizosphere, leading to changes in microbial dynamics, nutrient uptake, and competitive interactions in the plant community. One of the important mechanisms behind the modification of the rhizosphere by invasive species is the secretion of root exudates with allelochemicals. Native flora is adversely affected by the allelochemicals secreted by the invasive species, which is the reason for the success of the invasive species in that community. For example, in the western grasslands of North America, the invasion of *Centaurea diffusa* negatively affected native grasses, which further facilitated its invasion (Callaway and Aschehoug 2000). One more important side of rhizosphere interactions involves the release from pathogens or other hostile soil biota, which can assist the establishment and spread of invasive plants (Bever et al. 1997). Invasive plants can affect the functioning of an invaded ecosystem by shifting the time of nutrient uptake as a result of its differential phenology, and they can also influence fuel loads in fire-prone ecosystems (Levine et al. 2003).

Soil organisms govern the productivity (Moore et al. 2003) and structure (Grime et al. 1987) of a plant community and are more reactive to any external stress (Panikov 1999). Soil community can exert negative feedback, which is an important determinant of plant growth and a key mechanism by which plant species diversity can be maintained in several communities. Disturbances in soil lead to changes in ecosystem properties. Several studies have observed that native soil can have a negative effect on some native plants while having a positive effect on an invasive plant species (Reinhart et al. 2003, Callaway et al. 2004), which helps in the successful selective establishment of the invasive species. Introduction of larger herbivores, such as sheep, cattle, and goat, to the AG subsystem negatively affects soil organisms through changes in litter and root exudates, which results in soil disturbance and erosion (Yates et al. 2000, Wardle et al. 2001, Wardle and Bardgett 2004). Eldridge and Simpson (2002) in their study in Australia observed that rabbit burrows affected the soil surface, which favored the expansion of invasive species and enhanced the spread of fire. Similarly, it was observed that grazing by large, nonnative herbivores in the montane grasslands of central Argentina facilitated the expansion of nonnative plant species (Alejandro et al. 2010). Invasion by nonnative herbivorous insects also exerts various negative effects on ecosystems (Wallner 1996), for example, in the forests of eastern North America, invasion by gypsy moth and other insect herbivores directly or indirectly disturbed and affected various ecological interactions and processes (Gandhi and Herms 2010).

5.5.2 *Belowground biological invasion: Ecological effects*

Studies on invasion are mostly focused on AG invasion and its effects on various ecological parameters; much less attention has been paid to the invasion of BG organisms (Bohlen 2006). It is noted that BG invasion is equally important in consideration of ecosystem processes. Study of soil faunal invasion is recently receiving much attention in terms of earthworm invasion (Bohlen et al. 2004a), as earthworms are "ecosystem engineers" and are capable of considerably changing the physical and chemical characteristics of the soil environment. In North American forests, the invasion of earthworms, which are native to Europe and Asia, have greatly affected the nutrient storage and nutrient availability to the native floral and faunal community, which are largely dependent on the soil nutrients (Bohlen et al. 2004b, Zhang et al. 2010).

Ants are another important invader, which considerably impact both AG and BG subsystems and affect flora and fauna of the invaded ecosystem (Bohlen 2006). It is noted that red imported fire ants, *Solenopsis invicta*, adversely affect invaded ecosystems (Sanders et al. 2003, Rao and Vinson 2004) of the Southern United States; one new ant species, *Myrmica specioides* Bondroit, was discovered very recently in the state of Washington (Jansen and Radchenko 2009), which is considered to be an important invader of dry garden meadow in Olympia, Washington. Ants also help in seed dispersal because of their small size (Ness et al. 2004) and facilitate the invasion of a few plant species through selective predation of seeds of native flora (Seaman and Marino 2003). The aforementioned studies give evidence that ants interfere with the ecological functioning of invaded ecosystems at both AG and BG levels.

Other important BG invaders are termites: The Formosan termite has caused severe destruction in the Southern United States (Fairfax 2009). The invasive *Planaria Bipalium adventitium* is also a matter of concern these days as the species is invading various ecosystems; the most adverse effect of this invasion is the species' predation on earthworms (Fiore et al. 2004), which are very important in nutrient turnover and, ultimately, causing shifts in ecosystem functioning. Invasions by pathogens have dreadful effects on ecosystems as they commonly attack the dominant species of an ecosystem's plant community and cause its decline (Bohlen 2006), which further disturbs the BG subsystem (litter inputs, decomposition rates, nutrient cycling, etc.). For example, the invasive pathogen *Phytophthora kernoviae* causes bleeding stem lesions on forest trees and foliar necrosis of ornamental shrubs in the United Kingdom (Brasier et al. 2005). *Phytophthora cinnamomi*, another pathogen, causes dieback of eucalyptus trees and alters the plant and animal communities of ecosystems at both AG and BG levels (Weste et al. 2002). Similarly, in the west coast of the United States and in seminatural areas in Europe, an invasive pathogen, *Phytophthora ramorum*, has caused the sudden death of oaks (*Quercus* spp.) and tanoak (*Lithocarpus densiflorus*). This pathogen caused lethal stem infection resulting in the decline of oaks, which are the dominant tree species of the area, and had important effects on forest ecosystems (Condeso and Meentemeyer 2007). Alterations in soil biota, mainly pathogens, may promote invasion when interactions with exotic plants are less negative than those with native plants (Callaway et al. 2004, Mangla et al. 2008). Study of rhizosphere soil of a very destructive weed, *Chromolaena odorata*, showed accumulation in high concentrations of the generalist soilborne fungi *Fusarium semitectum*, which exerted a negative feedback for native plant species (Mangla et al. 2008). Deterioration of dominant trees leads to alteration in natural structure and functioning of the ecosystem through shifts in litter quality, litter quantity, and altered microbial composition. From the discussion in this section, it is clear that BG invaders play a key role in configuring the structure and function of ecosystems.

5.6   Research needs

Studies have observed that interactions between AG and BG subsystems are potentially the main mechanism that maintains the structure and function of ecosystems (Wardle 2002, Bardgett and Wardle 2010, Mangan et al. 2010). Several changes in either AG subsystem or BG subsystem can govern the ecological property of an ecosystem and may result in its alteration. Global climatic change affects the ecosystem in various ways, especially by invoking AG and BG invasions (Bardgett and Wardle 2010). Responses of any ecosystem to these changes are the result of AG–BG feedbacks (Vitousek and Walker 1989, Korner 2000, Hendrix and Bohlen 2002, Batten et al. 2006, Bradley et al. 2006, Addison 2009). Biological invasion is not the extreme, but it is the start that invokes further changes in the ecosystem to make it susceptible to further destruction. A review of the literature illustrates that AG alteration through invasive plants and herbivores affects the soil community, which in turn influences the plant community; similarly, BG alteration through invaders competitively affects soil community composition, interferes with their normal functioning, and also affects the plant community of the area. This chapter also gives insights that invasive pathogens play a major role in disturbing the ecosystem properties of both AG and BG subsystems. Hence, a better understanding of AG–BG interactions is required to understand the consequences of various alterations in ecosystems (Wardle et al. 2004).

Interactions among AG and BG subsystems are highly context dependent; therefore, to evaluate the effect of invasion on ecosystem processes and properties, multiapproach studies are required to investigate the intricate coupling of AG and BG subsystems (Zavaleta et al. 2003). Robust greenhouse experiments are needed with a follow up in natural and seminatural conditions through manipulation of AG and BG biodiversity at the level of communities and functiona, or taxonomic groups (De Deyn and van der Putten 2005, van der Putten et al. 2009). A feedback study between AG and BG subsystems also requires the consideration of the food web perspective (Moore et al. 2004), and a metabolomic approach can broaden current knowledge regarding this interaction (Bezemer and van Dam 2005). All these hypothetical approaches in the aforementioned prospective ecological studies, which include manipulation experiments with invasive species, should be tested at spatial and temporal scales. Insights from this chapter will enhance the predictive capacity of AG–BG interactions with one component as invasion; this could be useful in formulating future biodiversity management and conservation strategies for various ecosystems. We believe that designing management practices for an invaded ecosystem and for community reorganization after various AG–BG alterations will potentially generate information for the sustainable delivery of ecosystem services.

References

Addison, J.A. 2009. Distribution and impacts of invasive earthworms in Canadian forest ecosystems. *Biological Invasions* 11:59–79.

Alejandro, L., R.A. Distel, and S.M. Zalba. 2010. Large herbivore grazing and non-native plant invasions in montane grasslands of central Argentina. *Natural Areas Journal* 30:148–155.

Allen, M.E., E.B. Allen, and C.F. Friese. 1989. Responses of the nonmycotrophic *Salsola kali* to invasion by vesicular-arbuscular mycorrhizal fungi. *New Phytologist* 111:45–49.

Ball, B.A., M.A. Bradford, D.C. Coleman, and M.D. Hunter. 2009. Linkages between below and aboveground communities: Decomposer responses to simulated tree species loss are largely additive. *Soil Biology and Biochemistry* 41:1155–1163.

Bardgett, R.D., and A. Shine. 1999. Linkages between plant litter diversity, soil microbial biomass, and ecosystem function in temperate grasslands. *Soil Biology and Biochemistry* 31:317–321.

Bardgett, R.D., and D.A. Wardle. 2003. Herbivore mediated linkage between aboveground and belowground communities. *Ecology* 84:2258–2268.

Bardgett, R.D., and D.A. Wardle. 2010. *Aboveground-Belowground Linkages: Biotic Interactions, Ecosystem Processes, and Global Change*. Oxford Series in Ecology and Evolution. Oxford, UK: Oxford University Press.

Batten, K.M., K.M. Scow, K.F. Davies, and S.P. Harrison. 2006. Two invasive plants alter soil microbial community composition in serpentine grasslands. *Biological Invasions* 8:217–230.

Batten, K.M., J. Six, K.M. Scow, and M.C. Rillig. 2005. Plant invasion of native grassland on serpentine soils has no major effects upon selected physical and biological properties. *Soil Biology and Biochemistry* 37:2277–2282.

Bever, J.D. 2002. Host-specificity of AM fungal population growth rates can generate feedback on plant growth. *Plant and Soil* 244:281–290.

Bever, J.D. 2003. Soil community feedback and the coexistence of competitors: Conceptual frameworks and empirical tests. *New Phytologist* 157:465–473.

Bever, J.D., K.M. Westover, and J. Antonovics. 1997. Incorporating the soil community into plant population dynamics: The utility of the feedback approach. *Journal of Ecology* 85:561–573.

Bezemer, T.M., and N.M. van Dam. 2005. Linking aboveground and belowground interactions via induced plant defences. *Trends in Ecology and Evolution* 20:618–624.

Bezemer, T.M., R. Wagenaar, N.M. van Dam, and F.L. Wackers. 2002. Interactions between root and shoot feeding insects are mediated by primary and secondary plant compounds. *Proceedings of the Section Experimental and Applied Entomology of the Netherlands Entomological Society (NEV)* 13:117–121.

Blank, R.R., and J.A. Young. 2002. Influence of the exotic invasive crucifer, *Lepidium latifolium*, on soil properties and elemental cycling. *Soil Science* 167:821–829.

Bohlen, P.J. 2006. Biological invasions: Linking aboveground and belowground consequences. *Applied Soil Ecology* 32:1–5.

Bohlen, P.J., P.M. Groffman, T.J. Fahey, E. Suárez, D.M. Pelletier, and R.T. Fahey. 2004a. Ecosystem consequences of exotic earthworm invasion of north temperate forests. *Ecosystems* 7:1–12.

Bohlen, P.J., S. Scheu, C.M. Hale, M.A. McLean, S. Migge, P.M. Groffman, and D. Parkinson. 2004b. Non-native invasive earthworms as agents of change in northern temperate forests. *Frontier in Ecology and Environment* 2:427–435.

Bonkowski, M., I. Geoghegann, A. Nicholas, E. Birch, and B. Griffiths. 2001. Effects of soil decomposer invertebrates (protozoa and earthworms) on above-ground phytophagous insects cereal aphid) mediated through changes in the host plant. *Oikos* 95:441–450.

Bradley, K.L., R. Drijber, and J. Knops. 2006. Increased N availability in grassland soils modifies their microbial communities and decreases the abundance of arbuscular mycorrhizal fungi. *Soil Biology and Biochemistry* 38:1583–1595.

Brasier, C.M., P.A. Beales, S.A. Kirk, S. Denman, and J. Rose. 2005. *Phytophthora kernoviae* sp. nov., an invasive pathogen causing bleeding stem lesions on forest trees and foliar necrosis of ornamentals in Britain. *Mycological Research* 109:853–859.

Brown, V.K., and A.C. Gange. 2002. Tritrophic below- and above-ground interactions in succession. In *Multitrophic Level Interactions*, eds. T. Tscharntke, and B. Hawkins, pp. 197–222. Cambridge, UK: Cambridge University Press.

Brussaard, L., V. Behan-Pelletier, D. Bignell, V. Brown, W. Didden, P. Folgarait, C. Fragoso, et al. 1997. Biodiversity and ecosystem functioning in soil. *Ambio* 26:563–570.

Callaway, R.M., and E.T. Aschehoug. 2000. Invasive plants versus their new and old neighbors: A mechanism for exotic invasion. *Science* 290:521–523.

Callaway, R.M., G.C. Thelen, A. Rodriguez, and W.E. Holben. 2004. Soil biota and exotic plant invasion. *Nature* 427:731–733.

Cavigelli, M.A., and G.P. Robertson. 2000. The functional significance of denitrifier community composition in a terrestrial ecosystem. *Ecology* 81:1402–1414.

Chen, H., Y. Li, B. Li, J. Chen, and J. Wu. 2005. Impacts of exotic plant invasions on soil biodiversity and ecosystem processes. *Biodiversity Science* 13:555–565.

Christian, J.M., and S.D. Wilson. 1999. Long-term ecosystem impacts of an introduced grass in the northern Great Plains. *Ecology* 80:2397–2407.

Condeso, T.E., and R.K. Meentemeyer. 2007. Effects of landscape heterogeneity on the emerging forest disease sudden oak death. *Journal of Ecology* 95:364–375.

Daneshgar, P., and S. Jose. 2009. *Imperata cylindrica*, an alien invasive grass, maintains control over nitrogen availability in an establishing pine forest. *Plant and Soil* 320:209–218.

De Deyn, G.B., and W.H. van der Putten. 2005. Linking aboveground and belowground diversity. *Trends in Ecology and Evolution* 11:625–633.

DeWalt, S.J., J.S. Denslow, and K. Ickes. 2004. Natural-enemy release facilitates habitat expansion of the invasive tropical shrub *Clidemia hirta*. *Ecology* 85:471–483.

Domenech, R., M. Vila, J. Gesti, and I. Serrasolses. 2006. Neighbourhood association of *Cortaderia selloana* invasion, soil properties and plant community structure in Mediterranean coastal grasslands. *Acta Oecologica* 29:171–177.

Drenovsky, R.E., and K.M. Batten. 2007. Invasion by *Aegilops triuncialis* (barb goatgrass) slows carbon and nutrient cycling in a serpentine grassland. *Biological Invasions* 9:107–116.

Dubey, P., G.P. Sharma, A.S. Raghubanshi, and J.S. Singh. 2011. Leaf traits and herbivory as indicators of ecosystem function: A review. *Current Science* 100:313–320.

Ehrenfeld, J.G. 2003. Effect of exotic plant invasions on soil nutrient cycling processes. *Ecosystems* 6:503–523.

Ehrenfeld, J.G., P. Koutev, and W. Huang. 2001. Changes in soil functions following invasions of exotic understory plants in deciduous forests. *Ecological Applications* 11:1287–1300.

Ehrenfeld, J.G., and N. Scott. 2001. Invasive species and the soil: Effects on organisms and ecosystem processes. *Ecological Applications* 11:1259–1260.

Eldridge, D.J., and R. Simpson. 2002. Rabbit (*Oryctolagus cuniculus* L.) impacts on vegetation and soils, and implications for management of wooded rangelands. *Basic and Applied Ecology* 3:19–29.

Enloe, S.F., J.M. Di Tomaso, S.B. Orloff, and D.J. Drake. 2004. Soil water dynamics differ among rangeland plant communities dominated by yellow star thistle (*Centaurea solstitialis*), annual grasses, or perennial grasses. *Weed Science* 52:929–935.

Eviner, V.T., and F.S. Chapin III. 2003. Functional matrix: A conceptual framework for predicting multiple plant effects on ecosystem processes. *Annual Review of Ecology, Evolution, and Systematics* 34:455–485.

Fairfax, V.A. 2009. *Across the country, invasive pest species are causing problems*. National Pest Management Association, Inc., September 24.

Fiore, C., J.L. Tull, S. Zehner, and P.K. Ducey. 2004. Tracking and predation on earthworms by the invasive terrestrial planarian *Bipalium adventitium* (Tricladida, Platyhelminthes). *Behavioural Processes* 67:327–334.

Frey, S.D., M. Knorr, J. Parrent, and R.T. Simpson. 2004. Chronic nitrogen enrichment affects the structure and function of the soil microbial community in a forest ecosystem. *Forest Ecology and Management* 196:159–171.

Gandhi, K.J.K., and D.A. Herms. 2010. Direct and indirect effects of alien insect herbivores on ecological processes and interactions in forests of eastern North America. *Biological Invasions* 12:389–405.

Gange, A.C., and V.K. Brown. 1989. Effects of root herbivory by an insect on a foliar feeding species, mediated through changes in the host plant. *Oecologia* 81:38–42.

Griffiths B.S., R. Welschen, J.J.C.M. van Arendonk, and H. Lambers. 1992. The effect of nitrate-nitrogen supply on bacteria and bacterial-feeding fauna in the rhizosphere of different grass species. *Oecologia* 91:253–259.

Grime, J.P. 1973. Control of species density in herbaceous vegetation. *Journal of Environment Management* 1:157–167.

Grime, J.P. 2001. *Plant Strategies, Vegetation Processes and Ecosystem Properties.* 2nd. ed. Chichester, UK: Wiley.

Grime, J.P., J.M.L. Mackey, S.H. Hillier, and D.J. Read. 1987. Floristic diversity in a model system using experimental microcosms. *Nature* 328:420–422.

Hamilton, E.W.I., and D.A. Frank. 2001. Can plants stimulate soil microbes and their own nutrient supply? Evidence from a grazing tolerant grass. *Ecology* 82:2397–2404.

Hawkes, C.V., I.F. Wren, D.J. Herman, and M.K. Firestone. 2005. Plant invasion alters nitrogen cycling by modifying the soil nitrifier community. *Ecology Letters* 8:976–985.

Hendrix, P.F., and P.J. Bohlen. 2002. Exotic earthworm invasions in North America: Ecological and policy implications. *Bioscience* 52: 801–811.

Hooper, D.U., D.E. Bignell, V.K. Brown, L. Brussaard, J.M. Dangerfield, D.H. Wall, D.A. Wardle, et al. 2000. Interactions between aboveground and belowground biodiversity in terrestrial ecosystems: Patterns, mechanisms, and feedbacks. *Bioscience* 50:1049–1061.

Hutchinson, G.E. 1961. The paradox of the plankton. *American Naturalist* 153:137–145.

Inderjit, and W.H. van der Putten. 2010. Impacts of soil microbial communities on exotic plant invasion. *Trends in Ecology and Evolution* 25:512–519.

Jackson, R.B., J.L. Banner, E.G. Jobbagy, W.T. Pockman, and D.H. Wall. 2002. Ecosystem carbon loss with woody plant invasion of grasslands. *Nature* 418:623–626.

Jansen, G., and A. Radchenko. 2009. *Myrmica specioides* Bondroit: A new invasive ant species in the USA? *Biological Invasions* 11:253–256.

Jordan, N.R., D.L. Larson, and S.C. Huerd. 2008. Soil modification by invasive plants: Effect on native and invasive species of mixed-grass prairies. *Biological Invasions* 10:177–190.

Klironomos, J.N. 2002. Feedback with soil biota contributes to plant rarity and invasiveness in communities. *Nature* 417:67–70.

Korner, C. 2000. Biosphere responses to CO_2 enrichment. *Ecological Applications* 10:1590–1619.

Kourtev, P.S., J.G. Ehrenfeld, and M. Haggblom. 2002. Exotic plant species alter the microbial community structure and function in the soil. *Ecology* 83:3152–3166.

Lee, J.A. 1998. Unintentional experiments with terrestrial ecosystems: Ecological effects of sulphur and nitrogen pollutants. *Journal of Ecology* 86:1–12.

Lessard, J.P., J.A. Fordyce, N.J. Gotelli, and N.J. Sanders. 2009. Invasive ants alter the phylogenetic structure of ant communities. *Ecology* 90:2664–2669.

Levine, J.M., M. Vilà, C.M. D'Antonio, J.S. Dukes, K. Grigulis, and S. Lavorel. 2003. Mechanisms underlying the impacts of exotic plant invasions. *Proceedings of the Royal Society of London* 270:775–781.

MacArthur, R.H., and E.O. Wilson. 1967. *The Theory of Island Biogeography.* Princeton, NJ: Princeton University Press.

Mack, M.C., and C.M. D'Antonio. 2003. Exotic grasses alter controls over soil nitrogen dynamics in a Hawaiian woodland. *Ecological Applications* 13:154–166.

Mack, R.N., D. Simberloff, W.M. Lonsdale, H. Evans, M. Clout, and F.A. Bazzaz. 2000. Biotic invasions: Causes, epidemiology, global consequences, and control. *Ecological Applications* 10, 689–710.

Mangan, S.A., S.A. Schnitzer, E.A. Herre, K.M.L. Mack, M.C. Valencia, E.I. Sanchez, and J.D. Bever. 2010. Negative plant-soil feedback predicts tree-species relative abundance in a tropical forest. *Nature* 466:752–755.

Mangla, S., Inderjit, and R.M. Callaway. 2008. Exotic invasive plant accumulates native soil pathogens which inhibit native plants. *Journal of Ecology* 96:58–67.

Mason, T.J., K. Frencha, and K.G. Russell. 2007. Moderate impacts of plant invasion and management regimes in coastal hind dune seed banks. *Biological Conservation* 134:428–439.

Masters, G.J. 1995. The impact of root herbivory on aphid performance: field and laboratory evidence. *Acta Oecologia* 16:135–142.

Masters, G.J., and V.K. Brown. 1997. Host-plant mediated interactions between spatially separated herbivores: Effects on community structure. In *Multitrophic Interactions in Terrestrial Systems*, eds. A.C. Gange, and V.K. Brown, pp. 217–237. Oxford, UK: Blackwell Science.

Masters, G.J., V.K. Brown, and A.C. Gange. 1993. Plant mediated interactions between above- and below-ground insect herbivores. *Oikos* 66:148–151.

Masters, G.J., H. Jones, and M. Rogers. 2001. Host-plant mediated effects of root herbivory on insect seed predators and their parasitoids. *Oecologia* 127:246–250.

Mitchell, C.E., D. Tilman, and J.V. Groth. 2002. Effects of grassland plant species diversity, abundance, and composition on foliar fungal disease. *Ecology* 83:1713–1726.

Mooney, H.A., and J.A. Drake. 1986. *Ecology of Biological Invasions of North America and Hawaii.* New York: Springer.

Moore, J.C., E.L. Berlow, D.C. Coleman, P.C. de Ruiter, D. Quan, A. Hastings, N.C. Johnson, et al. 2004. Detritus, trophic dynamics and biodiversity. *Ecology Letters* 7:584–600.

Moore, J.C., K. McCann, H. Setala, and P.C. de Ruiter. 2003. Top-down is bottom-up: Does predation in the rhizosphere regulate aboveground dynamics? *Ecology* 84:846–857.

Ness, J.H., J.L. Bronstein, A.N. Andersen, and J.N. Holland. 2004. Ant body size predicts dispersal distance of ant-adapted seeds: Implications of small-ant invasions. *Ecology* 85:1244–1250.

Panikov, N.S. 1999. Understanding and prediction of soil microbial community dynamics under global change. *Applied Soil Ecology* 11:161–176.

Pastor, J., B. Dewey, R.J. Naiman, P.F. McInnes, and Y. Cohen. 1993. Moose browsing and soil fertility in the boreal forests of Isle Royale National Park. *Ecology* 74:467–480.

Peltzer, D.A., R.B. Allen, G.M. Lovett, D. Whitehead, and D.A. Wardle. 2010. Effects of biological invasions on forest carbon sequestration. *Global Change Biology* 16:732–746.

Porazinska, D.L., R.D. Bardgett, M.B. Blaauw, H.W. Hunt, A.N. Parsons, T.R. Seastedt, and D.H. Wall. 2003. Relationships at the aboveground-belowground interface: Plants, soil biota and soil processes. *Ecological Monographs* 73:377–395.

Prater, M.R., and E.H. DeLucia. 2006. Non-native grasses alter evapotranspiration and energy balance in Great Basin sagebrush communities. *Agricultural and Forest Meteorology* 139:154–163.

Raizada, P., A.S. Raghubanshi, and J.S. Singh. 2008. Invasive plant species impact on soil processes. *Proceedings of the National Academy of Sciences, India* 78:288–298.

Rao, A., and S.B. Vinson. 2004. Ability of resident ants to destruct small colonies of *Solenopsis invicta* (Hymenoptera: Formicidae). *Environmental Entomology* 33:587–598.

Reinhart, K.O., A. Packer, W.H. van der Putten, and K. Clay. 2003. Plant-soil biota interactions and spatial distribution of black cherry in its native and invasive ranges. *Ecological Letters* 6:1046–1050.

Rosenzweig, M.L. 1995. *Species Diversity in Space and Time.* Cambridge, UK: University Press.

Ruiter De, P.C., A.M. Neutel, and J.C. Moore. 1995. Energetics, patterns of interaction strengths, and stability in real ecosystems. *Science* 269:1257–1260.

Saetre, P., and E. Baath. 2000. Spatial variation and pattern of soil microbial community structure in a mixed spruce-birch stand. *Soil Biology and Biochemistry* 32:909–917.

Sanders, N.J., N.J. Gotelli, N.E. Heller, and D.M. Gordon. 2003. Community disassembly by an invasive species. *Proceedings of Natural Academy of Sciences, U.S.A.* 100:2474–2477.

Scheu, S., and H. Setala. 2002. Multitrophic interactions in decomposer food webs. In *Multitrophic Level Interactions*, eds. T. Tscharntke, and B. Hawkins, pp. 223–264. Cambridge, UK: Cambridge University Press.

Seaman, R.E., and P.C. Marino. 2003. Influence of mound building and selective seed predation by the red imported fire ant (*Solenopsis invicta*) on an old-field plant assemblage. *Journal of the Torrey Botanical Society* 130:193–201.

Sharma, G.P., S.A. Muhl, S.J. Milton, and K.J. Esler. 2010. Nitrogen enrichment and competitive interaction between *Avena fatua* and herbaceous indigenous plants of South African Renostervelds. *Biological Invasions* 12:3371–3378.

Sharma, G.P., and A.S. Raghubanshi. 2009. Lantana invasion alters the soil pools and processes: A case study with special reference to nitrogen dynamics in the tropical dry deciduous forest of India. *Applied Soil Ecology* 42:134–140.

Sharma, G.P., J.S. Singh, and A.S. Raghubanshi. 2005. Plant invasions: Emerging trends and future implications. *Current Science* 85:726–734.

Singh, H. 1993. Effect of crop residue management on microbial biomass accumulation in the soil. *Current Science* 65:487–488.

Singh, S.P., Y.S. Rawat, and Y.D. Bhatt. 2005. *Lantana camara* in Kumaun Himalaya: Invasive trait and ecosystem properties. In *Proceedings of the workshop on invasive alien species in India*, eds. L.C. Rai, and J.P. Gaur, pp. 150–155. Varanasi, India: Banaras Hindu University.

Smith, S.E., and D.J. Read. 1997. *Mycorrhizal Symbiosis*. London: Academic Press.

Sperry, L.J., J. Belnap, and R.D. Evans. 2006. *Bromus tectorum* invasion alters nitrogen dynamics in an undisturbed arid grassland ecosystem. *Ecology* 87: 603–615.

Stark, J.M., and M.K. Firestone. 1996. Kinetic characteristics of ammonium-oxidizer communities in a California oak woodland annual grassland. *Soil Biology and Biochemistry* 28:1307–1317.

Strickland, M.S., J.L. Devore, J.C. Maerz, and M.A. Bradford. 2010. Grass invasion of a hardwood forest is associated with declines in belowground carbon pools. *Global Change Biology* 16:1338–1350.

Swift, M.J., O.W. Heal, and J.M. Anderson. 1979. *Decomposition in Terrestrial Ecosystems*. Oxford, UK: Blackwell.

Tilman, D. 1982. *Resource Competition and Community Structure*. Princeton, New Jersey: Princeton University Press.

Tilman, D., and S. Pacala. 1993. The maintenance of species richness in plant communities. In *Species diversity in ecological communities: Historical and geographical perspectives*, eds. R.E. Richlefs, and D. Schluter, pp. 13–25. Illinois: University of Chicago Press.

van der Heijden, M.G.A., J.N. Klironomos, M. Ursic, P. Moutoglis, R. Streitwolf-Engel, T. Boller, A. Wiemken, and I.R. Sanders. 1998. Mycorrhizal fungal diversity determines plant biodiversity, ecosystem variability and productivity. *Nature* 396:69–72.

van der Putten, W.H. 2003. Plant defense belowground and spatiotemporal processes in natural vegetation. *Ecology* 84:2269–2280.

van der Putten, W.H., C. Ban Dijk, and B.A.M. Peters. 1993. Plant-specific soil-borne diseases contribute to succession in foredune vegetation. *Nature* 362:53–56.

van der Putten, W.H., R.D. Bardgett, P.C. de Ruiter, W.H. Hol, K.M. Meyer, T.M. Bezemer, M.A. Bradford, et al. 2009. Empirical and theoretical challenges in aboveground-belowground ecology. *Oecologia* 161:1–14.

van der Putten, W.H., L.E.M. Vet, J. Harvey, and F. Wackers. 2001. Linking above- and belowground multitrophic interactions of plants, herbivores, pathogens, and their antagonists. *Trees* 16:547–554.

Verma, P., and R. Sagar. 2011. Effects of N-application and resource heterogeneity on the herbaceous species diversity in permanent experimental plots established in the campus of Banaras Hindu University. *Environmentalist* Varanasi, India. In press.

Virginia, R.A., W.M. Jarrell, W.G. Whitford, and D.W. Freckman. 1992. Soil biota and soil properties in the surface rooting zone of mesquite (*Prosopis glandulosa*) in historical and recently desertified Chihuahuan desert habitats. *Biology and Fertility of Soils* 14:90–98.

Vitousek, P.M., C.M. D'Antonio, L.L. Loope, and R. Westbrooks. 1996. Biological invasions as global environmental change. *American Scientist* 84:218–228.

Vitousek, P.M., H.A. Mooney, J. Lubchenco, and J.M. Melillo. 1997. Human domination of Earth's ecosystem. *Science* 277:494–499.

Vitousek, P.M., and L.P. Walker. 1989. Biological invasions by *Myrica faya* in Hawaii: Plant demography, nitrogen fixation and ecosystem effects. *Ecological Monographs* 59:247–265.

Vivrette, N.J., and C.H. Muller. 1977. Mechanism of invasion and dominance of coastal grassland by *Mesembryanthemum crystallinum*. *Ecological Monographs* 47:301–318.

Wallner, W.E. 1996. Invasive pests ('biological pollutants') and US forests: Whose problem, who pays? *EPPO Bulletin* 26:167–180.

Wardle, D.A. 2002. *Communities and Ecosystems: Linking the Aboveground and Belowground Components*. Princeton, New Jersey: Princeton University Press.

Wardle, D.A., and R.D. Bardgett. 2004. Human-induced changes in densities of large herbivorous mammals: Consequences for the decomposer subsystem. *Frontiers in Ecology and the Environment* 2:145–153.

Wardle, D.A., R.D. Bardgett, J.N. Klironomos, H. Setala, W.H. van der Putten, and D.H. Wall. 2004. Ecological linkages between aboveground and belowground biota. *Science* 304:1629–1633.

Wardle, D.A., G.M. Barker, G.W. Yeates, K.I. Bonner, and A. Ghani. 2001. Introduced browsing mammals in New Zealand natural forests: Aboveground and belowground consequences. *Ecological Monographs* 71:587–614.

Wardle, D.A., K.E. Giller, and G.M. Barker. 1999. The regulation and functional significance of soil biodiversity in agroecosystems. In *Agrobiodiversity: Characterization, Utilization and Management*, eds. D. Wood, and J.M. Lenne, pp. 87–121. Bristol, UK: CABI Publishing.

Wardle, D.A., K.S. Nicholson, M. Ahmed, and A. Rahman. 1994. Interference effects of the invasive plant *Carduus nutans* L. against the nitrogen fixation ability of *Trifolium repens* L. *Plant and Soil* 163:287–297.

Weste, G., K. Brown, J. Kennedy, and T. Walshe. 2002. *Phytophthora cinnamomi* infestation— a 24-year study of vegetation change in forests and woodlands of the Grampians, Western Victoria. *Australian Journal of Botany* 50:247–274.

Windham, L., and J.G. Ehrenfeld. 2003. Net impact of a plant invasion on nitrogen-cycling processes within a brakish tidal marsh. *Ecological Applications* 13:883–896.

Yates, D.N., T.G.F. Kittel, and R.F. Cannon. 2000. Comparing the correlative Holdridge model to mechanistic biogeographical models for assessing vegetation distribution response to climate change. *Climate Change* 44:59–87.

Zak, D.R., D. Tilman, R.R. Parmenter, C.W. Rice, F.M. Fisher, J. Vose, D. Milchunas, and W.W. Martin. 1994. Plant production and soil microorganisms in late-successional ecosystems: A continental study. *Ecology* 75:2333–2347.

Zavaleta, E., M.R. Shaw, N.R. Chiariello, H.A. Mooney, and C.B. Field. 2003. Additive effects of simulated climate changes, elevated CO2, and nitrogen deposition on grassland diversity. *Proceedings of Natural Academy of Sciences, U.S.A.* 100:7650–7654.

Zhang, W., P.F. Hendrix, B.A. Snyder, M. Molina, J. Li, X. Rao, E. Siemann, and S. Fu. 2010. Dietary flexibility aids Asian earthworm invasion in North American forests. *Ecology* 91:2070–2079.

chapter six

From species coexistence to genotype coexistence

What can we learn from invasive plants?

Alexandra Robin Collins and Jane Molofsky

Contents

6.1 Introduction

A central question in plant ecology is how diversity is maintained. A number of theories have been proposed to explain coexistence including niche partitioning (Tilman 1982), lottery models (Chesson and Warner 1981), feedbacks (Antonovics and Kareiva 1988, Molofsky et al. 2001, 2002, Molofsky and Bever 2002), and neutral theory (Hubbell 2001) (Table 6.1). However, empirical and theoretical work examining coexistence has often regarded species as single taxonomic units despite the increasing recognition that a fundamental level of variation exists at the gene level (Harper 1977, Antonovics 2003). Genotypic diversity unmasks a whole new level of variation within species, but our understanding of how this diversity is maintained and the implications of these interactions remains poor (Fridley et al. 2007). A number of parallels exist between species diversity and genetic diversity (Van Valen 1965, Kassen 2002, Chase and Knight 2003, Vellend 2005, Vellend and Geber 2005, Johnson and Stinchcombe 2007), and many of the same mechanisms that maintain species diversity can be applied at the genotype level (Antonovics 2003) (Table 6.1). Furthermore, genotypic diversity is expected to have its biggest effect when species are habitat providing or competitively dominant (Whitham et al. 2006, Hughes et al. 2008); therefore, invasive species serve as prime experimental tools to study and evaluate the relative roles of different types of genotype interactions in determining community structure and composition.

Table 6.1 Four Different Theories or Models Commonly Used to Explain Species Coexistence and How They Can Be Extended to the Genotype Level

Theory or model influencing species/genotype coexistence	Description with regard to species diversity	Description with regard to genotypic diversity
Niche partitioning	• Species diversity is maintained if different species have different resources that are most limiting • Diversity will be maintained if the intensity of intraspecific competition for resources is greater than that of interspecific competition	• Genotypic diversity is maintained if different genotypes have different resources that are most limiting • Diversity will be maintained if the intensity of intragenotypic competition for resources is greater than that of intergenotypic competition
Lottery models	• Species diversity is maintained through random allocation of space • Assumes that communities are saturated and that species occupy space on a first-come, first-served basis	• Genotypic diversity is maintained through the random allocation of space • Assumes that the population is saturated and that genotypes occupy space on a first-come, first-served basis
Feedbacks (with respect to frequency and density)	• Frequency: Species diversity is maintained when a species' performance is dependent on its local abundance in the population relative to other species • Density: Species diversity is maintained when a species' performance is dependent on its density in the population	• Frequency: Genotypic diversity is maintained when a genotype's performance is dependent on its local abundance in the population relative to other genotypes of the same species • Density: Genotypic diversity is maintained when a genotype's performance is dependent on its density in the population
Neutral theory	• Nonequilibrium theory that predicts that species are functionally equivalent and that species diversity is maintained through stochastic extinction and dispersal	• Nonequilibrium theory that predicts that genotypes are functionally equivalent and that genotypic diversity is maintained through stochastic extinction and dispersal

In this chapter, we first briefly summarize what is known about the genetic variations of invasive species and how small-scale genotype interactions could be important for invasion dynamics. Second, we outline how different theories of coexistence that are typically applied at the species level can be applied at the genotype level and potentially serve as a means of evaluating invasion risk. Third, we use the invasive grass *Phalaris arundinacea* as a case study to explore how intraspecific feedbacks and/or neutral processes may structure diversity at the genotype level and influence invasion success.

6.2   Genetic variation of invasive plant populations

Numerous studies have compared the molecular genetic variation of invasive species with that of their conspecifics in the native range (Merila et al. 1996, Lee et al. 2004, Maron et al. 2004, Genton et al. 2005, Williams et al. 2005, Meimberg et al. 2006, Kang et al. 2007, Lavergne and Molofsky 2007, Taylor and Keller 2007, Hufbauer and Sforza 2008). Within invasive plant populations, it was previously thought that they would exhibit a reduction in genetic diversity during founding events; however, despite numerous studies that support this prediction (see the review by Dlugosch and Parker [2008]), several other studies have found that invasive populations have similar, if not more, nuclear genetic diversity compared to native populations (Molina-Freaner and Jain 1992, Lavergne and Molofsky 2007).

A common mechanism to explain the maintenance and/or increase of genetic diversity in introduced ranges is multiple introductions from genetically differentiated locations (Ellstrand and Schierenbeck 2000, Novak and Mack 2001, Kolbe et al. 2004, Bossdorf et al. 2005, Lockwood et al. 2005, Lavergne and Molofsky 2007). Here, repeated introductions may both increase the chance of introducing preadapted genotypes (Le Roux et al. 2008) and bring together genotypes from disparate regions, which may create novel genotype combinations (Lavergne and Molofsky 2007). Greater genetic diversity for invasive populations can be advantageous for two reasons: (1) Greater genetic diversity may lead to greater phenotypic variation and adaptability, and (2) greater overall heterozygosity may increase population fitness and reduce genetic bottlenecks that can limit the adaptive evolution of fitness-related traits (Reed and Frankham 2003, Dlugosch and Parker 2008). Therefore, invasive populations with greater genetic diversity may be at an evolutionary advantage, and there are a number of examples of invasive species that have evolved adaptive traits in their introduced ranges (Thompson 1998, Blair and Wolfe 2004, Rogers and Siemann 2004).

The maintenance of genetic variation at small spatial scales may also have implications for invasion dynamics (see the review by Ward [2006]). At these scales, the main determinant of spatial genetic structure is predicted to be restricted gene flow, with microhabitat selection being a possible secondary factor (Chung and Epperson 1999, Vekemans and Hardy 2004). The extent to which a plant population exhibits small-scale spatial genetic structure is primarily determined by seed dispersal and the mating system, with the size of the spatial structure being determined by seed dispersal distance and plant density (Dyer 2007, Pardini and Hamrick 2008). Although small-scale genetic structure has been examined for a number of herbaceous species (Jacquemyn et al. 2005, Premoli and Kitzberger 2005, Jones et al. 2006, Troupin et al. 2006, Williams et al. 2007), studies examining the small-scale genetic structure of invasive species are few (Roux et al. 2010). The latter will be important for understanding the breeding behavior, recruitment history (Premoli and Kitzberger 2005, Jones et al. 2006, Williams et al. 2007, Pardini and Hamrick 2008), and small-scale genotypic structure of invasive populations. Moreover, the intraspecific genetic diversity of plant populations is an important factor shaping the diversity and structure of communities (Whitham et al. 2003) and, thus, is also important for assessing the impact of invasive species on higher trophic levels.

When the predictable effects of genes within a species have extended consequences for communities or ecosystems, these are referred to as extended phenotypes (Whitham et al. 2006). For example, crosses of *Populus angustifolia* and *P. fremontii* identified a single quantitative trait locus (QTL) that was correlated with the phenotypic variation in condensed tannins (Whitham et al. 2003). Furthermore, variability in condensed-tannin levels produced significant community- and ecosystem-level phenotypes. The variation in condensed-tannin levels was associated with different terrestrial arthropod communities in tree canopies (Whitham et al. 2006) and was negatively associated with nitrogen mineralization rates (Schweitzer et al. 2004).

Even when precise QTLs are not determined, genotypic diversity alone has been identified as an important factor determining community productivity. Here, a greater number of genotypes may increase the chances of including phenotypes that are favorable to environmental conditions or allow genotypes to share resources (niche complementarity) and produce more biomass. Crutsinger et al. (2006) found that increasing the genotypic diversity of *Solidago altissima* populations led to increased predator richness, herbivore richness, and aboveground net primary productivity. In this chapter, we discuss how changing population genotypic diversity can influence population productivity.

For an invasive species, greater genotypic diversity may be an important factor allowing the species to outcompete native vegetation and increase its net primary productivity; however, studies examining how genotypic diversity influences invasion success remain limited (but see the studies by Collins et al. [2010] and Vellend et al. [2010]). While examining how genotypic diversity influenced the productivity of *Taraxacum officinale*, Vellend et al. (2010) found that genotypic diversity did not increase productivity but rather genotype identity was more important for plant performance. Thus, genotypic diversity may still be important in order to increase the chances of including genotypes that are favorable to invasion conditions.

Understanding the nature of intraspecific interactions within an invasive plant at small spatial scales will allow us to determine how genotypic diversity influences invasion success. Elton (1958) first hypothesized that the susceptibility of a community to invasion may be controlled by the composition and diversity of its resident species (diversity–invasibility hypothesis). A number of empirical tests of the diversity–invasibility relationship have supported the hypothesis (Tilman 1997, Knops et al. 1999, Naeem et al. 2000, Dukes 2002, Kennedy et al. 2002, Fargione et al. 2003, Fargione and Tilman 2005), rejected the hypothesis (Lonsdale 1999, Stohlgren et al. 1999, McKinney 2001, Pyšek et al. 2002, Stohlgren et al. 2003), or found no relationship (Crawley et al. 1999, Collins et al. 2007) between species diversity and invasibility. Similarly, increased genotypic diversity of the invader may also act to increase invasion success via similar mechanisms (Vellend et al. 2010). For example, if we consider genotypes interacting over small spatial scales, invasion potential may increase if plants exhibit increased intragenotypic competition due to niche complementarity. In other words, genotypes compete more against identical genotypes than they do against different genotypes. The reduction of competitive interactions among different genotypes due to differences in resource use could allow invasive genotypes to grow more densely, produce more biomass, and increase their invasion potential. In contrast, invasion potential could also increase if plants exhibit reduced intragenotypic competition for key fitness traits. In other words, genotypes have greater fitness when surrounded by identical genotype neighbors. For example, reduced intragenotypic competition may help overcome potential Allee effects (Allee 1931). An Allee effect occurs when an individual's ability to survive and reproduce is influenced by population density (Stephens et al. 1999) and is a step function whereby below a critical threshold a population will not establish itself and above this threshold the population will experience density-dependent growth. For invasive species, low density at the edge of an initial population can lead to Allee effects that can ultimately slow the rate of spread, increase lag times, and decrease the probability of establishment (Taylor and Hastings 2005).

Both the aforementioned scenarios (increased intragenotypic competition and reduced intragenotypic competition) assume that competitive differences among genotypes are important for determining invasion outcomes. However, an alternative scenario is that populations are structured stochastically and genotypes are functionally equivalent (or neutral). In this case, genotypes may interact, but because they are functionally equivalent their interactions are identical irrespective of genotype composition and identity. The debate as to whether stochastic or deterministic processes maintain diversity has been fiercely discussed for well over a decade and empirical tests have been equivocal. For invasive species, if genotypes are functionally equivalent, then genotypic diversity may be less important for invasion success but may still be maintained as long as there is a constant source of incoming propagules.

6.3 Applying theories of species coexistence to the genotype level

To date, theories that explain species coexistence have in general fallen into two categories: (1) Models that explain the distribution and abundance of species on the basis of niche differentiation (Tilman 1982), and (2) models of community dynamics where coexistence is unstable and diversity is maintained by ecological drift balanced by migration (Bell 2001, Hubbell 2001). For simplicity, we can think of these two categories as including mechanisms of coexistence where genotypes are considered functionally different (e.g., they compete for different resources) (category 1) or where genotypes are functionally equivalent (neutral) (category 2). In Sections 6.3.1 and 6.3.2, we outline two specific examples of how these two categories of models can also be applied at the genotype level and how doing so may help us to better understand invasion.

6.3.1 Category 1: Feedbacks

When neighbor identity and proximity influence performance, such that genotypes and/or species are not functionally equivalent, both frequency dependence and density dependence can be potential outcomes responsible for the maintenance of diversity (Chesson 2000). Here, we focus primarily on frequency dependence. We define frequency dependence as when a species performance is dependent on its local abundance in the population. Frequency dependence can be either positive or negative, where negative frequency dependence maintains diversity if different species and/or genotypes have different limiting resources (Harpole and Suding 2007) and positive frequency dependence maintains diversity through the formation of semistable species or genotype clusters (Molofsky and Bever 2002). Therefore, within a species, increased performance of a genotype when surrounded by same-genotype neighbors demonstrates positive frequency dependence (or reduced intragenotypic competition) and increased performance of a genotype when surrounded by different-genotype neighbors demonstrates negative frequency dependence (or increased intragenotypic competition). A number of different mechanisms function to promote positive and negative frequency-dependent interactions. Negative frequency dependence has been shown to occur through competition and niche partitioning (Antonovics and Kareiva 1988), parasitism (May and Anderson 1983), or mutualism through an asymmetry that is beneficial to mutualists (Bever 1999), whereas positive frequency dependence has been shown to occur by preferential predation on rare individuals (Futuyma and Wasserman 1980) or host specificity among mutualists (Bever 1999) through, for example, increased mycorrhizal hyphal connections among like genotypes (Ronsheim and Anderson 2001).

Intraspecific frequency-dependent feedbacks have been recognized as important mechanisms maintaining genotype diversity (Antonovics and Ellstrand 1984, Aarssen and Turkington 1985, Antonovics and Kareiva 1988, Ronsheim 1996, Bennington and Stratton 1998). For invasive species, genotypic diversity of the resident community can be an important factor buffering further invasion (Weltzin et al. 2003, Crutsinger et al. 2008). Crutsinger et al. (2008) found that the biomass of invading species was negatively correlated with the genotypic diversity of *S. altissima* and that plots with greater genotypic diversity have significantly greater stem density. Although it has received less attention, increased genotypic diversity of the invader may also act to increase invasion success through similar mechanisms and by favoring individual genotypes that are adapted to the invasion conditions (Vellend et al. 2010, Collins et al. 2010).

6.3.2 *Category 2: Neutral theory*

The neutral model draws from both island biogeography theory (MacArthur and Wilson 1963, 1967) and neutral theory in population genetics (Kimura 1983). Neutral theory assumes that communities comprise functionally equivalent individuals (equal fitness) and diversity is maintained through stochastic extinction and dispersal. Neutral theory explains community diversity as being limited by dispersal with dispersal being the key determinant defining community assembly. Although neutral theory is typically used to explain coexistence among species, if different genotypes have equal fitness then genotypic diversity can also be maintained due to ecological drift balanced by migration.

Neutral theory has been criticized for the unrealistic assumption of equal average fitness among species (Chesson 2000) and for how well it actually describes the patterns and properties of community assembly (Whitfield 2002, Gaston and Chown 2005). Despite these criticisms, neutral theory has been shown to be able to successfully predict community diversity in a number of different habitats (Bell 2001, Hubbell 2001, Volkov et al. 2003, He 2005). For studies that do not support neutral theory, results are often intermediate between purely stochastic and purely deterministic, and there is an increasing attempt to reconcile neutral and niche-based models into a comprehensive framework (Uriarte and Reeve 2003, Chave 2004, Gravel et al. 2006, Adler et al. 2007). For example, using a model of recruitment probabilities that combined both neutral and niche-based mechanisms, Gravel et al. (2006) presented coexistence as a continuum hypothesis where both neutral and niche-based processes can act simultaneously rather than at stochastic and competitive extremes of the spectrum.

For invasive species, several patterns such as the relationship between native and exotic species richness and the relationship between propagule release and invasion success have been shown to follow the assumptions of the neutral model (Daleo et al. 2009). However, other patterns such as how invasiveness is determined by species traits significantly contrast with the assumptions of the neutral theory. Thus, invasion success may be contingent on both neutral and nonneutral processes and the extent to which one category influences invasion over the other may serve as an important tool for determining appropriate management plans.

6.4   Case study: Exploring mechanisms of coexistence using genotypes of an invasive grass

In an attempt to explore the two categories of coexistence models outlined in Sections 6.3.1 and 6.3.2 and to discuss their implications for invasion success, we present the results of two experiments using the invasive grass *P. arundinacea* (Poaceae). *P. arundinacea* or reed canary grass is native to temperate zones of the Northern Hemisphere and is widely distributed throughout Eurasia (Lavergne and Molofsky 2007). Growing 1–2 m tall in dense stands, *P. arundinacea* is a C_3 perennial grass with an extensive underground rhizome network (Lavergne and Molofsky 2006). *P. arundinacea* has a high annual seed set as well as a high rate of outcrossing due to self-sterility (Ostrem 1988); it can also reproduce asexually by producing clonal tillers. Germination usually requires moist conditions (Lindig-Cisneros and Zedler 2001, 2002); however, seeds can have long dormancy periods and can therefore constitute a large component of the seed bank (Vose 1962). Tillers often form dense underground networks that can outcompete native vegetation.

Disturbance favors *P. arundinacea* invasion in areas where flood intensity and addition of sediments are high (Kercher and Zedler 2004) and, therefore, invasion by *P. arundinacea* is often found in highly disturbed areas such as stream banks subject to agricultural runoff, roadsides, and wetlands. Due to *P. arundinacea*'s aggressive growth and extensive belowground rhizome network, eradication and control strategies can be difficult. Currently, integrated management, consisting of a combination of physical, chemical, and biological methods, is considered to be the best means of control for *P. arundinacea* (Lavergne and Molofsky 2006).

P. arundinacea strains were originally introduced to the United States from Europe shortly after 1850 and have since spread throughout North America (Merigliano and Lesica 1998, Galatowitsch et al. 1999) (Figure 6.1). It has been proposed that European strains that were planted for cattle forage and stream bank erosion control likely hybridized with native genotypes to create more aggressive clones. *P. arundinacea* has been introduced multiple times to North America, which may lend to its increased invasibility (Lavergne and Molofsky 2007). Purposes of introduction include the following: *P. arundinacea* has

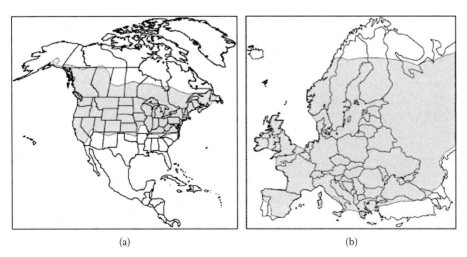

(a)   (b)

Figure 6.1 Current distribution map of *Phalaris arundinacea* in (a) North America and (b) Europe. (From Lavergne, S., and J. Molofsky, *Crit. Rev. Plant Sci.*, 23, 415–429, 2004.)

been introduced as a pure or a mixed forage crop (Ostrem 1988, Sheaffer and Marten 1992, Buxton et al. 1998), a restoration tool for degraded soils and waters (Vymazal 2001), and a number of bioenergy crops (Lewandowski et al. 2003).

   P. arundinacea is a good study species for experiments examining genotypic differences because individuals can be easily genotyped through allozyme screening and rapidly cloned through repeated vegetative tillering. Previous collections of genotypes from both native and invasive ranges of *P. arundinacea* have found that genotypic differences among plants translate into differences in physiological and morphological characteristics (Morrison and Molofsky 1999, Lavergne and Molofsky 2007, Brodersen et al. 2008, Lavergne et al. 2010). Furthermore, differences in these characteristics result in differences in competitive ability and survival (Morrison and Molofsky 1998, 1999, Lavergne and Molofsky 2007). For example, Morrison and Molofsky (1999) grew three genotypes of *P. arundinacea* under 10 artificial environments and found that two genotypes responded differently to competitors: One allocated more biomass to roots when competitors were sparse, whereas one allocated more biomass to roots when neighborhood competition was more intense. The differential response of genotypes to their local neighborhood may in turn facilitate invasion into new habitats.

   Invasive *P. arundinacea* populations have higher genetic diversity and heritable phenotypic variation when compared with native populations (Lavergne and Molofsky 2007). Multiple introduction events from the native range have been proposed to have led to recombination events between native genotypes from disparate regions in the invasive range. These novel genotypes have greater colonization ability and phenotypic plasticity. Furthermore, because invasive populations have greater heritable variation than native populations, this may have led to the rapid selection of genotypes with increased invasion potential (Lavergne and Molofsky 2007).

6.5   Experiment 1: Feedbacks

To understand if and/or how invasive genotypes interact at small spatial scales, we present data from an experiment conducted by Collins et al. (2010) that examined the variability of individual genotype responses to neighbor identity for multiple growth measures using genotypes of *P. arundinacea*. Collins et al. (2010) experimentally manipulated the intraspecific frequency of genotypes using hexagonal arrays with one "target" plant in the center of the array surrounded by six neighbor plants to test for competition-mediated frequency-dependent feedbacks. A feedback simply refers to a sequence of interactions that determine either a positive or a negative effect on a system. The experiment consisted of three treatments: (1) The target plant with no neighbors, (2) the target plant surrounded by six same-genotype neighbors, and (3) the target plant surrounded by six different-genotype neighbors, with each of the six neighbor genotypes representing a different genotype. Thus, increased performance of the same-neighbor treatment provides support for positive frequency-dependent feedback and increased performance of the different-neighbor treatment provides support for negative frequency dependent feedback. The experiment used seven unique genotypes; each of the seven genotypes served as the target plant for the three treatments and was replicated three times for a total of 63 plots.

   Interestingly, the authors (Collins et al. 2010) found that genotypes experienced significant positive frequency-dependent feedbacks for several important growth traits including stem height and aboveground biomass (Figure 6.2). In other words, genotypes exhibited

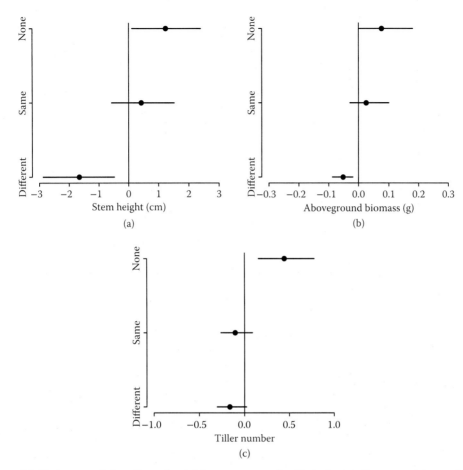

Figure 6.2 Estimates of the effect of neighbor treatment for *Phalaris arundinacea* on (a) stem height, (b) aboveground biomass, and (c) tiller number: "Same" refers to the same-neighbor treatment, "Different" refers to the different-neighbor treatment, and "None" refers to the no-neighbor treatment. Each estimate is the effect relative to the grand mean of the trait. Thin lines represent 95% credible intervals, and the center point is the median of the posterior distribution. The significance of an effect is determined by whether the 95% credible interval bounds 0; if an effect bounds 0, it is not considered significant. (From Collins et al., *Oecol.*, 164, 959–969, 2010.)

greater performance when planted next to identical genotype clones. Furthermore, when the authors examined the extent of frequency dependence for the seven unique genotypes used in the experiment, they found that the extent of frequency dependence was not consistent. Genotypes experienced a range of frequency-dependent responses with some genotypes showing evidence of negative frequency-dependent feedbacks and others showing evidence of positive frequency-dependent feedbacks (although only the genotypes with positive frequency-dependent feedbacks were statistically significant). The authors concluded that positive feedbacks may be important for overcoming potential Allee effects and that a range of frequency-dependent responses may increase the likelihood of including genotypes that are favorable to the invasion conditions and allow genotypes to quickly adapt to and invade a range of novel environments.

6.6   Experiment 2: Neutral processes

To examine the second category of coexistence models, we present an experiment by Collins and Molofsky (2010) that evaluated which processes, stochastic versus deterministic, were stronger for maintaining the coexistence of *P. arundinacea* genotypes. Our objectives were twofold: (1) Determine whether genotypes exhibit equal fitness when grown alone to determine if genotypes could be considered functionally equivalent, and (2) determine rank abundance curves for genotypes when grown alone and under mixed conditions to elucidate the primary processes driving coexistence (competitive interactions vs. stochastic drift). Rank abundance curves are a common way for ecologists to display the relative abundance of species and/or genotypes in a community. The *x* axis represents abundance rank, where the most abundant species and/or genotype is given a rank of 1, the second most abundant is given a rank of 2, and so on, and the *y* axis represents relative abundance. Relative abundance is measured as the abundance of a species and/or a genotype relative to the abundance of all other species in a plot.

We used eight unique genotypes of the invasive grass *P. arundinacea*, and the genotypes were planted both alone and in plots consisting of one-, two-, four-, or eight-genotype mixtures (although we examine only the one- and eight-genotype plots here). Density was kept constant at eight tillers per plot, and the experiment consisted of three blocks with each one-genotype treatment replicated eight times per block (a total of 24 plots) and each eight-genotype plot replicated four times per block (a total of 12 plots). The experiment was established at the Biological Research Complex (BRC), Burlington, Vermont (44°27′N; 73°11′W), and all plots were removed of native vegetation prior to planting and were weeded every two weeks. We predicted that if coexistence was determined by deterministic processes, genotypes would have functional differences when grown alone and a predictable competitive hierarchy of genotypes when planted in mixtures. In contrast, we predicted that if stochastic processes were responsible for coexistence, genotypes would exhibit functional equivalence when grown alone and an unpredictable competitive hierarchy when grown in mixtures.

When genotypes were planted alone, there was no significant effect of genotype on total individual biomass ($F = 1.07$, $p = 0.41$) (Figure 6.3). Therefore, in the absence of

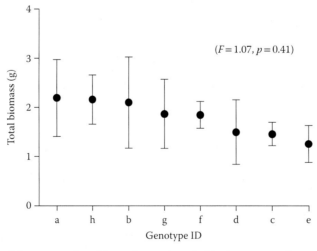

Figure 6.3 Mean total biomass produced by each genotype (a–h) when grown alone under disturbed conditions: All genotypes are ranked according to the total biomass produced over the course of one field season. Inset F-statistic is the effect of genotype in the model and error bars represent +/− 1 standard error.

competitors, genotypes were functionally equivalent. In the eight-genotype mixtures, the rank abundance curves for the proportion of total biomass produced by each genotype for each of the 12 replicate plots showed that in each plot several genotypes always outperformed the others; however, the competitive hierarchy of genotypes was not consistent across plots. The rank abundance curves for the mean proportion of total biomass produced by each genotype across all eight-genotype mixture plots had no significant effect of genotype under disturbed conditions ($F = 0.58$, $p = 0.77$) (Figure 6.4).

When genotypes were planted in monocultures, we found a similar pattern for the rank abundance curves, where certain individuals grew well and others grew poorly or died before the end of the experiment. However, the total biomass produced by each genotype in monoculture did not differ significantly ($F = 0.03$, $p = .86$). Furthermore, the total biomass produced by one-genotype plots did not significantly differ from the total biomass produced by eight-genotype plots, indicating that greater genotypic diversity did not increase *P. arundinacea* productivity.

Interestingly, we showed that genotypes of *P. arundinacea* exhibit ecological equivalence when grown alone and that equal fitness may lead to stochastic processes determining the abundance of genotypes when grown in mixtures (ecological drift). We found that no one genotype consistently outperformed the others. If it is assumed that immigration exists from local and/or regional genotype pools, then stochastic processes may maintain diversity at small scales. We also validated our assumption of ecological equivalence by showing similar genotype rank abundance curves in monocultures and no significant difference in productivity between mixture and monoculture plots. This is interesting, as complementarity among genotypes has been shown to increase plant productivity (Crutsinger et al. 2008) and ecosystem processes (Crutsinger et al. 2006) for native species. Similarly, our results contrasted those of Whitlock et al. (2007) who found that the rank abundance curves of genotypes for several different species showed evidence for deterministic processes such that certain genotypes consistently outperformed others across homogenous environmental conditions (Whitlock et al. 2007).

Previous work with *P. arundinacea* has found that the genotypic neighborhood in which a plant is growing significantly influences plant performance (Collins et al. 2010).

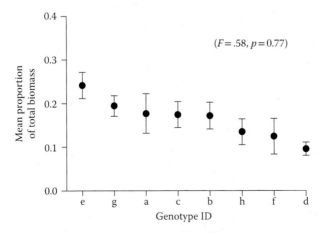

Figure 6.4 Mean proportion of the total biomass produced by each genotype (a–h) when grown in eight-genotype mixtures under disturbed conditions: Inset F-statistic is the effect of genotype in the model and error bars represent +/− 1 standard error.

Here, however, eight genotypes were introduced to a plot and the most abundant geno-type was determined stochastically. We cannot, however, discount that productivity dif-ferences between plots may be explained by how specific genotypes interact with their surrounding neighbors and/or microenvironment. For example, the eight genotypes that were used in our experiments were collected from three different populations and, there-fore, the stochastic processes we observed may be due to the lack of coevolutionary his-tory among them. If genotypes share a common ancestry, then complementarity among genotypes may be more important. However, for invasive plant species, propagules that are released may not share a coevolutionary history.

These complicated relationships may be difficult to specify deterministically, and the inability to quantify these higher-order interactions has been considered a fault of sto-chastic models (Clark 2009). However, in an attempt to reconcile neutral and niche-based processes, many models have assumed that recruitment is a lottery function of differences in competitive ability among species and/or genotypes (niche processes) and of local and long-distance dispersal limitation (neutral processes) (Gravel et al. 2006). Therefore, geno-types introduced to a community through dispersal may be neutral. However, once intro-duced, niche differentiation exists between genotypes such that distinct relationships with environmental factors will increase or decrease abundance. For *P. arundinacea*, the initial introduction and establishment of propagules may occur at random; however, intraspecific feedbacks and differences in competitive ability may become more important over time.

6.7   Summary

The links between species diversity and genetic diversity have been well documented in the literature (see Section 6.1); however, it is only recently that empirical work has begun to unmask the interactions that exist at the genotype level. Luckily, many of the same eco-logical theories used to explain species coexistence can be applied at the genotype level. Given that genotypic diversity is predicted to be of great importance for dominant species (Hughes et al. 2008), the role of genotypic diversity in invasion success presents an exciting new direction for future research. In this chapter, we highlight the fact that interactions among genotypes can have extended consequences for communities and ecosystems; how-ever, the nature of such interactions is poorly understood. We provide two detailed examples of how well-established theories of coexistence at the species level can be applied to the genotype level using the invasive grass *P. arundinacea*. Our case study shows that determin-ing the mechanisms maintaining genotypic diversity can be complicated as genotypes do not behave in a purely stochastic or a purely deterministic manner. Instead, it is most likely that maintenance of diversity at the genotype level involves some interplay between the two.

References

Aarssen, L. W., and R. Turkington. 1985. Biotic specialization between neighboring genotypes in *Lolium perenne* and *Trifolium repens* from a permanent pasture. *Journal of Ecology* 73:605–614.

Adler, P. B., J. Hille Ris Lambers, and J. M. Levine. 2007. A niche for neutrality. *Ecology Letters* 10:95–104.

Allee, W. C. 1931. *Animal Aggregations*. Chicago, IL: University of Chicago Press.

Antonovics, J. 2003. Toward community genomics? *Ecology* 84:598–601.

Antonovics, J., and N. C. Ellstrand. 1984. Experimental studies of the evolutionary significance of sexual reproduction. 1. A test of the frequency-dependent selection hypothesis. *Evolution* 38:103–115.

Antonovics, J., and P. Kareiva. 1988. Frequency-dependent selection and competition: Empirical approaches. *Philosophical Transactions of the Royal Society of London Series B-Biological Sciences* 319:601–613.

Bell, G. 2001. Ecology—Neutral macroecology. *Science* 293:2413–2418.

Bennington, C. C., and D. A. Stratton. 1998. Field tests of density- and frequency-dependent selection in *Erigeron annuus* (Compositae) *American Journal of Botany* 85:540–545.

Bever, J. D. 1999. Dynamics within mutualism and the maintenance of diversity: Inference from a model of interguild frequency dependence. *Ecology Letters* 2:52–62.

Blair, A. C., and L. M. Wolfe. 2004. The evolution of an invasive plant: An experimental study with *Silene latifolia*. *Ecology* 85:3035–3042.

Bossdorf, O., H. Auge, L. Lafuma, W. E. Rogers, E. Siemann, and D. Prati. 2005. Phenotypic and genetic differentiation between native and introduced plant populations. *Oecologia* 144:1–11.

Brodersen, C., S. Lavergne, and J. Molofsky. 2008. Genetic variation in photosynthetic characteristics among invasive and native populations of reed canary grass (*Phalaris arundinacea*). *Biological Invasions* 10:1317–1325.

Buxton, D. R., I. C. Anderson, and A. Hallam. 1998. Intercropping sorghum into alfalfa and reed canarygrass to increase biomass yield. *Journal of Production Agriculture* 11:481–486.

Chase, J. M., and T. M. Knight. 2003. Community genetics: Toward a synthesis. *Ecology* 84:580–582.

Chave, J. 2004. Neutral theory and community ecology. *Ecology Letters* 7:241–253.

Chesson, P. 2000. Mechanisms of maintenance of species diversity. *Annual Review of Ecology and Systematics* 31:343–366.

Chesson, P. L., and R. P. Warner. 1981. Environmental variability promotes coexistence in lottery competitive systems. *The American Naturalist* 117:923–943.

Chung, M. G., and B. K. Epperson. 1999. Spatial genetic structure of clonal and sexual reproduction in populations of *Adenophora grandiflora* (Campanulaceae). *Evolution* 53:1068–1078.

Clark, J. S. 2009. Beyond neutral science. *Trends in Ecology and Evolution* 24:8–15.

Collins, A., E. M. Hart, and J. Molofsky. 2010. Differential response to frequency-dependent interactions: An experimental test using genotypes of an invasive grass. *Oecologia* 164:959–969.

Collins, A. R., S. Jose, P. Daneshgar, and C. L. Ramsey. 2007. Elton's hypothesis revisited: An experimental test using cogongrass. *Biological Invasions* 9:433–443.

Crawley, M. J., S. L. Brown, M. S. Heard, and G. R. Edwards. 1999. Invasion-resistance in experimental grassland communities: Species richness or species identity? *Ecology Letters* 2:140–148.

Crutsinger, G. M., M. D. Collins, J. A. Fordyce, Z. Gompert, C. C. Nice, and N. J. Sanders. 2006. Plant genotypic diversity predicts community structure and governs an ecosystem process. *Science* 313:966–968.

Crutsinger, G. M., L. Souza, and N. J. Sanders. 2008. Intraspecific diversity and dominant genotypes resist plant invasions. *Ecology Letters* 11:16–23.

Daleo, P., J. Alberti, and O. Iribarne. 2009. Biological invasions and the neutral theory. *Diversity and Distributions* 15:547–553.

Dlugosch, K. M., and I. M. Parker. 2008. Founding events in species invasions: Genetic variation, adaptive evolution, and the role of multiple introductions. *Molecular Ecology* 17:431–449.

Dukes, J. S. 2002. Species composition and diversity affect grassland susceptibility and response to invasion. *Ecological Applications* 12:602–617.

Dyer, R. J. 2007. Powers of discerning: Challenges to understanding dispersal processes in natural populations. *Molecular Ecology* 16:4881–4882.

Ellstrand, N. C., and K. A. Schierenbeck. 2000. Hybridization as a stimulus for the evolution of invasiveness in plants? *Proceedings of the National Academy of Sciences of the United States of America* 97:7043–7050.

Elton, C. S. 1958. *The Ecology of Invasions by Animals and Plants*. London, UK: Chapman and Hall.

Fargione, J., C. S. Brown, and D. Tilman. 2003. Community assembly and invasion: An experimental test of neutral versus niche processes. *Proceedings of the National Academy of Sciences of the United States of America* 100:8916–8920.

Fargione, J. E., and D. Tilman. 2005. Diversity decreases invasion via both sampling and complementarity effects. *Ecology Letters* 8:604–611.

Fridley, J. D., J. P. Grime, and M. Bilton. 2007. Genetic identity of interspecific neighbours mediates plant responses to competition and environmental variation in a species-rich grassland. *Journal of Ecology* 95:908–915.

Futuyma, D. J., and S. S. Wasserman. 1980. Resource concentration and herbivory in oak forests. *Science* 210:920–922.

Galatowitsch, S. M., N. O. Anderson, and P. D. Ascher. 1999. Invasiveness in wetland plants in temperate North America. *Wetlands* 19:733–755.

Gaston, K. J., and S. L. Chown. 2005. Neutrality and the niche. *Functional Ecology* 19:1–6.

Genton, B. J., J. A. Shykoff, and T. Giraud. 2005. High genetic diversity in French invasive populations of common ragweed, *Ambrosia artemisiifolia*, as a result of multiple sources of introduction. *Molecular Ecology* 14:4275–4285.

Gravel, D., C. D. Canham, M. Beaudet, and C. Messier. 2006. Reconciling niche and neutrality: The continuum hypothesis. *Ecology Letters* 9:399–409.

Harper, J. L. 1977. *Population Biology of Plants*. New York, NY: Academic Press.

Harpole, W. S., and K. N. Suding. 2007. Frequency-dependence stabilizes competitive interactions among four annual plants. *Ecology Letters* 10:1164–1169.

He, F. 2005. Deriving a neutral model of species abundance from fundamental mechanisms of population dynamics. *Functional Ecology* 19:187–193.

Hubbell, S. P. 2001. *A Unified Theory of Biodiversity and Biogeography*. Princeton, NJ: Princeton University Press.

Hufbauer, R. A., and R. Sforza. 2008. Multiple introductions of two invasive *Centaurea* taxa inferred from cpDNA haplotypes. *Diversity and Distributions* 14:252–261.

Hughes, A. R., B. D. Inouye, M. T. J. Johnson, N. Underwood, and M. Vellend. 2008. Ecological consequences of genetic diversity. *Ecology Letters* 11:609–623.

Jacquemyn, H., R. Brys, O. Honnay, M. Hermy, and I. Roldan-Ruiz. 2005. Local forest environment largely affects below-ground growth, clonal diversity and fine-scale spatial genetic structure in the temperate deciduous forest herb *Paris quadrifolia*. *Molecular Ecology* 14:4479–4488.

Johnson, M. T. J., and J. R. Stinchcombe. 2007. An emerging synthesis between community ecology and evolutionary biology. *Trends in Ecology & Evolution* 22:250–257.

Jones, F. A., J. L. Hamrick, C. J. Peterson, and E. R. Squiers. 2006. Inferring colonization history from analyses of spatial genetic structure within populations of *Pinus strobus* and *Quercus rubra*. *Molecular Ecology* 15:851–861.

Kang, M., Y. M. Buckley, and A. J. Lowe. 2007. Testing the role of genetic factors across multiple independent invasions of the shrub Scotch broom (*Cytisus scoparius*). *Molecular Ecology* 16:4662–4673.

Kassen, R. 2002. The experimental evolution of specialists, generalists, and the maintenance of diversity. *Journal of Evolutionary Biology* 15:173–190.

Kennedy, T. A., S. Naeem, K. M. Howe, J. M. H. Knops, D. Tilman, and P. Reich. 2002. Biodiversity as a barrier to ecological invasion. *Nature* 417:636–638.

Kercher, S. M., and J. B. Zedler. 2004. Multiple disturbances accelerate invasion of reed canary grass (*Phalaris arundinacea* L.) in a mesocosm study. *Oecologia* 138:455–464.

Kimura, M. 1983. *The Neutral Theory of Molecular Evolution*. Cambridge, UK: Cambridge University Press.

Knops, J. M. H., D. Tilman, N. M. Haddad, S. Naeem, C. E. Mitchell, J. Haarstad, M. E. Ritchie, et al. 1999. Effects of plant species richness on invasion dynamics, disease outbreaks, insect abundances and diversity. *Ecology Letters* 2:286–293.

Kolbe, J. J., R. E. Glor, L. R. G. Schettino, A. C. Lara, A. Larson, and J. B. Losos. 2004. Genetic variation increases during biological invasion by a Cuban lizard. *Nature* 431:177–181.

Lavergne, S., and J. Molofsky. 2004. Reed canary grass (*Phalaris arundinacea*) as a biological model in the study of plant invasions. *Critical Reviews in Plant Sciences* 23:415–429.

Lavergne, S., and J. Molofsky. 2006. Control strategies for the invasive reed canarygrass (*Phalaris arundinacea* L.) in North American wetlands: The need for an integrated management plan. *Natural Areas Journal* 26:208–214.

Lavergne, S., and J. Molofsky. 2007. Increased genetic variation and evolutionary potential drive the success of an invasive grass. *Proceedings of the National Academy of Sciences of the United States of America* 104:3883–3888.

Lavergne, S., N. J. Muenke, and J. Molofsky. 2010. Genome size reduction can trigger rapid phenotypic evolution in invasive plants. *Annals of Botany* 105:109–116.

Lee, P. L. M., R. M. Patel, R. S. Conlan, S. J. Wainwright, and C. R. Hipkin. 2004. Comparison of genetic diversities in native and alien populations of hoary mustard (*Hirschfeldia incana* [L.] Lagreze-Fossat). *International Journal of Plant Sciences* 165:833–843.

Le Roux, J. J., A. M. Wieczorek, and J. Y. Meyer. 2008. Genetic diversity and structure of the invasive tree *Miconia calvescens* in Pacific islands. *Diversity and Distributions* 14:935 948.

Lewandowski, I., J. M. O. Scurlock, E. Lindvall, and M. Christou. 2003. The development and current status of perennial rhizomatous grasses as energy crops in the US and Europe. *Biomass & Bioenergy* 25:335–361.

Lindig-Cisneros, R., and J. Zedler. 2001. Effect of light on seed germination in *Phalaris arundinacea* L. (reed canary grass). *Plant Ecology* 155:75–78.

Lindig-Cisneros, R., and J. B. Zedler. 2002. *Phalaris arundinacea* seedling establishment: Effects of canopy complexity in fen, mesocosm, and restoration experiments. *Canadian Journal of Botany* 80:617–624.

Lockwood, J. L., P. Cassey, and T. Blackburn. 2005. The role of propagule pressure in explaining species invasions. *Trends in Ecology & Evolution* 20:223–228.

Lonsdale, W. M. 1999. Global patterns of plant invasions and the concept of invasibility. *Ecology* 80:1522–1536.

MacArthur, R. H., and E. O. Wilson. 1963. An equilibrium theory of insular zoogeography. *Evolution* 17:1373–1387.

MacArthur, R. H., and E. O. Wilson. 1967. *The Theory of Island Biogeography*. Princeton NJ: Princeton University Press.

Maron, J. L., M. Vila, R. Bommarco, S. Elmendorf, and P. Beardsley. 2004. Rapid evolution of an invasive plant. *Ecological Monographs* 74:261–280.

May, R. M., and R. M. Anderson. 1983. Epidemiology and genetics in the coevolution of parasites and hosts *Proceedings of the Royal Society of London Series A-Mathematical Physical and Engineering Sciences* 390:219.

McKinney, C. L. 2001. Effects of human population, area, and time on non-native plant and fish diversity in the United States. *Biological Conservation* 100:243–252.

Meimberg, H., J. I. Hammond, C. M. Jorgensen, T. W. Park, J. D. Gerlach, K. J. Rice, and J. K. McKay. 2006. Molecular evidence for an extreme genetic bottleneck during introduction of an invading grass to California. *Biological Invasions* 8:1355–1366.

Merigliano, M. F., and P. Lesica. 1998. The native status of reed canary grass (*Phalaris arundinacea* L.) in the inland Northwest, USA. *Natural Areas Journal* 18:223–230.

Merila, J., M. Bjorklund, and A. J. Baker. 1996. The successful founder: Genetics of introduced *Carduelis chloris* (greenfinch) populations in New Zealand. *Heredity* 77:410–422.

Molina-Freaner, F., and S. K. Jain. 1992. Isozyme variation in Californian and Turkish populations of the colonizing species *Trifolium hirtum*. *Journal of Heredity* 83:423–430.

Molofsky, J., and J. D. Bever. 2002. A novel theory to explain species diversity in landscapes: Positive frequency dependence and habitat suitability. *Proceedings of the Royal Society of London Series B-Biological Sciences* 269:2389–2393.

Molofsky, J., J. D. Bever, and J. Antonovics. 2001. Coexistence under positive frequency dependence. *Proceedings of the Royal Society of London Series B-Biological Sciences* 268:273–277.

Molofsky, J., J. D. Bever, J. Antonovics, and T. J. Newman. 2002. Negative frequency dependence and the importance of spatial scale. *Ecology* 83:21–27.

Morrison, S. L., and J. Molofsky. 1998. Effects of genotypes, soil moisture, and competition on the growth of an invasive grass, *Phalaris arundinacea* (reed canary grass). *Canadian Journal of Botany-Revue Canadienne De Botanique* 76:1939–1946.

Morrison, S. L., and J. Molofsky. 1999. Environmental and genetic effects on the early survival and growth of the invasive grass *Phalaris arundinacea*. *Canadian Journal of Botany* 77:1447–1453.

Naeem, S., J. M. H. Knops, D. Tilman, K. M. Howe, T. Kennedy, and S. Gale. 2000. Plant diversity increases resistance to invasion in the absence of covarying extrinsic factors. *Oikos* 91:97–108.

Novak, S. J., and R. N. Mack. 2001. Tracing plant introduction and spread: Genetic evidence from *Bromus tectorum* (Cheatgrass). *Bioscience* 51:114–122.

Ostrem, L. 1988. Studies on genetic variation in reed canary grass *Phalaris arundinacea* L. III. Seed yield and seed yield components. *Hereditas* 107:15–168.

Pardini, E. A., and J. L. Hamrick. 2008. Inferring recruitment history from spatial genetic structure within populations of the colonizing tree *Albizia julibrissin* (Fabaceae). *Molecular Ecology* 17:2865–2879.

Premoli, A. C., and T. Kitzberger. 2005. Regeneration mode affects spatial genetic structure of *Nothofagus dombeyi* forests. *Molecular Ecology* 14:2319–2329.

Pyšek, P., V. Jarošik, and T. Kučera. 2002. Patterns of invasion in temperate nature reserves. *Biological Conservation* 104:13–24.

Reed, D. H., and R. Frankham. 2003. Correlation between fitness and genetic diversity. *Conservation Biology* 17:230–237.

Rogers, W. E., and E. Siemann. 2004. Invasive ecotypes tolerate herbivory more effectively than native ecotypes of the Chinese tallow tree *Sapium sebiferum*. *Journal of Applied Ecology* 41:561–570.

Ronsheim, M. L. 1996. Evidence against a frequency-dependent advantage for sexual reproduction in *Allium vineale*. *American Naturalist* 147:718–734.

Ronsheim, M. L., and S. E. Anderson. 2001. Population-level specificity in the plant-mycorrhizae association alters intraspecific interactions among neighboring plants *Oecologia* 128:77–84.

Roux, J. J., A. M. Wieczorek, C. T. Tran, and A. E. Vorsino. 2010. Disentangling the dynamics of invasive fireweed (*Senecio madagascariensis* Poir. species complex) in the Hawaiian Islands. *Biological Invasions* 12:2251–2264.

Schweitzer, J. A., J. K. Bailey, B. J. Rehill, G. D. Martinsen, S. C. Hart, R. L. Lindroth, P. Keim, and T. G. Whitham. 2004. Genetically based trait in a dominant tree affects ecosystem processes. *Ecology Letters* 7:127–134.

Sheaffer, C. C., and G. C. Marten. 1992. Seeding patterns affect grass and alfalfa yield in mixtures. *Journal of Production Agriculture* 5:328–332.

Stephens, P. A., W. J. Sutherland, and R. P. Freckleton. 1999. What is the Allee effect? *Oikos* 87:185–190.

Stohlgren, T. J., D. T. Barnett, and J. Kartesz. 2003. The rich get richer: Patterns of plant invasions in the United States. *Frontiers in Ecology and the Environment* 1:11–14.

Stohlgren, T. J., D. Binkley, G. W. Chong, M. A. Kalkhan, L. D. Schell, K. A. Bull, Y. Otsuki, G. Newman, M. Bashkin, and Y. Son. 1999. Exotic plant species invade hot spots of native plant diversity. *Ecological Monographs* 69:25–46.

Taylor, C. M., and A. Hastings. 2005. Allee effects in biological invasions. *Ecology Letters* 8:895–908.

Taylor, D. R., and S. R. Keller. 2007. Historical range expansion determines the phylogenetic diversity introduced during contemporary species invasion. *Evolution* 61:334–345.

Thompson, J. N. 1998. Rapid evolution as an ecological process. *Trends in Ecology & Evolution* 13:329–332.

Tilman, D. 1982. Resource competition and community structure. *Monographs in Population Biology* 17:1–296.

Tilman, D. 1997. Community invasibility, recruitment limitation, and grassland biodiversity. *Ecology* 78:81–92.

Troupin, D., R. Nathan, and G. G. Vendramin. 2006. Analysis of spatial genetic structure in an expanding *Pinus halepensis* population reveals development of fine-scale genetic clustering over time. *Molecular Ecology* 15:3617–3630.

Uriarte, M., and H. K. Reeve. 2003. Matchmaking and species marriage: A game-theory model of community assembly. *Proceedings of the National Academy of Sciences of the United States of America* 100:1787–1792.

Van Valen, L. 1965. Morphological variation and width of ecological niche. *American Naturalist* 99:377–389.

Vekemans, X., and O. J. Hardy. 2004. New insights from fine-scale spatial genetic structure analyses in plant populations. *Molecular Ecology* 13:921–935.

Vellend, M. 2005. Species diversity and genetic diversity: Parallel processes and correlated patterns. *American Naturalist* 166:199–215.

Vellend, M., E. B. M. Drummond, and H. Tomimatsu. 2010. Effects of genotype identity and diversity on the invasiveness and invasibility of plant populations. *Oecologia* 162:371–381.

Vellend, M., and M. A. Geber. 2005. Connections between species diversity and genetic diversity. *Ecology Letters* 8:767–781.

Volkov, I., J. R. Banavar, S. P. Hubbell, and A. Maritan. 2003. Neutral theory and relative species abundance in ecology. Nature 424:1035–1037.

Vose, P. B. 1962. Delayed germination in reed canary-grass *Phalaris arundinacea* L. *Annals of Botany* 26:197–206.

Vymazal, J. 2001. Constructed wetlands for wastewater treatment in the Czech Republic. *Water Science and Technology* 44:369–374.

Ward, S. 2006. Genetic analysis of invasive plant populations at different spatial scales. *Biological Invasions* 8:541–552.

Weltzin, J. F., N. Z. Muth, B. Von Holle, and P. G. Cole. 2003. Genetic diversity and invasibility: A test using a model system with a novel experimental design. *Oikos* 103:505–518.

Whitfield, J. 2002. Ecology: Neutrality versus the niche. *Nature* 417:480–481.

Whitham, T. G., J. K. Bailey, J. A. Schweitzer, S. M. Shuster, R. K. Bangert, C. J. Leroy, E. V. Lonsdorf, et al. 2006. A framework for community and ecosystem genetics: From genes to ecosystems. *Nature Reviews Genetics* 7:510–523.

Whitham, T. G., W. P. Young, G. D. Martinsen, C. A. Gehring, J. A. Schweitzer, S. M. Shuster, G. M. Wimp, et al. 2003. Community and ecosystem genetics: A consequence of the extended phenotype. *Ecology* 84:559–573.

Whitlock, R., J. P. Grime, R. Booth, and T. Burke. 2007. The role of genotypic diversity in determining grassland community structure under constant environmental conditions. *Journal of Ecology* 95:895–907.

Williams, D. A., E. Muchugu, W. A. Overholt, and J. P. Cuda. 2007. Colonization patterns of the invasive Brazilian peppertree, *Schinus terebinthifolius*, in Florida. *Heredity* 98:284–293.

Williams, D. A., W. A. Overholt, J. P. Cuda, and C. R. Hughes. 2005. Chloroplast and microsatellite DNA diversities reveal the introduction history of Brazilian peppertree (*Schinus terebinthifolius*) in Florida. *Molecular Ecology* 14:3643–3656.

chapter seven

Mycorrhizae and alien plants

Donald L. Hagan and Shibu Jose

Contents

7.1 Introduction

Invasions of alien plants have become one of the greatest threats to ecosystem function and timber and crop production worldwide. In the United States alone, some 5000 invasive alien (IA) plant species have become established, with an estimated economic impact (mostly in the form of control costs and lost productivity) in billions of dollars annually (Pimentel et al. 2005). Although the ecological ramifications of invasions are more difficult to quantify monetarily, it is generally accepted that IA species (including plants) have become a leading cause of biodiversity loss, second only to habitat destruction (Simberloff 2005). Largely for these reasons, the field of invasion ecology has grown significantly in recent years, and increased emphasis is placed on predicting, preventing, and mitigating plant invasions and their impacts.

What are the causes and effects of invasions, and why do some alien species become invasive and others do not? These are the primary questions that have motivated invasion ecology research over the last few decades. Williamson and Fitter (1996), in an effort to develop a predictive framework for biological invasions, proposed the "tens rule," which states that 1/10 of all introduced species escape, 1/10 of those that escape become established, and 1/10 of those that become established become invasive. Although this rule is more of a generalization than an actual scientific rule, it illustrates that invasion is a multistep process, with many barriers that must be overcome in order for an alien plant to become invasive. Moreover, it implies that interactions between an alien plant and its new environment are a major determinant of whether or not the plant establishes and becomes invasive. Indeed, it is believed that a primary factor that affects the performance of alien species is the degree of resistance that they face in their new environment, and this is arguably the basis of most contemporary invasion hypotheses (Daneshgar and Jose 2008 and citations therein). The biotic resistance hypothesis (Maron and Vilà 2001), for example, states that an alien plant is more likely to establish and spread (i.e., become invasive) when negative interactions with native biota (i.e., herbivores, parasites, interspecific competition) are minimal or, at least, less deleterious than they are in its native range.

Unfortunately, most studies of interspecific interactions in invaded ecosystems focus on the primary producers (i.e., the plants themselves), typically viewing invasion as the end result of a plant-to-plant interaction in which an introduced alien species successfully outcompetes established natives to become invasive. Furthermore, most studies are "aboveground centric," with only a handful of researchers attempting to elucidate the complex suite of interspecific interactions—many of which are mediated by soil microbiota—that take place in the rhizosphere (Ehrenfeld 2003, Wolfe and Klironomos 2005). As more studies are conducted, however, the role of belowground mutualisms in plant invasions becomes clearer, as do the changes to soil community that occur following invasions (Richardson et al. 2000a, Reynolds et al. 2003, Pringle et al. 2009).

The plant–mycorrhizal fungi symbiosis (mycorrhiza [singular], mycorrhizae [plural]) is one of the most important mutualisms on Earth. It has been estimated that some 90% of all plant families and 82% of all plant species depend, to some extent, on mycorrhizae for nutrient and water uptake (Wang and Qiu 2006) and pathogen defense (Pozo and Azcón-Aguilar 2007). Mycorrhizal fungi provide these benefits to the host plant in exchange for carbon, which the plant acquires via photosynthesis (Smith and Read 2008). Enhanced phosphorus uptake is thought to be the most important benefit that the plant receives from the symbiosis, since this nutrient is generally much less mobile in the soil compared to most other macronutrients (Bolan 1991). According to van der Heijden et al. 2008, mycorrhizae are responsible for up to 75% of all phosphorus that is taken up annually by terrestrial plants worldwide.

Although mycorrhizae can be classified into several functional groups, the bulk of mycorrhizal plants form associations with either arbuscular mycorrhizal (AM) or ectomycorrhizal (EM) fungi (Figure 7.1). Certain plants (e.g., eucalypts) have been shown

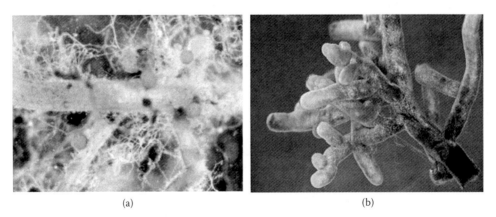

(a)   (b)

Figure 7.1 Mycorrhizae consist of specialized intraradical and extraradical structures that aid in resource uptake/transfer and pathogen defense: (a) Photograph of a plant root colonized by arbuscular mycorrhizal fungi, showing extraradical hyphae and spores, and (b) photograph of a plant root colonized by ectomycorrhizal fungi, showing characteristic extraradical hyphal sheath. (From Rosendahl, S., *New Phytol.*, 178, 253–266, 2008; Nilsson et al., *BMC Bioinformatics*, 6, 178, 2008. With permission.)

to associate with both AM and EM fungi (dos Santos et al. 2001). The two groups of fungi are ecologically quite distinct, as AM fungi are thought to derive the majority of their carbon from the plant host (Schwab et al. 1991), whereas some EM fungi have substantial saprophytic (decomposer) ability (Koide et al. 2008). There are some notable exceptions, but the majority of the world's most problematic IA plant species are mycorrhizal (Richardson et al. 2000a). Some are facultative mycotrophs, which means they benefit from the symbiosis but do not require it, while others are obligate (Richardson et al. 2000a).

In this chapter, we provide a broad overview of the interactions between alien plant species and AM and EM fungal communities. An attempt is made to outline the role of plant–fungi interactions within the framework of current invasion ecology theory, with particular emphasis on the biotic resistance hypothesis and its related hypotheses. We also highlight several areas where further research is needed. The body of this chapter is divided into three main parts: In Section 7.2, we generalize and describe the ways in which alien plant species interact with local populations of mycorrhizal fungi, focusing on how these interactions might affect their establishment. In Section 7.3, we scale up and address the mycorrhizal fungal community–level changes that occur directly or indirectly as the result of plant invasion. Finally, in Section 7.4, we discuss mycorrhizal "legacy effects" and the implications they might have for the reestablishment of desirable native plants following the local eradication of IA species.

7.2   Interactions between alien plants and mycorrhizal fungi

If an alien plant is successful in overcoming the many barriers to dispersal, the next hurdle is establishment (Richardson et al. 2000b). For an obligate mycorrhizal plant species, this likely means, among other things, that it must form mycorrhizal associations with the local mycorrhizal fungal community soon after seeds or vegetative propagules break dormancy (note, however, that some alien plants are introduced intentionally and are often preinoculated with the appropriate fungi). Perhaps the most straightforward way in which this occurs is via direct interaction between the root system of an alien plant and the native mycorrhizal fungal inocula in the soil (Pringle et al. 2009). In sites that currently are or recently were vegetated with mycorrhizal plants, for example, viable fungal propagules (spores, infected roots and hyphal fragments) should be present in the soil (Smith and Read 2008). In general, the greater the flexibility a mycorrhizal alien plant has in its ability to form a mutualism with locally available mycorrhizal propagules, the greater the chance it has to establish and become invasive (Pringle et al. 2009). This would seemingly make communities with diverse populations of mycorrhizal fungi more amenable to the establishment of mycorrhizal alien plants, but this has yet to be demonstrated.

The introduction of alien mycorrhizal fungal propagules into a plant community can greatly facilitate the establishment of alien plant species. This is particularly true for EM fungi, which are thought to be more host specific in their associations and have historically been more limited in their global distribution compared with AM fungi (Wang and Qiu 2006). Vellinga et al. (2009) determined that at least 200 species of EM fungi have been transported (usually on the roots of inoculated plants) outside their native ranges (Figure 7.2). The movement, and subsequent establishment, of these fungal species has greatly aided in the establishment of alien EM plant species. The lack of compatible EM fungi in many parts of the Southern Hemisphere, for example, was long a major barrier to

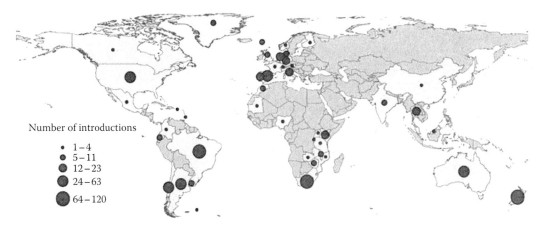

Figure 7.2 Map of global distribution of ectomycorrhizal (EM) fungal introductions: Countries in light gray are those that have at least one introduction, and the sizes of circles in the figure are proportional to the number of EM species that have been introduced. (From Vellinga et al., *New Phytol.*, 181, 960–973, 2009. With permission.)

the establishment of commercially important obligate EM tree species, particularly members of the Pinaceae family. The introduction of EM fungi has alleviated this recruitment limitation and has sometimes allowed these plant species to establish themselves well outside their initial area of introduction (Richardson et al. 1994, but see Nuñez et al. 2009). Similarly, the introduction of several Australian strains of EM fungi has facilitated the establishment of nonnative eucalypts across many parts of Europe (Díez 2005).

The effects of mycorrhizal fungi on their host plants can range from highly beneficial to highly parasitic (Smith and Read 2008). If an alien plant benefits more from mycorrhizal fungi in its new habitat than it does in its old habitat, or if it benefits more strongly from the association than do the native plants of the new habitat, it has an increased likelihood of becoming established. This has been referred to as the "enhanced mutualisms" hypothesis (Reinhart and Callaway 2006) and, although its empirical and theoretical basis remains weak, it does have some supporting evidence in the literature. Nijjer et al. (2008) reported that the Chinese tallow tree (*Triadica sebifera*) derived greater benefit from native (Texas) soil biota than did various native mycorrhizal tree species. High levels of AM fungal root colonization for *T. sebifera*, relative to the native species, led the researchers to speculate that enhanced nutrient uptake could explain the observed differences in species performance. Baynes et al. (2012) propose that a novel association with fire-associated AM fungi results in increased biomass production and fecundity for alien cheatgrass (*Bromus tectorum*) in the western United States. Higher stand density for black cherry (*Prunus serotina*) in its invaded range (Europe) compared to its native range (North America) may also be evidence of an enhanced mutualism (Reinhart et al. 2003, Reinhart and Callaway 2006).

The relatively non-host-specific nature of most AM and EM fungi permits them to link multiple species simultaneously in a common mycorrhizal network. This not only results in root inoculation but also may allow the movement of carbohydrates from one plant to another through their mycelia (Simard and Durall 2004). This in turn could be a mechanism by which an alien plant gains a competitive advantage over established natives. In a greenhouse study, Carey et al. (2004) demonstrated that the alien plant spotted knapweed (*Centaurea maculosa*) can obtain up to 15% of its carbon from the native plant Idaho fescue (*Fetescua idahoensis*) via a common mycorrhizal network. Similar studies have yet to be conducted with alien EM plants. However, in native EM forests, isotopic evidence suggests that substantial carbon transfer can occur between different tree species via mycorrhizal networks (Simard et al. 1997, Selosse et al. 2006). It deserves mention, however, that for mycorrhizae-mediated interplant carbohydrate transfer to facilitate alien plant establishment, it would have to be in a large enough quantity to influence the fitness of the donor and/or recipient plant. To date, definitive evidence of fluxes of this magnitude has not been reported (Simard and Durall 2004, Selosse et al. 2006).

Interspecific carbohydrate transfer is not the only way by which a common mycorrhizal network can potentially mediate interactions between native and alien plant species. Significant one-way nitrogen transfer via EM and AM mycorrhizal networks, for example, has been observed between alien N_2-fixing plants and nonfixers. Interestingly, nitrogen fixers (e.g., *Casuarina* sp.) often serve as the N sink in this relationship (He et al. 2005), although they can also be the source (He et al. 2004). Phosphorus transfer has been reported with agronomic species (Johansen and Jensen 1996), but transfer between native and alien species has yet to be studied. As is the case with carbohydrates, the biological significance of these nutrient fluxes remains debatable.

Enhanced nutrition from mycorrhizal networks may not, however, be limited to that which occurs via plant-to-plant resource transfer. In undisturbed or relatively undisturbed soil, for example, the intact mycorrhizal network represents a major carbon investment, which is "paid for" by the established plant community (Simard and Durall 2004). An alien plant that taps into this infrastructure could, therefore, gain access to a large pool of soil resources without first having to establish an extensive root and mycorrhizae system. For these reasons, mycorrhizal networking may have significant implications for the establishment of an alien plant species, particularly in sites with healthy populations of mycorrhizal plants. Zabinski et al. (2002), for example, reported that *C. maculosa* had increased access to soil phosphorus when joined to native species via an established mycorrhizal network. The effect, however, was highly species specific, occurring only when *C. maculosa* was grown with a grass (rather than a forb) neighbor.

7.3 When alien plants become invasive: Implications for mycorrhizal fungal communities

As mentioned in Section 7.1, only a fraction of introduced alien plant species that become established eventually become invasive. It is these plants, however, that are of most concern, as the transition from alien to IA often represents the point at which a plant begins to significantly transform ecosystem structure and function (Richardson et al. 2000b). Although researchers have long been studying the effects of invasion on aboveground communities, only recently have they paid much attention to how IA plants affect the soil. It is now generally accepted, however, that IA plants can significantly alter soil microbial communities, including mycorrhizal fungi (Richardson et al. 2000a, Ehrenfeld 2003, Wolfe and Klironomos 2005, Reinhart and Callaway 2006, Pringle et al. 2009).

There is a growing theoretical basis that describes how IA plants alter soil community structure and function. In general, the extent to which an IA plant species affects soil communities is believed to depend on how different the suite of traits possessed by it (e.g., competitiveness, growth rate, phenology, rooting habit, mutualisms) is from that possessed by the displaced native species (Ehrenfeld 2003). Since below- and aboveground processes are tightly linked, any significant change in plant community assembly that occurs with invasion is ultimately caused by a combination of top-down and bottom-up controls (Hausmann and Hawkes 2010). Although they do not focus solely on mycorrhizal fungi, Wolfe and Klironomos (2005) propose that IA plants disrupt these controls through their effects on plant community composition, soil community composition, and ecosystem processes and properties. Invasive alien plants likely affect one of these three components directly (Figure 7.3), with subsequent alterations occurring as the result of feedbacks and indirect effects. Plants that affect mycorrhizal fungal communities directly are often those that exude phytotoxic and/or antimicrobial allelochemicals (Roberts and Anderson 2001, Vivanco et al. 2004). In the case of IA plants, the lack of coevolution among their native neighbors may augment the effects of these compounds (see "novel weapons hypothesis" in the studies by Callaway and Ridenour [2004] and Vivanco et al. [2004]). Indirect plant–fungi interactions, however, may be more common; this is likely the case when an invader outcompetes native species, in the process creating a novel mycorrhizal fungal community (Pringle et al. 2009), or when a species alters soil chemical properties (e.g., soil nutrient status) in a manner that increases or decreases its dependence on the mycorrhizal symbiosis.

The effects of IA plant species on mycorrhizal inoculum availability are variable, and likely highly dependent on context. For nonmycorrhizal or weakly mycorrhizal plants, which are frequent invaders of early successional and non-phosphorus-limited sites (Richardson et al. 2000b), invasion often results in a decrease in mycorrhizal inoculum availability over time. This can create a positive feedback that shifts the competitive balance further in favor of the invader (Wolfe et al. 2008). Roberts and Anderson (2001), for example, found AM inoculum potential to be inversely related to the cover of garlic mustard (*Alliaria petiolata*) (Figure 7.4), a nonmycorrhizal allelopathic IA species of the Brassicaceae family. These alterations significantly impeded the performance of seedling AM sugar maple (*Acer saccharum*), red maple (*A. rubrum*), and white ash (*Fraxinus americana*) (Stinson et al. 2006). Reductions in AM fungal density have also been observed in sites invaded by AM *B. tectorum* in the semiarid western United States (Busby et al. 2012). Shifts in inoculum availability may be less likely when obligate mycorrhizal species invade sites with healthy populations of compatible mycorrhizal fungi. In longleaf pine (*Pinus palustris*) sandhill stands in central Florida, for example, similar AM fungal spore

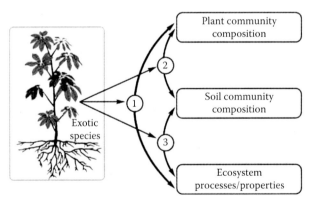

Figure 7.3 Simplified conceptual diagram outlining the three points at which an invasive alien (IA) plant species can directly affect plant communities, soil communities (including mycorrhizal fungi), and ecosystem processes and properties: (1) The IA plant interferes with the linkage between native plant community composition and important ecosystem processes and properties. (2) Invasion-induced changes in plant community composition alter the composition of the soil community (or vice versa). (3) Invasion modifies the soil community, which subsequently alters ecosystem processes and properties (or vice versa). (From Wolfe, B.E., and J.N. Klironomos, *Bioscience*, 55, 477–487, 2005. With permission.)

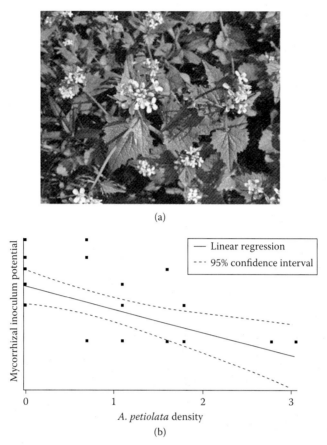

(a)

(b)

Figure 7.4 (a) Garlic mustard (*Alliaria petiolata*), a problematic nonmycorrhizal invader of North American forest ecosystems. (Courtesy of Frank Mayfield, Encyclopedia of Life Images.) (b) Relationship between *A. petiolata* density (plants [dm²];) and AM fungal inoculum potential in hardwood forests in Illinois. Natural log transformations were performed on all data. (From Roberts, K.J., and R.C. Anderson, *Am. Midl. Nat.*, 146, 146–152, 2001. With permission.)

counts were observed between sites heavily invaded by AM cogongrass (*Imperata cylindrica*) and those where native understory species predominated (Hagan et al. unpublished data). Similarly, no reductions in inoculum potential were observed in mountain grassland sites invaded by AM leafy spurge (*Euphorbia esula*) and Canada thistle (*Cirsium arvense*) (Pritekel et al. 2006). In some cases, the establishment and spread of mycorrhizal IA plant species may result in an increase in the inoculum potential of one mycorrhizal functional group relative to another (e.g., increases in EM inoculum following pine establishment in AM-dominated communities) (Thiet and Boerner 2007, Salomón et al. 2011).

Inoculum potential assays and spore counts provide valuable insight into the effects of IA plants on mycorrhizal fungal populations. However, recent developments in molecular genetics approaches have enabled researchers to address more complex questions related to mycorrhizal community assembly. Well-conserved genes or internal transcribed spacer (ITS) sequences can be amplified from DNA extracted from the soil (Curlevski et al. 2010) or from infected roots (Nuñez et al. 2009, Shi et al. 2012), and powerful programs are readily available for sequence analysis and community profiling

(Altschul et al. 1990, Lozupone et al. 2007, Schloss et al. 2009, Tamura et al. 2011). Their relative lack of host specificity suggests that mycorrhizal fungal communities are resistant to change following invasion, but this has not always proven to be the case (Hausmann and Hawkes 2010). Even when no change in fungal inoculum availability occurs due to invasion, sequence analyses have shown that the composition of the resultant mycorrhizal fungal community can be considerably different from the composition of the original. In AM fungal clone libraries assembled from 18S ribosomal RNA sequences from the *I. cylindrica*–invaded pine stands, Hagan et al. (unpublished data) reported substantial alterations to AM fungal community assembly. In this study, uninvaded and invaded sites shared only about 77% of AM fungal OTUs (operational taxonomic units, sensu species). This scenario supports the "passenger" hypothesis (Hart et al. 2001), which states that changes in AM fungal community properties are a top-down by-product of changes in the plant community. Interestingly, no differences in OTU richness or diversity were observed. The latter trends are quite intriguing, since *I. cylindrica* invasion results in a dramatic decline in plant species richness (Brewer 2008), and there is typically a positive relationship between above- and belowground richness (Kernaghan 2005). Indeed, when diverse, semiarid steppe communities became dominated by IA cheatgrass, substantial reductions in AM fungal richness and diversity were shown to occur (Busby 2011).

In an invaded system, feedback between plants, soil, and ecosystem processes may continue to alter community structure and function until a new equilibrium is reached (Reinhart and Callaway 2006). Once this quasi-stable state is reached, the plant and mycorrhizal fungal communities become increasingly novel and may further shift the competitive balance further in favor of the IA species. This has been proposed as a mechanism that is responsible for the dominance of spotted knapweed (Mummey and Rillig 2006) and numerous other weedy IA plant species (Vogelsang and Bever 2009) in North American grassland ecosystems. Vogelsang and Bever (2009) refer to this as the "degraded mutualist" hypothesis, highlighting the role of the mycorrhizal fungal community as an important mediator of biotic resistance in invaded plant communities. However, the concept of equilibrium in natural systems is a bit tenuous, and the cascade of above- and belowground alterations brought about by plant invasion certainly results in much short-term, if not long-term, instability. Therefore, while the novel mycorrhizal characteristics of an invaded plant community may select against the native species, they likely facilitate the establishment and codominance of other invaders (see "invasional meltdown hypothesis" in the study by Simberloff and Von Holle [1999]), in the process resulting in further alterations to the mycorrhizal fungal community.

7.4   Posteradication legacy effects

The body of knowledge on the interactions between IA plant species and mycorrhizal fungi is limited, but it has increased significantly in recent years. However, comparatively less attention has been paid to the legacy that invasives leave behind once they are locally eradicated. This clearly is an area that deserves more consideration, as many desirable native plant species are obligately mycorrhizal and the restoration of native plant communities following the eradication of IA plants is a top priority for land managers (Hartman and McCarthy 2004). If soil properties (including mycorrhizal fungi) are altered by an IA plant and/or by control efforts, novel characteristics may persist for some time after the invader is gone. Arbuscular mycorrhizal fungal populations may be particularly susceptible to transformative impacts, since they have only limited ability to persist without a compatible plant host (Smith and Read 2008). A lag in the recovery of the mycorrhizal fungal community could significantly impair the establishment and early growth of a desired native species (Anderson et al. 2010). This would be especially true in infertile soils, where dependence on mycorrhizal symbiosis is typically high.

Fortunately, the few studies addressing legacy effects have shown that mycorrhizal fungal communities recover quite rapidly following the local eradication of IA plants. Using a 7-year recovery chronosequence, which also included invaded and uninvaded native pine sandhill reference sites, Hagan et al. (unpublished data) found that significant modifications to AM fungal community assembly (based on analyses of polymerase chain reaction [PCR] clone libraries) develop in the years following the chemical eradication of *I. cylindrica* monocultures. While soils from sites where *I. cylindrica* was eradicated 3 years prior shared only approximately 55% of AM fungal OTUs with the native reference community, a principal components analysis (PCA) of AM fungal small-subunit ribosomal RNA sequences indicated that community assembly approached the native reference state over time (Figure 7.5). These same authors observed that native plant communities were much slower to recover than AM fungal communities. The fact that AM fungal communities recover to a preinvasion state before native plant communities do is intriguing, and it suggests that bottom-up control from AM fungi may play a role in determining plant successional trajectories following *I. cylindrica* eradication (see "driver hypothesis" in the study by Hart et al. [2001]). Unfortunately, the relatively short length of the chronosequence makes it difficult to ascertain the implications that these temporary alterations to

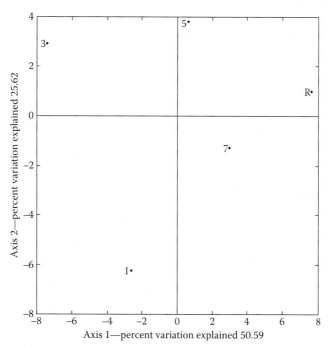

Figure 7.5 Principal components analysis (PCA) biplot illustrating separation in variable space between the AM fungal community assembly of longleaf pine (*Pinus palustris*) sandhill sites where cogongrass (*Imperata cylindrica*) was eradicated 3, 5, and 7 years prior, native reference (R) sites, and currently invaded (I) sites: Distance between points is an indication of compositional similarity. Each point represents the mean of 2 replicates. Each replicate was a composite of soil samples from 4 sites.

the fungal community might have for plant communities in the longer term. Anderson et al. (2010) also found that mycorrhizal inoculum potential rebounds rapidly following IA plant (*A. petiolata*) removal, but the reestablishment of mycorrhizal species lagged behind that of nonmycorrhizal species. Posteradication legacy effects on mycorrhizal fungal communities, even short-lived ones, may, however, not be a universal phenomenon. Although they did not assess differences in community assembly, Pritekel et al. (2006) reported no significant difference in AM fungal inoculum potential between sites formerly invaded by *E. esula* or *C. arvense* and diverse native grassland sites. Postulated explanations included the relatively small sizes of invaded patches and the rapid dispersal of new AM fungal propagules.

7.5   Conclusions and future directions

Research conducted over the last few years has provided valuable insight into mycorrhizal fungal dynamics in ecosystems impacted by alien plants. After a review of the literature, we were able to make some generalizations about these complex and poorly understood interactions. However, it is clear that much more research is needed in this domain as experimental evidence for many ideas is limited, often having weak theoretical foundations or being based on just a few species or ecosystems. This is particularly true for interactions involving EM fungi (as well as other fungi, e.g., ericoid mycorrhizal fungi), which have received comparatively less research attention in invaded ecosystems than AM fungi and present unique methodological challenges.

Additional questions remain. An inherent problem with IA plant research, for example, is that although many studies have been conducted to address the drivers and consequences of successful plant invasions, little is known about the causes of failed invasions (Reinhart and Callaway 2006). Do certain characteristics of mycorrhizal fungal communities (e.g., diversity, species identity) impart resistance or susceptibility to the establishment and invasion of some alien plant species? Decreased dependence on mycorrhizal symbiosis in alien plant populations (relative to their native range) was reported recently (Seifert et al. 2009); the implications that this finding may have for native plant and mycorrhizal fungal communities should be thoroughly evaluated. Clearly, hypothesis-testing approaches that couple modern molecular techniques with the manipulation of fungal and plant assemblages will be of great benefit in these types of studies. The intentional introduction of potentially invasive alien plants is usually not feasible (or recommended), which means that greenhouse studies will be required. Whenever possible, however, corroborating evidence should be provided from field studies, as controlled experiments cannot adequately simulate the complex suite of direct and indirect interactions that occurs in the field between plant communities, soil communities, and ecological processes.

References

Altschul, S.F., W. Gish, W. Miller, E.W. Myers, and D.J. Lipman. 1990. Basic local alignment search tool. *Journal of Molecular Biology* 215:403–410.

Anderson, R.C., M.R. Anderson, J.T. Bauer, M. Slater, J. Herold, P. Baumhardt, and V. Borowicz. 2010. Effect of removal of garlic mustard (*Alliaria petiolata*, Brassicaceae) on arbuscular mycorrhizal fungi inoculum potential in forest soils. *The Open Ecology Journal* 3:41–47.

Baynes, M., G. Newcombe, L. Dixon, L. Castlebury, and K. O'Donnell. 2012. A novel plant-fungal mutualism associated with fire. *Fungal Biology* 116:133–144.

Bolan, N.S. 1991. A critical review on the role of mycorrhizal fungi in the uptake of phosphorus by plants. *Plant and Soil* 134:189–207.

Brewer, S. 2008. Declines in plant species richness and endemic plant species in longleaf pine savannas invaded by *Imperata cylindrica*. *Biological Invasions* 10:1257–1264.

Busby, R.R. 2011. Interactions between plants and soil microbes in Florida communities: Implications for invasion and ecosystem ecology. Dissertation, Colorado State University.

Busby, R.R., M.W. Paschke, M.E. Stromberger, and D.L. Gebhart. 2012. Seasonal variation in arbuscular mycorrhizal fungi root colonization of cheatgrass (*Bromus tectorum*) an invasive winter annual. *Journal of Ecosystem and Ecography* http://dx.doi.org/10.4172/2157 7625.S8-001.

Callaway, R.M., and W.M. Ridenour. 2004. Novel weapons: Invasive success and the evolution of increased competitive ability. *Frontiers in Ecology and the Environment* 2:436–443.

Carey, E.V., M.J. Marler, and R.M. Callaway. 2004. Mycorrhizae transfer carbon from a native grass to an invasive weed: evidence from stable isotopes and physiology. *Plant Ecology* 172:133–141.

Curlevski, N.J.A., Z. Xu, I.C. Anderson, and J.W.G. Cairney. 2010. Soil fungal communities differ in native mixed forest and adjacent *Araucaria cunninghamii* plantations in subtropical Australia. *Journal of Soils and Sediments* 10:1278–1288.

Daneshgar, P., and S. Jose. 2008. Mechanisms for plant invasion: a review. In: R.K. Kohli, S. Jose and H.P. Singh (Eds). *Invasive Plants and Forest Ecosystems*. Boca Raton, FL: CRC Press.

Díez, J. 2005. Invasion biology of Australian ectomycorrhizal fungi introduced with eucalypt plantations into the Iberian Peninsula. *Biological Invasions* 7:3–15.

dos Santos, V.L., R.M. Muchovej, A.C. Borges, J.C.L. Neves, and M.C.M. Kasuya. 2001. Vesicular arbuscular-/ecto-mycorrhiza succession in seedlings of *Eucalyptus* spp. *Brazilian Journal of Microbiology* 32:81–86.

Ehrenfeld, J.G. 2003. Effects of exotic plant invasions on soil nutrient cycling processes. *Ecosystems* 6:503–523.

Hart, M.M., R.J. Reader, and J.N. Klironomos. 2001. Life-history strategies of arbuscular mycorrhizal fungi in relation to their successional dynamics. *Mycologia* 93:1186–1194.

Hartman, K.M., and B.C. McCarthy. 2004. Restoration of a forest understory after the removal of an invasive shrub, Amur honeysuckle (*Lonicera maackii*). *Restoration Ecology* 12:154–165.

Hausmann, N.T., and C.V. Hawkes. 2010. Order of plant host establishment alters the composition of arbuscular mycorrhizal communities. *Ecology* 91:2333–2343.

He, X., C. Critchley, H. Ng, and C. Bledsoe. 2005. Nodulated N_2-fixing *Casuarina cunninghamiana* is the sink for net N transfer from non-N_2-fixing *Eucalyptus maculata* via an ectomycorrhizal fungus *Pisolithus* sp. using $^{15}NH_4^+$ or $^{15}NO_3^-$ supplied as ammonium nitrate. *New Phytologist* 167:897–912.

He, X.H., C. Critchley, H. Ng, and C.S. Bledsoe. 2004. Reciprocal N ($^{15}NH_4^+$ or $^{15}NO_3$) transfer between nonN_2-fixing *Eucalyptus maculata* and N_2-fixing *Casuarina cunninghamiana* linked by the ectomycorrhizal fungus *Pisolithus* sp. *New Phytologist* 163:629–640.

Johansen, A., and E.S. Jensen. 1996. Transfer of N and P from intact or decomposing roots of pea to barley interconnected by an arbuscular mycorrhizal fungus. *Soil Biology and Biochemistry* 28:73–81.

Kernaghan, G. 2005. Mycorrhizal diversity: Cause and effect? *Pedobiologia* 49:511–520.

Koide, R.T., J.N. Sharda, J.R. Herr, and G.M. Malcolm. 2008. Ectomycorrhizal fungi and the biotrophy–saprotrophy continuum. *New Phytologist* 178:230–233.

Lozupone, C.A., M. Hamady, S.T. Kelley, and R. Knight. 2007. Quantitative and qualitative β diversity measures lead to different insights into factors that structure microbial communities. *Applied and Environmental Microbiology* 73:1576–1585.

Maron, J.L., and M. Vilà. 2001. When do herbivores affect plant invasion? Evidence for the Natural Enemies and Biotic Resistance hypotheses. *Oikos* 95:361–373.

Mummey, D.L., and M.C. Rillig. 2006. The invasive plant species *Centaurea maculosa* alters arbuscular mycorrhizal fungal communities in the field. *Plant and Soil* 288:81–90.

Nijjer, S., W.E. Rogers, C.A. Lee, and E. Siemann. 2008. The effects of soil biota and fertilization on the success of *Sapium sebiferum*. *Applied Soil Ecology* 38:1–11.

Nilsson, R.H., E. Kristiansson, M. Ryberg, and K.H. Larsson. 2005. Approaching the taxonomic affiliation of unidentified sequences in public databases – an example from the mycorrhizal fungi. *Bioinformatics* 6:178.

Nuñez, M.A., T.R. Horton, and D. Simberloff. 2009. Lack of belowground mutualisms hinders Pinaceae invasions. *Ecology* 9:2352–2359.

Pozo, M.J., and C. Azcón-Aguilar. 2007. Unraveling mycorrhiza-induced resistance. *Current Opinion in Plant Biology* 10:393–398.

Pimentel, D., R. Zuniga, and D. Morrison. 2005. Update on the environmental and economic costs associated with alien-invasive species in the United States. *Ecological Economics* 52:273–288.

Pringle, A., J.D. Bever, M. Gardes, J.L. Parrent, M.C. Rillig, and J.N. Klironomos. 2009. Mycorrhizal symbioses and plant invasions. *Annual Review of Ecology, Evolution and Systematics* 40:699–715.

Pritekel, C., A. Whittemore-Olson, N. Snow, and J.C. Moore. 2006. Impacts from invasive plant species and their control on the plant community and belowground ecosystem at Rocky Mountain National Park, USA. *Applied Soil Ecology* 32:132–141.

Reinhart, K.O., and R.M. Callaway. 2006. Soil biota and invasive plants. *New Phytologist* 170:445–457.

Reinhart, K.O., A. Packer, W.H. van der Putten, and K. Clay. 2003. Plant-soil biota interactions and spatial distribution of black cherry in its native and invasive ranges. *Ecology Letters* 6:1046–1050.

Reynolds, H.L., A. Packer, A. Bever, and K. Clay. 2003. Grassroots ecology: Plant-microbe-soil interactions as drivers of plant community structure and dynamics. *Ecology*, 84:2281–2291.

Richardson, D.M., N. Allsopp, C.M. D'Antonio, S.J. Milton, and M. Rejmanek. 2000a. Plant invasions—the role of mutualisms. *Biological Reviews* 75:65–93.

Richardson, D.M., P. Pysek, M. Rejmanek, M.G. Barbour, F.D. Panetta, and C.J. West. 2000b. Naturalization and invasion of alien plants: concepts and definitions. *Diversity and Distributions* 6:193–207.

Richardson, D.M., P.A. Williams, and R.J. Hobbs. 1994. Pine invasions in the Southern Hemisphere: determinants of spread and invadability. *Journal of Biogeography* 21:511–527.

Roberts, K.J., and R.C. Anderson. 2001. Effect of garlic mustard [*Alliaria petiolata* (Beib. Cavara & Grande)] extracts on plants and arbuscular mycorrhizal (AM) fungi. *American Midland Naturalist* 146:146–152.

Rosendahl, S. 2008. Communities, populations and individuals of arbuscular mycorrhizal fungi. *New Phytologist* 178:253–266.

Salomón, M.E.S., C. Barrotaveña, and M. Rajchenberg. 2011. Do pine plantations provide mycorrhizal inocula for seedlings establishment in grasslands from Patagonia, Argentina? *New Forests* 41:191–205.

Schloss, P.D., S.L. Westcott, T. Ryabin, J.R. Hall, M. Hartmann, et al. 2009. Introducing mothur: Open-source, platform-independent, community-supported software for describing and comparing microbial communities. *Applied and Environmental Microbiology* 75:7537–7541.

Schwab, S.M., J.A. Menge, and P.B. Tinker. 1991. Regulation of nutrient transfer between host and fungus in vesicular-arbuscular mycorrhizas. *New Phytologist* 117:387–398.

Seifert, E.K., J.D. Bever, and J.L. Maron. 2009. Evidence for the evolution of reduced mycorrhizal dependence during plant invasion. *Ecology* 90:1055–1062.

Selosse, M.A., F. Richard, X. He, and S.W. Simard. 2006. Mycorrhizal networks: des liaisons dangereuses? *Trends in Ecology and Evolution* 21:579–582.

Shi, P., L.K. Abbot, N.C. Banning, and B. Zhao. 2012. Comparison of morphological and molecular genetic quantification of relative abundance of arbuscular mycorrhizal fungi within roots. *Mycorrhiza* doi:10.1007/s00572-011-0425-8.

Simard, S.W., and D.M. Durall. 2004. Mycorrhizal networks: A review of their extent, function and importance. *Canadian Journal of Botany* 82:1140–1165.

Simard, S.W., D.A. Perry, M.D. Jones, D.D. Myrold, D.M. Durall, and R. Molina. 1997. Net transfer of carbon between ectomycorrhizal tree species in the field. *Nature* 388:579–582.

Simberloff, D. 2005. Non-native species DO threaten the natural environment! *Journal of Agricultural and Environ Ethics* 18:595–607.

Simberloff, D., and B. Von Holle. 1999. Positive interactions of nonindigenous species: Invasional meltdown? *Biological Invasions* 1:21–32.

Smith, S.E., and D.J. Read. 2008. *Mycorrhizal Symbiosis*. Elsevier Academic Press, St. Louis.

Stinson, K.A., S.A. Campbell, J.R. Powell, B.E. Wolfe, R.M. Callaway, et al. 2006. Invasive plant suppresses the growth of native tree seedlings by disrupting belowground mutualisms. *PLoS Biology* 4:727–731.

Tamura, K., D. Peterson, N. Peterson, G. Stecher, M. Nei, S. Kumar. 2011. MEGA5: Molecular evolutionary genetics analysis using maximum likelihood evolutionary distance, and maximum parsimony methods. *Molecular Biology and Evolution* 28:2731–2739.

Thiet, R.K., and R.E.J. Boerner. 2011. Spatial patterns of ectomycorrhizal fungal inoculum in arbuscular mycorrhizal barrens communities: Implications for controlling invasion by *Pinus virginiana*. *Mycorrhiza* 17:507–517.

Van der Heijden, M.G.A., R.D. Bardgett, and N.M. van Straalen. 2008. The unseen majority: soil microbes as drivers of plant diversity and productivity in terrestrial systems. *Ecology Letters* 11:296–310.

Vellinga, E.C., B.E. Wolfe, and A. Pringle. 2009. Global patterns of ectomycorrhizal introductions *New Phytologist* 181:960–973.

Vivanco, J.M., H.P. Bais, F.R. Stermitz, G.C. Thelen, R.M. Callaway. 2004. Biogeographical variation in community response to root allelochemistry: Novel weapons and exotic invasion. *Ecology Letters* 7:285–292.

Vogelsang, K.M., and J.D. Bever. 2006. Mycorrhizal densities decline in association with nonnative plants and contribute to plant invasion. *Ecology* 90:399–407.

Wang, B., and Y.L. Qiu. 2006. Phylogenetic distribution and evolution of mycorrhizas in land plants. *Mycorrhiza* 16:299–363.

Willamson, M., and A. Fitter. 1996. The varying success of invaders. *Ecology* 77:1661–1666.

Wolfe, B.E., and J.N. Klironomos. 2005. Breaking new ground: soil communities and exotic plant invasion. *Bioscience* 55:477–487.

Wolfe, B.E., V.L. Rodgers, K.A. Stinson, and A. Pringle. 2008. The invasive plant *Alliaria petiolata* (garlic mustard) inhibits ectomycorrhizal fungi in its introduced range. *Journal of Ecology* 96:777–783.

Zabinski, C.A., L. Quinn, and R.M. Callaway. 2002. Phosphorus uptake, not carbon transfer, explains arbuscular mycorrhizal enhancement of *Centaurea maculosa* in the presence of native grassland species. *Functional Ecology* 16:758–765.

chapter eight

Exotic plant response to forest disturbance in the western United States

Carl E. Fiedler, Erich K. Dodson, and Kerry L. Metlen

Contents

8.1   Introduction

Concerns about exotic plant invasions in forests of the western United States are recent, even though many exotics were introduced into the region more than a century ago (Holechek et al. 1998, Parks et al. 2005). Several factors account for this delayed concern. Principal among these are that invasion levels have been low as forests in the West are generally remote from population centers, cover huge contiguous areas in single ownerships—primarily federal (Smith et al. 2008), and have relatively few heavily traveled roads. Another factor is that western forests have historically been managed for wood, and the low levels of exotic plant invasion extant in the region have posed little threat to this resource. However, the increased focus on ecological values in forests has sparked concern about exotic plant invasions, which can alter fire regimes (Mack et al. 2000, Arno and Fiedler 2005), nutrient cycles (Vitousek et al. 1987, Hawkes et al. 2005), soil biota (Belnap et al. 2005, Grman and Suding 2010), native species richness (Yurkonis et al. 2005, Hejda et al. 2009), forage quality and use (Trammell and Butler 1995, DiTomaso 2000), and tree regeneration (Randall and Rejmánek 1993, Reinhart et al. 2005).

Forest invasion in the West generally decreases with increasing elevation (Keeley et al. 2003, Fisher and Fulé 2004), with the greatest threat in lower-lying forests adjacent to invaded grasslands, woodlands, and human developments. Lower-elevation forests are also commonly grazed by livestock and receive greater recreational and off-road vehicle use as well. However, substantial forest areas occupy mid- to high-elevation sites that have been relatively invasion-free. Regardless of elevation, some areas of the Colorado Front

Range, interior Southwest, and Pacific Northwest still exhibit surprisingly little exotic plant invasion (Fornwalt et al. 2003, Laughlin et al. 2004, Anzinger and Radosevich 2008, respectively). As recently as a decade ago, Withers et al. (1999) reported that exotics comprised a lower proportion of total plant species in the mountainous Northwest than in any other region of North America.

The current status of exotic plant invasions in western U.S. forests can be described as early stage and not yet well defined. However, the primary natural disturbances, environmental conditions, and human-related disturbances that influence invasibility are all changing. The scale and intensity of wildfires and biotic disturbances have increased sharply (Arno and Fiedler 2005); warming temperatures and drought have dominated the climate in recent years (Westerling et al. 2006); and broad-scale implementation of thinning and burning (forest restoration) treatments has reduced canopy cover and disturbed understories.

The convergence of larger and more intense natural disturbances, warm dry climatic conditions, and extensive forest treatments suggests elevated levels of invasion in the future. It also points to the need for an integrated examination of exotic species characteristics, the invasion process, and management strategies to slow invasions or reduce habitat invasibility. This chapter provides such a synthesis, including broad guidance to forest managers for minimizing, reducing, or living with exotic plant invasions in the future.

In this chapter, we address exotic plant invasiveness and habitat invasibility and their interrelationship with forest treatments and wildfire. Specifically, we will (1) assess the generalized invasive threats posed by exotic species to dry, moist, and subalpine coniferous forests in the western United States (Table 8.1); (2) characterize known and problematic exotic species as to invasive potential, habitat occurrence, and response to forest disturbance; (3) review invasive plant ecology and identify factors that may interact to resist or promote invasion; (4) interpret the above information to identify knowledge gaps and high-priority research needs; and (5) synthesize existing knowledge into management recommendations for reducing or neutralizing invasion following forest disturbance.

Table 8.1 General Characteristics of Dry, Moist, and Subalpine Conifer Forests in the Western United States By Forest Type, Primary Natural Disturbance Regime, Typical Recovery Pattern After Disturbance, and Landscape Location

	Dry conifer forests[a]	**Moist conifer forests**	**Subalpine conifer forests**
Major forest types	Ponderosa pine Interior Douglas-fir White fir	Coastal conifer forests (Douglas-fir, redwood, true fir/ spruce/cedar/ hemlock) Red fir and white fir Mixed conifer (Interior Northwest, Pacific Northwest, Sierra Nevada, Southwestern)	Engelmann spruce/ subalpine fir Lodgepole pine True fir/mountain hemlock

Table 8.1 (continued) General Characteristics of Dry, Moist, and Subalpine Conifer Forests in the Western United States By Forest Type, Primary Natural Disturbance Regime, Typical Recovery Pattern After Disturbance, and Landscape Location

	Dry conifer forests[a]	Moist conifer forests	Subalpine conifer forests
Primary natural disturbance regime	Low-severity fires at 0–35 year intervals, with relatively little overstory killed. Secondary disturbances include bark beetles and fungal pathogens	Mixed-severity fires at 35–100+ year intervals (<75% of overstory killed). Secondary disturbances include windthrow, defoliators, and fungal pathogens	High-severity fires at 35–200+ year intervals (>75% of overstory killed). However, biotic agents comprise the primary disturbance regime in some Rocky Mountain and Pacific Northwest subalpine forests
Typical recovery pattern after primary disturbance	Recovery rapid due to minor effects of disturbance on structure and species composition	Variable rates of recovery due to disturbance effects ranging from minor to stand-replacement at multiple spatial scales	Slow recovery of forest structure from stand-replacement disturbance
Landscape location	Low elevation	Low and middle elevation	High elevation

[a] Dry or seasonally dry.

8.2   *Exotic species invasive potential following disturbance*

The term "invasive" has been defined by numerous authors in numerous ways (e.g., Alpert et al. 2000, Richardson et al. 2000, Colautti and MacIsaac 2004, Huston 2004, MacDougall and Turkington 2005). For the purposes of this chapter, we will define invaders as *exotic species* that increase over time and negatively affect native species. We will further refine this definition by degree to address the differential invasive threats posed by different exotic species. For example, some invaders are capable of fundamentally altering conditions to favor self-perpetuation in the community, and we will use the term "transformer" *sensu* Richardson et al. (2000) to differentiate these aggressive invaders from relatively benign exotic species.

We have integrated the preceding concepts to describe three types of exotic invaders (or invasive potentials) based on how the species respond to forest disturbance and to the differential threats they pose to native plant communities (Figure 8.1). *Responders-decreasers* (such as the forb *Cirsium vulgare*) are either absent or present at very low abundances in the undisturbed plant community but respond rapidly to disturbance. They are soon outcompeted, however, and sharply decline in abundance. *Responders-transformers* (such as the forb *Centaurea diffusa*) are also absent or present in the undisturbed community at very low levels, but disturbance allows their populations to increase rapidly to a critical threshold. Once these species are broadly established, the community can enter an alternative, exotic-dominated state. *Invaders-transformers* (such as the forb *Alliaria petiolata* in eastern deciduous forests) are species capable of invading undisturbed native plant communities and may be present at moderate to high levels. These species further increase in relative abundance after disturbance and permanently alter biotic and abiotic conditions.

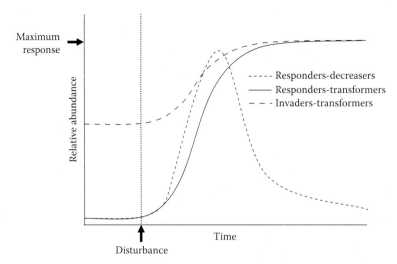

Figure 8.1 Differential responses ("invasive potentials") of exotic plant species to forest disturbance. *Responders-decreasers* are absent or present at low levels in the undisturbed community, spike sharply after disturbance, but then decline rapidly. *Responders-transformers* are also absent or present at very low abundances in undisturbed conditions, but increase rapidly after disturbance and transform the community. *Invaders-transformers* are present at moderate to high abundances in undisturbed conditions, then increase further after disturbance to transform and dominate the community.

Hundreds of exotic plant species can be found in the West, and these species vary greatly in their capacity to invade under different kinds and intensities of forest disturbances, habitat conditions, existing plant communities, and environmental conditions. Yet, only a modest subset of species raises concerns about disturbance-induced forest invasions (Table 8.2). The species identified in Table 8.2 were selected from the Fire Effects Information System (FEIS) invasive plants list if they met the following two criteria: (1) the plant is exotic to the United States, and (2) the plant occurs in upland western coniferous forests. Additional species were identified from recently published in-depth summaries of invasive plants by western U.S. bioregion (Anzinger and Radosevich 2008, Rice et al. 2008) and from a synthesis of understory response studies conducted under the Fire and Fire Surrogate national study network (Bartuszevige and Kennedy 2009).

Selected species were further designated as to "invasive potential" (Table 8.2) based on the strongest invasive response observed for the species, which may not reflect how the species would respond to disturbance in forest environments. The assignment of "invasive potential" is necessarily subjective, but it provides a comparative framework based on current knowledge of the relative threat posed by each species following disturbance. A given species' designation as to invasive potential (e.g., "responder-decreaser" or "responder-transformer") may change over time and vary at a given time in different forest ecosystems across the region. For example, *Centaurea stoebe* is a perennial forb that can invade dry conifer forests with or without disturbance, moist conifer forests only with disturbance, but does not invade subalpine forests regardless of disturbance (Table 8.2).

Nearly all of the species listed in Table 8.2 appear on the noxious plant list of at least one western state. However, we did not use federal or state noxious species lists to identify invasives because these lists are typically derived from agricultural or range management perspectives. A designation as noxious does not translate into a species' invasive potential in forested environments, or indicate its potential impacts on native plant communities or ecological services from forests.

Table 8.2 Characteristics and Invasive Potential of the Primary Exotic Plant Species Found in Coniferous Forests of the Western United States

Species[a]	Invasive Potential[b]	Dry Conifer forest[c]	Moist Conifer forest	Subalpine forest	Life-form[d]	Shade tolerance	Reproductive strategy[e]	Response to fire[f]
Acer platanoides	IT	I	I	N	PT	High	V, S	R, SD
Agrostis gigantea	RD	N	D	N	PG	Moderate	V, S	R, SD, SB
Albizia julibrissin	RD	D	N	N	PT	Low	V, S	R, SD, SB
Amaranthus retroflexus	RD	D	D	N	AF	Low	S	SD, SB
Artemisia absinthium	RD	D	D	N	PF	Moderate	S	R, SD, SB
Brachypodium sylvaticum	IT	N	I	N	PG	High	V, S	R, SD
Bromus inermis	RT	D	D	D	PG	Moderate	V, S	R, SD
Bromus japonicus	RD	D	N	N	AG	Low	S	SD, SB
Bromus tectorum	IT	D	N	N	AG	Low	S	SD, SB
Cardaria spp.	RD	D	D	N	PF	Low	V, S	R, SD, SB
Carduus nutans	RD	D	D	D	BF	Low	S	SD, SB
Centaurea diffusa	RT	I	D	D	PF	Low	S	R, SD, SB
Centaurea solstitialis	RD	D	D	N	AF	Low	S	SD, SB
Centaurea stoebe	IT	I	D	N	PF	Low	S	R, SD, SB
Chenopodium album	RD	D	D	N	AF	Low	S	R, SD
Chondrilla juncea	RD	D	N	N	PF	Low	V, S	R, SD
Cirsium arvense	RD	D	D	N	PF	Moderate	V, S	R, SD
Cirsium vulgare	RD	D	D	N	BF	Low	S	SD, SB
Convolvulus arvensis	RD	D	D	N	PF	Low	V, S	R, SD, SB

(Continued)

Table 8.2 (continued) Characteristics and Invasive Potential of the Primary Exotic Plant Species Found in Coniferous Forests of the Western United States

Species[a]	Invasive Potential[b]	Dry Conifer forest[c]	Moist Conifer forest	Subalpine forest	Life-form[d]	Shade tolerance	Reproductive strategy[e]	Response to fire[f]
Cynoglossum officinale	RD	D	D	N	BF	Low	S	SD
Cytisus spp.	RD	D	D	N	PS	Low-Mod	V, S	R, SD, SB
Dactylis glomerata	RD	D	D	D	PG	High	V, S	R, SD
Descurainia sophia	RD	D	D	D	AF	Low	S	SD, SB
Elymus repens	RD	N	D	D	PG	Low	V, S	R, SD
Erodium cicutarium	RD	D	D	N	AF	Low	S	SD, SB
Euphorbia esula	RD	D	D	N	PF	Low	V, S	R, SD, SB
Genista monspessulana	RD	D	D	N	PS	Mod-High	S	R, SD, SB
Heracleum mantegazzianum	RD	N	D	N	PF	Moderate	S	R, SD, SB
Hieracium spp.	RD	D	D	N	PF	Low	V, S	R, SD, SB
Holcus lanatus	RD	N	D	N	PG	Moderate	V, S	R, SD, SB
Hypericum perforatum	RD	D	D	N	PF	Low	V, S	R, SD, SB
Hypochaeris radicata	RD	N	D	N	PF	Low	V, S	R, SD, SB
Isatis tinctoria	RD	D	N	N	BF	Moderate	S	R, SD, SB
Lactuca serriola	RD	D	D	D	AF	Low	S	SD, SB
Linaria spp.	RT	D	D	N	PF	Moderate	V, S	R, SD, SB
Lolium perenne	RD	D	D	D	PG	Low	V, S	R, SD, SB
Medicago sativa	RD	D	D	N	PF	Low	V, S	R, SD, SB
Melilotus alba	RD	D	D	D	BF	Low	S	R, SD, SB

Species							
Melilotus officinalis	RD	D	D	D	BF	Low	S	R, SD, SB
Mycelis muralis	RD	N	D	N	AF	Moderate	S	SD, SB
Onopordum acanthium	RD	D	N	N	BF	Low	S	SD, SB
Phleum pratense	RD	D	D	D	PG	Moderate	V, S	R, SD, SB
Poa pratensis	RT	D	D	D	PG	Moderate	V, S	R, SD, SB
Potentilla recta	RT	I	N	N	PF	Moderate	S	R, SD, SB
Rosa multiflora	RD	D	D	N	PS	Moderate	V, S	R, SD, SB
Rubus discolor	RD	N	D	N	PS	Low	V, S	R, SD, SB
Rubus laciniatus	RD	N	D	N	PS	Low	V, S	R, SD, SB
Rumex acetosella	RD	D	D	N	PF	Moderate	V, S	R, SB
Salsola kali	RT	D	N	N	PF	Low	S	SD
Schedonorus arundinaceus	RT	D	D	D	PG	Moderate	V, S	R, SD, SB
Senecio jacobaea	RD	D	D	N	BF	Low	V, S	R, SD, SB
Senecio sylvaticus	RD	N	D	N	AF	Low	S	SD, SB
Sisymbrium altissimum	RD	D	D	N	AF	Low	S	SD, SB
Sonchus arvensis	RD	D	D	D	PF	Low	V, S	R, SD, SB
Spartium junceum	RD	D	D	N	PS	Low	S	R, SD, SB
Taeniatherum caput-medusae	IT	D	D	N	AG	Low	S	SD, SB
Taraxacum officinale	RD	D	D	D	PF	Low-Mod	V, S	R, SD, SB
Tragopogon dubius	RD	D	D	D	BF	Moderate	S	SD

(Continued)

Table 8.2 (continued) Characteristics and Invasive Potential of the Primary Exotic Plant Species Found in Coniferous Forests of the Western United States

Species[a]	Invasive Potential[b]	Dry Conifer forest[c]	Moist Conifer forest	Subalpine forest	Life-form[d]	Shade tolerance	Reproductive strategy[e]	Response to fire[f]
Ulex europaeus	RT	D	D	N	PS	Mod-High	V, S	R, SD, SB
Verbascum thapsus	RD	D	D	D	BF	Low	S	SD, SB
Vinca spp.	IT	N	I	N	PS	Mod-High	V	R
Vulpia myuros	RT	D	D	N	AG	Low	S	SD, SB

[a] Species included in the table were selected using the Fire Effects Information System (FEIS, www.fs.fed.us/database/feis) invasive plants list, summaries of invasive plants by western U.S. bioregion (Anzinger and Radosevich 2008, Rice et al. 2008), and a synthesis of western U.S. understory response studies (Bartuszevige and Kennedy 2009). This is an open rather than absolute or exhaustive list of exotic invaders, and will likely become longer in the future.

[b] NR = no significant response, RD = responders-decreasers, RT = responders-transformers, IT = invaders-transformers. "Invasive potential" is based on the strongest invasive response observed for the species, which may not reflect how the species would respond to disturbance in forest environments.

[c] For each forest type, N = not likely to establish high cover even with disturbance, D = invades with disturbance, I = invades with or without disturbance.

[d] A = annual, B = biennial, P = perennial; and G = graminoid, F = forb, S = shrub, T = tree.

[e] S = seed, V = vegetative.

[f] R = resprout, SD = seed (dispersed), SB = seed bank.

While there are not enough long-term data available to make definitive determinations as to which exotic responders will ultimately turn out to be transformers (Figure 8.1, Table 8.2), experimental evidence suggests that very few species in western forests currently meet this definition. Examination of Table 8.2 shows that most exotic responders in western coniferous forests are short-lived, shade-intolerant grasses or forbs that primarily regenerate from seed. However, these shared attributes provide few clues to help differentiate the benign short-term nuisance species (i.e., "responder-decreasers") from the long-term ecosystem transformers (i.e., "responder-transformers" and "invader-transformers").

Recent studies in the West that assess understory species' response to thinning and burning treatments (Griffis et al. 2001, Wienk et al. 2004, Kerns et al. 2006, Metlen and Fiedler 2006, Collins et al. 2007, Knapp et al. 2007, Dodson et al. 2008) provide little evidence that exotic invaders at these respective sites are displacing native species in the community. Instead, both exotic and native species are increasing after treatment (Metlen and Fiedler 2006, Dodson et al. 2007, Dodson et al. 2008, McGlone et al. 2009), suggesting that exotics are responding to increased resource availability and not transforming the community. While increases in invasive species abundances were often statistically significant, exotic cover seldom exceeded 2% in these studies. However, McGlone et al. (2009) reported 17% cover of the annual grass *Bromus tectorum* 5 years following restoration treatments in Arizona, indicating this species had become a major community component with the potential to increase abundance at the cost of native species (i.e., a responder-transformer). The bigger concern in all of these examples is what would happen in the future—5, 10, or more years after disturbance—as the previously cited studies report only short-term (≤5-year) responses. Will exotics stay at low levels or nearly disappear (Figure 8.2a) or will they increase and ultimately have transformative effects on the community (Figure 8.2b), and can we predict which species are likely to behave as transformers in particular

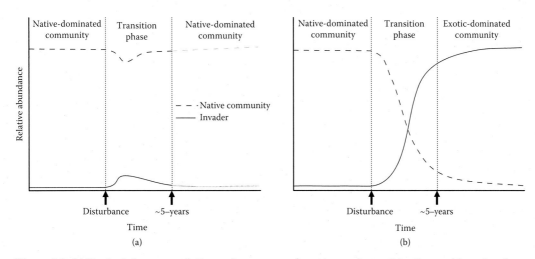

Figure 8.2 (a) Typical short-term (≤5 years) response of exotic species to thinning and burning (restoration) treatments, based on ongoing experiments at numerous locations across the West. These studies provide evidence that exotic invaders are not transforming native plant communities during the first few years after treatment, as exotic species cover averages <2% at nearly all sites. Dashed lines (in gray) extended after the fifth year indicate expected native and exotic relative abundances based on current trends. (b) Hypothesized transformation of the plant community by exotics if changing environment or propagule pressure allow exotic species to increase and dominate the community over longer time periods (i.e., >5 years).

environments? Even if exotic abundances stay low, a subsequent disturbance could induce a more ecologically significant response, since future invasion would be less dependent on external propagule pressure.

Many exotics in the western United States are dependent on disturbance to establish and remain in forested communities (Keeley 2006); indeed, nearly all of the species in Table 8.2 are expected to decline without further disturbance. Trees in particular may mitigate the competitive effects of even strongly invasive exotics such as *C. stoebe* (Metlen et al. 2012), making forests inhospitable habitat for many species that would be designated as transformers in other ecosystems.

Many exotic species can potentially establish in dry coniferous forests, especially following disturbance (Table 8.2). These forests typically border drier shrub/steppe and woodland ecosystems that are heavily invaded by exotic species and which may provide a substantial propagule source for invasion into forests. Dry coniferous forests are subject to frequent disturbance (fire) and have relatively low canopy cover, further contributing to their invasibility. A century of fire exclusion has removed the primary disturbance agent in these forests and allowed tree density to increase—a change that has likely resulted in reduced invasion by exotic species (Keeley 2006). However, fire exclusion has also contributed to increased likelihood of uncharacteristically severe wildfire in dry conifer forests, which can ultimately result in very high exotic species establishment (Crawford et al. 2001, Griffis et al. 2001).

The well-documented decline in exotics with increasing elevation (i.e., Forcella and Harvey 1983, Fisher and Fulé 2004, Pauchard and Alaback 2004, Laughlin et al. 2005, Klinger et al. 2006, Ferguson et al. 2007) can be seen in Table 8.2 by comparing the large number of exotic species found in lower-elevation dry conifer and moist conifer forests compared to higher-elevation subalpine forests. Few exotic species are considered invasive in subalpine forests without disturbance, and even with disturbance, the suite of exotic species that is likely to establish is much smaller than in other forest types. Recent studies have confirmed that even when subalpine forests burn in severe stand-replacing wildfires, exotics do not compose much of the recovering vegetation (Turner et al. 1999, Schoennagel et al. 2004, Ferguson et al. 2007). However, the threat of invasion in high-elevation ecosystems is likely to increase with ongoing globalization and climate change (Pauchard et al. 2009), which emphasizes the importance of monitoring programs to detect new threats as they arise.

At lower elevations, wildfire and treatment-induced forest disturbances are almost always followed by some level of exotic plant invasion. Yet exceptions occur. For example, Fornwalt et al. (2003) found virtually no evidence of exotic invasion following forest treatments in Colorado, and Laughlin et al. (2004) and Dodson and Peterson (2009) reported a similar lack of invasives following wildfires in Arizona and Washington, respectively. Identifying key factors, conditions, or treatment characteristics whose presence or absence short-circuited the invasion process in these examples could provide insights into strategies for avoiding or minimizing exotic invasions elsewhere.

8.3   *Exotic plant ecology*

Knowledge about exotic plant ecology and invasions has largely come from research in grassland and rangeland environments. This work has been driven by the negative effects exotic species have on native community composition, forage palatability and production, and associated economic values (DiTomaso 2000). Forest managers need to be aware of this work while a deeper knowledge base specific to forest invasions is being developed, and recently installed experiments are followed over longer time periods.

Unfavorable environmental conditions, negative interactions with native species, and chance events either resist or repel most exotic plant invasions (Mack et al. 2000); as a result, Williamson and Fitter (1996) estimate that only 0.1% of exotic plant invasions are successful. However, this seemingly inconsequential proportion of successful exotic invasions belies the considerable threat they pose to native biodiversity, particularly following disturbance. While biotic resistance—or the capacity to inhibit exotic invasion—is an emergent property of diverse native plant communities (Elton 1958, Kennedy et al. 2002, Maron and Marler 2007), disturbances that severely impact key native species can leave the entire community vulnerable to invasion.

Disturbance is an essential component of healthy and diverse native forest ecosystems (D'Antonio and Vitousek 1992, Hobbs and Huenneke 1992, Daehler 2003). For example, low- to moderate-intensity fire historically maintained many western forests in relatively open conditions with vigorous understory communities (Arno 1980, Everett et al. 2000). Over the last century, many of these forests have become dense and closed-canopied, with depauperate understories (Arno and Brown 1989, Covington and Moore 1994). Thinning and prescribed burning (restoration) treatments applied in these forests reduce fuel build-ups and increase native species richness—including richness of relatively rare native plants (Moore et al. 1999, Laughlin et al. 2004, Fiedler et al. 2006, Dodson et al. 2007), but they can also promote exotic plant invasion (Hobbs and Huenneke 1992, Dodson and Fiedler 2006, Metlen and Fiedler 2006).

Invasive exotic species often respond strongly to disturbance, and management attempts to simply return a natural disturbance (such as fire) may not be sufficient to restore a native community, particularly if the community has been degraded (D'Antonio and Meyerson 2002, Suding et al. 2004). Management activities such as thinning and burning provide a flush of available nutrients, especially nitrogen (Kaye and Hart 1998, DeLuca and Zouhar 2000, Gundale et al. 2005), which benefits natives and exotics alike, although these effects tend to be short-lived (Reich et al. 2001, Wan et al. 2001). Burning can also increase soluble phosphorus and aggressively promote exotic plant invasion (Gundale et al. 2008). In nutrient-poor ecosystems, increased nutrient availability especially benefits invasive exotics (Vitousek et al. 1987, Daehler 2003). A strong invasive response may push the system past a critical threshold to an exotic-dominated system, where feedbacks among exotics alter conditions and make it difficult to restore the native community (Figure 8.2b; D'Antonio and Meyerson 2002, Suding et al. 2004, Radosevich 2007).

Disproportionate exotic plant response to disturbance suggests that invasive exotics interact differently with their neighbors than native species do. The majority of plant species require disturbance to invade an established community (Table 8.2), but a few species are particularly successful at creating their own "disturbance" (e.g., "invaders-responders," Figure 8.1). Invasiveness can be promoted by novel plant adaptations such as potent allelochemicals (Callaway and Aschehoug 2000, He et al. 2009, Thorpe et al. 2009), unique abilities to alter resource availability (Vitousek et al. 1987, Ehrenfeld 2003), or association with arbuscular mycorrhizae (Marler et al. 1999, Callaway et al. 2004). Regardless of the specific mechanism, invading exotic species tend to exhibit strong competitive effects (D'Antonio and Mahall 1991, Bakker and Wilson 2001), particularly on species within their invaded range (Ridenour et al. 2008, He et al. 2009). Life history trade-offs generally explain invasive plant success (Blossey and Nötzold 1995, Feng et al. 2009), but a few strong invaders such as *C. stoebe* do not exhibit these trade-offs (Vilá et al. 2003, Ridenour et al. 2008). Instead, these select invaders have the ability to transform or "disturb" native systems, and their response may not conform to conventional expectations of "responder-decreaser" plant succession after disturbance (Figure 8.1). Such transformative exotic species (i.e., "responder-transformers" and "invader-transformers") may also facilitate the invasion of other exotics by altering disturbance regimes (D'Antonio and Vitousek 1992) or nutrient cycling (Vitousek et al. 1987, Sperry et al. 2006), eliminating natives that are strong competitors (Ortega and Pearson 2005, Kulmatiski 2006), or promoting soil fungi that favor exotics over natives (Jordan et al. 2008, Grman and Suding 2010).

Exotic invasions are a growing concern for broader application of forest restoration treatments in the West, yet transient exotics ("responder-decreasers") should be low priority for control (D'Antonio and Myerson 2002) because native plant communities neutralize or "eradicate" them relatively soon after disturbance (Figure 8.2a). Few plant traits are reliable indicators of effective invaders in western forests (i.e., species designated as either "responder-transformers" or "invader-transformers" in Table 8.2), supporting Daehler's (2003) contention that the characteristics that make a species particularly invasive are highly context-specific.

Although not yet observed in western forests, invasive exotics are capable of competitive exclusion. However, invasive monocultures are invariably patchy and heterogeneous at larger scales (Melbourne et al. 2007). This patchiness allows native populations to adapt to invasive species tolerances, making "experienced" populations more competitive against invasives than "naïve" populations (Callaway et al. 2005, Lau 2008, Leger 2008). Longer term, this implies that native plant communities can develop resistance to exotics; shorter term, it argues for avoiding treatments that damage native plant populations and for using "experienced" seed stock in any reseeding efforts. In summary, forest managers can learn from other systems, and these sources suggest that site- and species-specific ecological knowledge is critical for effective invasive control and native species restoration. The consequences of imprudent actions could be exotic-dominated communities in the future, where remedial treatments are neither effective nor affordable (Figure 8.3).

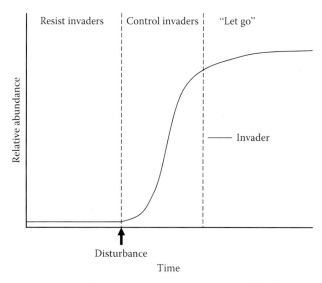

Figure 8.3 Management strategies (i.e., resisting invaders, controlling invaders, or "letting go") for addressing plant invasions, depending on current exotic species abundances and invasiveness. Measures used to resist invaders include promoting natives, spot-spraying, seeding, and education; measures to control invaders include herbicides, biocontrols, burning, and grazing; while simply "letting go" is recommended for advanced invasions on low priority sites.

8.4 Knowledge gaps and research needs

Perhaps the most pressing knowledge gap relative to exotic plant invasions in western forests is one whose solution cannot be accelerated or shortcut, and that is the need to ascertain longer-term responses of exotic species to disturbance. For example, long-term experiments across a range of forest environments can differentiate species that aggressively invade after disturbance—with potentially significant and long-term effects on the native community—from those that soon drop in abundance or disappear. Most randomized and replicated field experiments of invasive plant response in western U.S. forests report short-term (≤5-year) results. Two longer-term retrospective studies—one by Selmants and Knight (2003) working in Wyoming and another by Nelson et al. (2008) working in Washington—provide some insight into expected invasive species response to forest disturbances, although not at the species level. Both of these studies report the presence of exotic species decades after management treatments were applied (30–50 years after treatment in Wyoming and 3–19 years after treatment in Washington). However, no combination of cutting and burning treatments resulted in >1% cover of exotic species in the Wyoming study (Selmants and Knight 2003), and exotic cover averaged just 2% across the combined thinning/burning treatments in the Washington study (Nelson et al. 2008). Although long-term responses of exotic species to cutting and burning treatments were statistically significant in some cases, these responses were considered to have little ecological impact on the native plant community. These findings have limited geographic scope, but when combined with results from more broadly distributed short-term experiments, they suggest that under most conditions, restoration treatments have relatively low risk of transformative invasion. Furthermore, managers can select the location, intensity, and timing of treatments to avoid combinations with the highest invasion risk. Alternatively, not treating strategically located forests with high fire hazard may lead to more severe and extensive site impacts from wildfire in the future.

Additional critical research needs are to develop effective methods for restoring vigorous native plant communities as a means of resisting invasion and to develop affordable methods with minimal side effects for mitigating invasions once they have occurred. Common methods of preventing or neutralizing invasions—such as sowing seeds on disturbed sites, introducing biocontrols, or applying herbicides—have had mixed success, and in some cases even negative effects or outcomes. Sowing seed of supposedly innocuous exotic plant species may turn out to have unknown and unwanted side effects (Beyers 2004, Keeley 2006), and trace contamination of even certified "weed-free" seed lots (Robichaud et al. 2000) may be enough for exotics to establish high populations in disturbed environments (Hunter et al. 2006). The unintended consequences of biocontrols can at times have different but equally undesirable ecosystem impacts as the invasive species that were targeted by their release (Bush et al. 2007, Pearson and Callaway 2008). Herbicide applications have also demonstrated unexpected effects such as eradicating the target species, but allowing subsequent invasion by an even more aggressive invader (Ortega and Pearson 2010), or inducing more vigorous reinvasion by the same invader (Rinella et al. 2009).

Monitoring treatments over time is essential to identify unexpected and undesirable outcomes and is one strategy that managers can adopt immediately at relatively low cost. However, longer-term experiments that test both the efficacy and the collateral damage associated with alternative methods for controlling potential exotic transformers are a critical research need. It is also the only way to isolate specific relationships and to provide a solid base for building knowledge and framing hypotheses for further testing.

8.5   Synthesis—General relationships, management recommendations, and conclusions

8.5.1   General relationships

A short history of experimental research and management experience explains the limited knowledge base available for designing strategies to resist or control invasions. Despite these limitations, a number of important general principles or relationships have been identified:

Exotic plant invasions in western forests are not yet advanced, particularly at higher elevations, due to relatively low propagule pressure and stressful environments.

Invasiveness of exotic species varies across species, across the range of a given species, and with frequency, intensity, and type of disturbance.

Only a few exotics in the West have been identified as species capable of invading after disturbance and transforming the native community—or invading the undisturbed forest and transforming the community.

Climate change could itself become an invasion-inducing disturbance in the future, particularly at higher-elevation sites.

8.5.2   Management recommendations

The following recommendations are based on knowledge gained from short-term experiments (≤5 years), several longer-term retrospective studies, species-level ecological studies, and management observations and experiences. In way of qualification, these recommendations are based on responses and relationships documented for exotic species, climate, and disturbance conditions that prevailed in the recent past. The only certain knowledge is that the future will not exactly replicate the past, so management should consider potential climate change and associated shifts of species' ranges.

A growing body of evidence suggests that preventing or minimizing invasions is more ecologically and economically effective than trying to neutralize them after they have occurred (Klinger et al. 2006). For example, Figure 8.3 illustrates a staged management strategy for first resisting invasions before they occur and then employing more aggressive control measures depending on exotic species abundance and invasiveness. Preventative measures for resisting invaders include promoting vigorous native plant communities, seeding native species in areas where slash piles were burned, hand-pulling or spot-spraying exotics in road corridors and adjacent areas, cleaning equipment used in forest operations, and public education to reduce exotic dispersal. Applying herbicides, releasing biocontrol agents, sequential burning, and directed livestock grazing are measures used for controlling transformer species that are beginning to establish or are actively spreading. Recognizing the reality of "triage" situations points to simply "letting go" of advanced invasions in low priority or highly disturbed areas that overwhelm the resources available to address them (Hobbs and Humphries 1995).

A key first step for managers is to evaluate whether existing invaders, or invaders expected to establish after a disturbance, are species that can potentially transform the community—Table 8.2 is a useful resource for this purpose. Of the exotic invaders listed in Table 8.2, *B. tectorum* invasion of dry conifer forests presents the greatest immediate threat for transforming native plant communities. However, changing climatic patterns and the massive scale of recent disturbances may alter or influence the invasive potentials of all species listed in Table 8.2, raising the probability that some species currently classified as responders-decreasers may take on transformer status in the future. If problematic transformer species are identified, measured approaches are recommended before any action is taken, including seeking advice from forestry extension agents or forest ecologists. Caution is advised before applying chemical or biological control methods, because to date, few exotic species have been documented to have invasive potential beyond what the post-disturbance native plant community can neutralize or "eradicate." When eradication is deemed necessary, the means and methods selected to address the invaders should be carefully vetted to prevent a short-term solution from becoming a bigger, long-term problem (Rinella et al. 2009, Ortega and Pearson 2010).

The factor that managers have more direct control over at the project level is developing resistance to invasion by reducing existing plant community invasibility. One way of reducing invasibility is to weigh the location, intensity, and timing of planned treatments in light of known or suspected invaders. Specifically, when designing forest treatments to meet management objectives such as fuel reduction, forest restoration, wildlife habitat, or resource production, managers should modify prescriptions, consider seasonality of operations, and select equipment and techniques in ways that limit invasibility and benefit the native community. For example, the abundance of responders-transformers typically increases with increasing disturbance intensity, which suggests that lower intensity or incremental treatments (Dodson and Fiedler 2006) or longer intervals between treatments (Laughlin et al. 2008) may be preferred for projects with high invasion risk. Where intense single treatments or sequential thinning and burning treatments are necessary to meet management objectives, specific strategies such as seeding native understory species (Korb et al. 2004), restricting livestock grazing before but especially after treatment (Dodson and Fiedler 2006, Keeley 2006), or conducting forest thinning operations over a snowpack (Gundale et al. 2005), may limit exotic plant invasion.

8.5.3   Conclusions

This chapter synthesizes a fairly small and recent body of literature on exotic plant invasion of western coniferous forests. Most of the few extant experimental studies focus on short-term understory response (of which exotic species are a part) to various thinning and burning treatments. As longer-term data on exotic species invasion become available, current management recommendations will likely need modification or even replacement. In the interim, managers should stay observant for recurring invasion patterns and relationships, as well as for unexpected exotic species responses to disturbance, because local experience and site-specific ecological knowledge are the first lines of defense in effective invasion control.

References

Alpert, P., E. Bone, and C. Holzapfel. 2000. Invasiveness, invasibility and the role of environmental stress in the spread of non-native plants. *Perspectives in Plant Ecology, Evolution and Systematics* 3:52–66.

Anzinger, D., and S.R. Radosevich. 2008. Fire and nonnative invasive plants in the Northwest Coastal Bioregion. In *Wildland fire in ecosystems: Fire and nonnative invasive plants*, K. Zouhar, J.K. Smith, S. Sutherland, and M.L. Brooks (Eds.). USDA Forest Service General Technical Report RMRS-GTR-42. Washington, DC: U.S. Department of Agriculture, Forest Service, Washington Office.

Arno, S.F. 1980. Forest fire history in the northern Rockies. *Journal of Forestry* 78:460–465.

Arno, S.F., and J.K. Brown. 1989. Managing fire in our forests—Time for a new initiative. *Journal of Forestry* 87:44–46.

Arno, S.F., and C.E. Fiedler. 2005. *Mimicking nature's fire: Restoring fire-prone forests in the West*, Chapter 6, p. 242. Washington, DC: Island Press.

Bakker, J., and S. Wilson. 2001. Competitive abilities of introduced and native grasses. *Plant Ecology* 157:119–127.

Bartuszevige, A.M., and P.L. Kennedy. 2009. *Synthesis of knowledge on the effects of fire and thinning treatments on understory vegetation in U.S. dry forests.* Agricultural Experimental Station Special Report 1095. Oregon State University, Oregon: Agricultural Experimental Station.

Belnap, J., S.L. Phillips, S.K. Sherrod, and A. Moldenke. 2005. Soil biota can change after exotic plant invasion: Does this affect ecosystem processes? *Ecology* 86:3007–3017.

Beyers, J. 2004. Postfire seeding for erosion control: Effectiveness and impacts on native plant communities. *Conservation Biology* 18:947–956.

Blossey, B., and R. Nötzold. 1995. Evolution of increased competitive ability in invasive nonindigenous plants: A hypothesis. *Journal of Ecology* 83:887–889.

Bush, R.T., T.R. Seastedt, and D. Buckner. 2007. Plant community response to the decline of diffuse knapweed in a Colorado grassland. *Ecological Restoration* 25:169–174.

Callaway, R.M., and E.T. Aschehoug. 2000. Invasive plants versus their new and old neighbors: A mechanism for exotic invasion. *Science* 290:521–523.

Callaway, R.M., W.M. Ridenour, T. Laboski, T. Weir, and J.M. Vivanco. 2005. Natural selection for resistance to the allelopathic effects of invasive plants. *Journal of Ecology* 93:576–583.

Callaway, R.M., G.C. Thelen, S. Barth, P.W. Ramsey, and J.E. Gannon. 2004. Soil fungi alter interactions between the invader *Centaurea maculosa* and North American natives. *Ecology* 85:1062–1071.

Colautti, R.I., and H.J. MacIsaac. 2004. A neutral terminology to define "invasive" species. *Diversity and Distributions* 10:135–141.

Collins, B.M., J.J. Moghaddas, and S.L. Stephens. 2007. Initial changes in forest structure and understory plant communities following fuel reduction activities in a Sierra Nevada mixed conifer forest. *Forest Ecology and Management* 239:102–111.

Covington, W.W., and M.M. Moore. 1994. Southwestern ponderosa forest structure: Changes since Euro-American settlement. *Journal of Forestry* 92:39–47.

Crawford, J.A., C.-H.A. Wahren, S. Kyle, and W.H. Moir. 2001. Responses of exotic plant species to fires in *Pinus ponderosa* forests in northern Arizona. *Journal of Vegetation Science* 12:261–268.

Daehler, C.C. 2003. Performance comparisons of co-occurring native and alien invasive plants: Implications for conservation and restoration. *Annual Review of Ecology and Systematics* 34:183–211.

D'Antonio, C., and L.A. Meyerson. 2002. Exotic plant species as problems and solutions in ecological restoration: A synthesis. *Restoration Ecology* 10:703–713.

D'Antonio, C., and P.M. Vitousek. 1992. Biological invasions by exotic grasses, the grass/fire cycle, and global change. *Annual Review of Ecology and Systematics* 23:63–87.

D'Antonio, C.M., and B.E. Mahall. 1991. Root profiles and competition between the invasive exotic perennial, *Carpobrotus edulis*, and two native shrub species in California coastal scrub. *American Journal of Botany* 78:885–894.

DeLuca, T.H., and K.L. Zouhar. 2000. Effects of selection harvest and prescribed fire on the soil nitrogen status of ponderosa pine forests. *Forest Ecology and Management* 138:263–271.

DiTomaso, J.M. 2000. Invasive weeds in rangelands: Species, impacts, and management. *Weed Science* 48:255–265.

Dodson, E.K., and C.E. Fiedler. 2006. Impacts of restoration treatments on alien plant invasion in *Pinus ponderosa* forests, Montana, USA. *Journal of Applied Ecology* 43:887–897.

Dodson, E.K., K.L. Metlen, and C.E. Fiedler. 2007. Common and uncommon understory species differentially respond to restoration treatments in ponderosa pine/Douglas-fir forests, Montana. *Restoration Ecology* 15:696–708.

Dodson, E.K., and D.W. Peterson. 2009. Seeding and fertilization effects on plant cover and community recovery following wildfire in the eastern Cascade Mountains, USA. *Forest Ecology and Management* 258:1586–1593.

Dodson, E.K., D.W. Peterson, and R.J. Harrod. 2008. Understory vegetation response to thinning and burning restoration treatments in dry conifer forests of the eastern Cascades, USA. *Forest Ecology and Management* 255:3130–3140.

Ehrenfeld, J.G. 2003. Effects of exotic plant invasions on soil nutrient cycling processes. *Ecosystems* 6:503–523.

Elton, C.S. 1958. *The ecology of invasions by animals and plants.* London, UK: Methuen Ltd.

Everett, R.L., R. Schellhaas, D. Keenum, D. Spurbeck, and P. Ohlson. 2000. Fire history in the ponderosa pine/Douglas-fir forests on the east slope of the Washington Cascades. *Forest Ecology and Management* 129:207–225.

FEIS. Fire Effects Information System, [Online]. U.S. Department of Agriculture, Forest Service, Rocky Mountain Research Station, Fire Sciences Laboratory (Producer). Available: http://www.fs.fed.us/database/feis.

Feng, Y.-L., Y.-B. Lei, R.-F. Wang, R.M. Callaway, A. Valiente-Banuet, Inderjit, Y.-P. Li, and Y.-L. Zheng. 2009. Evolutionary tradeoffs for nitrogen allocation to photosynthesis versus cell walls in an invasive plant. *Proceedings of the National Academy of Sciences* 106:1853–1856.

Ferguson, D.E., C.L. Craig, and K.Z. Schneider. 2007. Spotted knapweed (*Centaurea biebersteinii*) response to forest wildfires on the Bitterroot National Forest, Montana. *Northwest Science* 81:138–146.

Fiedler, C.E., K.L. Metlen, and E.K. Dodson. 2006. Restoration/fuels reduction treatments differentially affect native and exotic understory species in a ponderosa pine forest (Montana). *Ecological Restoration* 24:44–46.

Fisher, M.A., and P.Z. Fulé. 2004 Changes in forest vegetation and arbuscular mycorrhizae along a steep elevation gradient in Arizona. *Forest Ecology and Management* 200:293–311.

Forcella, F., and S.J. Harvey. 1983. Eurasian weed infestation in western Montana in relation to vegetation and disturbance. *Madrono* 30:102–109.

Fornwalt, P.J., M.R. Kaufmann, L.S. Huckaby, J.M. Stoker, and T.J. Stohlgren. 2003. Non-native plant invasions in managed and protected ponderosa pine/Douglas-fir forests of the Colorado Front Range. *Forest Ecology and Management* 177:515–527.

Griffis, K.L., J.A. Crawford, M.R. Wagner, and W.H. Moir. 2001. Understory response to management treatments in northern Arizona ponderosa pine forests. *Forest Ecology and Management* 146:239–245.

Grman, E., and K.N. Suding. 2010. Within-year soil legacies contribute to strong priority effects of exotics on native California grassland communities. *Restoration Ecology* 18:664–670.

Gundale, M.J., T.H. DeLuca, C.E. Fiedler, P.W. Ramsey, M.G. Harrington, and J.E. Gannon. 2005. Restoration treatments in a Montana ponderosa pine forest: Effects on soil physical, chemical and biological properties. *Forest Ecology and Management* 213:25–38.

Gundale, M.J., S. Sutherland, and T.H. DeLuca. 2008. Fire, native species, and soil resource interactions regulate the spatio-temporal invasion pattern of *Bromus tectorum. Ecography* 31:201–210.

Hawkes, C.V., I.F. Wren, D.H. Herman, and M.K. Firestone. 2005. Plant invasion alters nitrogen cycling by modifying the soil nitrifying community. *Ecology Letters* 8:976–985.

He, W.-M., Y. Feng, W.M. Ridenour, G.C. Thelen, J.L. Pollock, A. Diaconu, and R.M. Callaway. 2009. Novel weapons and invasion: Biogeographic differences in the competitive effects of *Centaurea maculosa* and its root exudate (±)-catechin. *Oecologia* 159:803–815.

Hejda, M., P. Pyšek, and V. Jarošík. 2009. Impact of invasive plants on the species richness, diversity and composition of invaded communities. *Journal of Ecology* 97:393–403.

Hobbs, R.J., and L.F. Huenneke. 1992. Disturbance, diversity, and invasion: Implications for conservation. *Conservation Biology* 6:324–337.

Hobbs, R.J., and S.E. Humphries. 1995. An integrated approach to the ecology and management of plant invasions. *Conservation Biology* 9:761–770.

Holechek, J.L., R.D. Pieper, and C.H. Herbel. 1998. Range management history. In *Range management: Principles and practices,* pp. 29–39, 3rd ed. Upper Saddle River, NJ: Prentice Hall.

Hunter, M.E., P.N. Omi, E.J. Martinson, and G.W. Chong. 2006. Establishment of non-native plant species after wildfires: Effects of fuel treatments, abiotic and biotic factors, and post-fire grass seeding treatments. *International Journal of Wildland Fire* 15:271–281.

Huston, M.A. 2004. Management strategies for plant invasions: Manipulating productivity, disturbance, and competition. *Diversity and Distributions* 10:167–178.

Jordan, N.R., D.L. Larson, and S.C. Huerd. 2008. Soil modification by invasive plants: Effects on native and invasive species of mixed-grass prairies. *Biological Invasions* 10:177–190.

Kaye, J.P., and S.C. Hart. 1998. Ecological restoration alters nitrogen transformations in a ponderosa pine-bunchgrass ecosystem. *Ecological Applications* 8:1052–1060.

Keeley, J.E. 2006. Fire management impacts on invasive plants in the western United States. *Conservation Biology* 20:375–384.

Keeley, J.E., D. Lubin, and C.J. Fotheringham. 2003. Fire and grazing impacts on plant diversity and alien plant invasions in the southern Sierra Nevada. *Ecological Applications* 13:1355–1374.

Kennedy, T.A., S. Naeem, K.M. Howe, J.M.H. Knops, D. Tilman, and P. Reich. 2002. Biodiversity as a barrier to ecological invasion. *Nature* 417:636–628.

Kerns, B.K., W.G. Thies, and C.G. Niwa. 2006. Season and severity of prescribed burn in ponderosa pine forests: Implications for understory native and exotic plants. *Ecoscience* 13:44–55.

Klinger, R., E.C. Underwood, and P.E. Moore. 2006. The role of environmental gradients in nonnative plant invasion into burnt areas of Yosemite National Park, California. *Diversity and Distributions* 12:139–156.

Knapp, E.E., D.W. Schwilk, J.M. Kane, and J.E. Keeley. 2007. Role of burning season on initial understory response to prescribed fire in a mixed conifer forest. *Canadian Journal of Forest Research* 37:11–22.

Korb, J.E., N.C. Johnson, and W.W. Covington. 2004. Slash pile burning effects on soil biota and chemical properties and plant establishment: Recommendations for amelioration. *Restoration Ecology* 12:52–62.

Kulmatiski, A. 2006. Exotic plants establish persistent communities. *Plant Ecology* 187:261–275.

Lau, J.A. 2008. Beyond the ecological: Biological invasions alter natural selection on a native plant species. *Ecology* 89:1023–1031.

Laughlin, D.C., J.D. Bakker, M.L. Daniels, M.M. Moore, C.A. Casey, and J.D. Springer. 2008. Restoring plant species diversity and community composition in a ponderosa pine-bunchgrass ecosystem. *Plant Ecology* 197:139–151.

Laughlin, D.C., J.D. Bakker, and P.Z. Fulé. 2005. Understory plant community structure in lower montane and subalpine forests, Grand Canyon National Park, USA. *Journal of Biogeography* 32:2083–2102.

Laughlin, D.C., J.D. Bakker, M.T. Stoddard, M.L. Daniels, J.D. Springer, C.N. Gildar, A.M. Green, and W.W. Covington. 2004. Toward reference conditions: Wildfire effects on flora in an old-growth ponderosa pine forest. *Forest Ecology and Management* 199:137–152.

Leger, E.A. 2008. The adaptive value of remnant native plants in invaded communities: An example from the Great Basin. *Ecological Applications* 18:1226–1235.

MacDougall, A.S., and R. Turkington. 2005. Are invasive species the drivers or passengers of change in degraded ecosystems? *Ecology* 86:42–55.

Mack, R.N., D. Simberloff, W.M. Lonsdale, H. Evans, M. Clout, and F.A. Bazzaz. 2000. Biotic invasions: Causes, epidemiology, global consequences, and control. *Ecological Applications* 10:689–710.

Marler, M.J., C.A. Zabinski, and R.M. Callaway. 1999. Mycorrhizae indirectly enhance competitive effects of an invasive forb on a native bunchgrass. *Ecology* 80:1180–1186.

Maron, J., and M. Marler. 2007. Native plant diversity resists invasion at both low and high resource levels. *Ecology* 88:2651–2661.

McGlone, C.M., J.D. Springer, and D.C. Laughlin. 2009. Can pine forest restoration promote a diverse and abundant understory and simultaneously resist nonnative invasion? *Forest Ecology and Management* 258:2638–2646.

Melbourne, B.A., H.V. Cornell, K.F. Davies, C.J. Dugaw, S. Elmendorf, A.L. Freestone, R.J. Hall, S. Harrison, A. Hastings, M. Holland, M. Holyoak, J. Lambrinos, K. Moore, and H. Yokomizo. 2007. Invasion in a heterogeneous world: Resistance, coexistence or hostile takeover? *Ecology Letters* 10:77–94.

Metlen, K.L., and C.E. Fiedler. 2006. Restoration treatment effects on the understory of ponderosa pine/Douglas-fir forests in western Montana, USA. *Forest Ecology and Management* 222:355–369.

Metlen, K.L., E.K. Aschehoug, and R.M. Callaway. In press. Competitive outcomes between two exotic invaders are modified by direct and indirect effects of a native conifer. *Oikos*.

Moore, M.M., W.W. Covington, and P.Z. Fulé. 1999. Reference conditions and ecological restoration: A southwestern ponderosa pine perspective. *Ecological Applications* 9:1266–1277.

Nelson, C.R., C.B. Halpern, and J.K. Agee. 2008. Thinning and burning result in low-level invasion by non-native plants but neutral effects on natives. *Ecological Applications* 18:762–770.

Ortega, Y.K., and D.E. Pearson. 2005. Weak vs. strong invaders of natural plant communities: Assessing invasibility and impact. *Ecological Applications* 15:651–661.

Ortega, Y.K., and D.E. Pearson. 2010. Effects of picloram application on community dominants vary with initial levels of spotted knapweed invasion. *Invasive Plant Science and Management* 3:70–80.

Parks, C.G., S.R. Radosevich, B.A. Endress, B.J. Naylor, D. Anzinger, L.J. Rew, B. Maxwell, and K.A. Dwire. 2005. Natural and land-use history of the North west mountain ecoregions (USA) in relation to patterns of plant invasions. *Perspectives in Plant Ecology, Evolution and Systematics* 7:137–158.

Pauchard, A., and P.B. Alaback. 2004. Influence of elevation, land use, and landscape context on patterns of alien plant invasions along roadsides in protected areas of south–central Chile. *Conservation Biology* 18:238–248.

Pauchard, A., C. Kueffer, H. Dietz, C.C. Daehler, J. Alexander, P.J. Edwards, J.R. Arévalo, L.A. Cavieres, A. Guisan, S. Haider, G. Jakobs, K. McDougall, C.I. Millar, B.J. Naylor, C.G. Parks, L.J. Rew, and T. Seipel. 2009. Ain't no mountain high enough: Plant invasions reaching new elevations. *Frontiers in Ecology and the Environment* 7:479–486.

Pearson, D.E., and R.M. Callway. 2008. Weed-biocontrol insects reduce native-plant recruitment through second-order apparent competition. *Ecological Applications* 18:1489–1500.

Radosevich, S. 2007. Plant invasions and their management. In *Invasive plant management*, Chapter 3, CIPM (Ed.), CIPM online textbook. Bozeman, MT: Center for Invasive Plant Management.

Randall, J.M., and M. Rejmánek. 1993. Interference of bull thistle (*Cirsium vulgare*) with growth of ponderosa pine (*Pinus ponderosa*) seedlings in a forest plantation. *Canadian Journal of Forest Research* 23:1507–1513.

Reich, P.B., D.W. Peterson, D.A. Wedin, and K. Wrage. 2001. Fire and vegetation effects on productivity and nitrogen cycling across a forest-grassland continuum. *Ecology* 82:1703–1719.

Reinhart, K.O., E. Greene, and. R.M. Callaway. 2005. Effects of *Acer platanoides* invasion on understory plant communities and tree regeneration in the northern Rocky Mountains. *Ecography* 28:573–582.

Rice, P.M., G.R. McPherson, and L.J. Rew. 2008. Fire and nonnative invasive plants in the Interior West Bioregion. In *Wildland fire in ecosystems: Fire and nonnative invasive plants*, Chapter 8, Zouhar, K., J.K. Smith, S. Sutherland, and M.L. Brooks (Eds.). USDA Forest Service General Technical Report RMRS-GTR-42. Washington, DC: U.S. Department of Agriculture, Forest Service, Washington Office.

Richardson, D.M., P. Pyšek, M. Rejmánek, M.G. Barbour, F.D. Panetta, and C.J. West. 2000. Naturalization and invasion of alien plants: Concepts and definitions. *Diversity and Distributions* 6:93–107.

Ridenour, W.M., J.M. Vivanco, Y. Feng, J. Horiuchi, and R.M. Callaway. 2008. No evidence for trade-offs: *Centaurea* plants from America are better competitors and defenders. *Ecological Monographs* 78:369–386.

Rinella, M.J., B.D. Maxwell, P.K. Fay, T. Weaver, and R.L. Sheley. 2009. Control effort exacerbates invasive-species problem. *Ecological Applications* 19:155–162.

Robichaud, P.R., J.L. Beyers, and D.G. Neary. 2000. *Evaluating the effectiveness of postfire rehabilitation treatments*. USDA Forest Service General Technical Report RMRS-GTR-63. Washington, DC: U.S. Department of Agriculture, Forest Service, Washington Office.

Schoennagel, T., D.M. Waller, M.G. Turner, and W.H. Romme. 2004. The effect of fire interval on post-fire understorey communities in Yellowstone National Park. *Journal of Vegetation Science* 15:797–806.

Selmants, P.C., and D.H. Knight. 2003. Understory plant species composition 30–50 years after clearcutting in southeastern Wyoming coniferous forests. *Forest Ecology and Management* 185:275–289.

Smith, W.B., P.D. Miles, C.H. Perry, and S.A. Pugh. 2008. *Forest resources of the United States, 2007.* USDA Forest Service General Technical Report WO-78. Washington, DC: U.S. Department of Agriculture, Forest Service, Washington Office.

Sperry, L.J., J. Belnap, and R.D. Evans. 2006. *Bromus tectorum* invasion alters nitrogen dynamics in an undisturbed arid grassland ecosystem. *Ecology* 87:603–615.

Suding, K.N., J.R. Larson, E. Thorsos, H. Steltzer, and W.D. Bowman. 2004. Species effects on resource supply rates: Do they influence competitive interactions? *Plant Ecology* 175:47–58.

Thorpe, A.S., G.C. Thelen, A. Diaconu, and R.M. Callaway. 2009. Root exudate is allelopathic in invaded community but not in native community: Field evidence for the novel weapons hypothesis. *Journal of Ecology* 97:641–645.

Trammell, M.A., and J.L. Butler. 1995. Effects of exotic plants on native ungulate use of habitat. *Journal of Wildlife Management* 59:808–816.

Turner, M.G., W.H. Romme, and R.H. Gardner. 1999. Prefire heterogeneity, fire severity, and early postfire plant reestablishment in subalpine forests of Yellowstone National Park, Wyoming. *International Journal of Wildland Fire* 9:21–36.

Vilá, M., A. Gómez, and J.L. Maron. 2003. Are alien plants more competitive than their native conspecifics? A test using *Hypericum perforatum* L. *Oecologia* 137:211–215.

Vitousek, P.M., L.R. Walker, L.D. Whiteaker, D. Mueller-Dumbois, and P.A. Matson. 1987. Biological invasion by *Myrica faya* alters ecosystem development in Hawaii. *Science* 238:802–808.

Wan, S., D. Hui, and Y. Luo. 2001. Fire effects on nitrogen pools and dynamics in terrestrial ecosystems: A meta-analysis. *Ecological Applications* 11:1349–1365.

Westerling, A.L., H.G. Hidalgo, D.R. Cayan, and T.W. Swetnam. 2006. Warming and earlier spring increase western U.S. forest wildfire activity. *Science* 12:940–943.

Wienk, C.L., C.H. Sieg, and G.R. McPherson. 2004. Evaluating the role of cutting treatments, fire and soil seed banks in an experimental framework in ponderosa pine forests of the Black Hills, South Dakota. *Forest Ecology and Management* 192:375–393.

Williamson, M., and A. Fitter. 1996. The varying success of invaders. *Ecology* 77:1661–1666.

Withers, M.A., M.W. Palmer, G.L. Wade, P.S. White, and P.R. Neal. 1999. Changing patterns in the number of species in North American floras. In *Perspectives on the land-use history of North America: A context for understanding our changing environment*, pp. 23–32, T.D. Sisk (Ed.), Biological Science Report USGS/BRD/BSR-1998-0003 (Revised September 1999). U.S. Geological Survey, Biological Resources Division, Reston, VA.

Yurkonis, K.A., S.J. Meiners, and B.E. Wachholder. 2005. Invasion impacts diversity through altered community dynamics. *Journal of Ecology* 93:1053–1061.

chapter nine

Effects of silvicultural practices on invasive plant species abundance in the Missouri Ozark forests of the central United States

R.M. Muzika and S.J. Farrington

Contents

9.1 Introduction

Shifts in community composition with forest management activities have been well documented but also controversial. With increasing concern about nonnative plant species pressure and the recognition and use of a variety of silvicultural approaches, continued investigation is necessary. Since the seminal work of Elton (1958), repeated studies have demonstrated that disturbance promotes invasive species, a concept solidified by Lozon and MacIssacs's (1997) review showing that 86% of the exotic species they studied were associated with disturbance. Disturbance, however, stands distinct from management, the latter tending to be more systematic and at times more extensive. Furthermore, many forest management practices and contemporary silvicultural approaches strive to reduce impact on the ecosystems in contrast with those of previous decades.

Possibly more important than the direct effect of tree removal and opening of the canopy in forest operations is the associated changes from the operation. The construction of roads and the development of edge habitats can provide opportunity for colonization, particularly of ruderal species, or for range expansion of established species. Along with the associated effects, any intensive forest management may be assumed to promote invasive species. Unequivocal evidence does not exist to suggest that the generic forest practices will affect plant species (invasive or native) in a uniform, predictable manner. Rather, the wide ranging type of forest harvest, from small single-tree removal to extensive clear-cut, prevents a broad, all-encompassing assumption. Furthermore, because the suite of invasive plants is so great, that is, at least 1000 plant species have been documented in North America alone, each plant species will respond in a unique fashion to disturbance, including forest management practices.

Recent forest-specific investigations have revealed that invasive species' success has been found to be inversely related to successional age (Flory and Clay 2006) and plant species richness (Belote et al. 2008) and positively related to mesic site conditions (Meekins and McCarthy 2001). Belote et al. (2008) specifically evaluated the effect of disturbance intensity and species richness at various scales to provide a comprehensive examination of invasion potential. The comprehensive article by Martin et al. (2009) covers much of the primary hypothetical framework of invasive plant species ecology. However, the specific question of management activity and effects on invasibility was not presented. Despite a vast amount of research including experimentation, the role of disturbance and forest management on the invasion and development of nonnative species has remained elusive and inconclusive.

The goal of this chapter is to examine if specific forest management practices have influenced the presence and abundance of invasive plant species over an approximately 15-year time frame in the Missouri Forest Ecosystem Project (MOFEP). This project represents a unique opportunity to analyze repeated sampling of permanent plots in forests that had been harvested under varying silvicultural treatments.

9.2   Methods

MOFEP is a long-term, multidisciplinary research project in the Missouri Ozarks. This project is located in the Ozark Highlands of Missouri. It was designed to investigate the longer term and large-scale events of varying forest management. The project uses several silvicultural treatments to evaluate a range of effects of management on many facets of the ecosystem, including vegetation and wildlife abundance. The overall management objective of MOFEP is to sustainably regenerate oak-hickory forests and woodlands while maintaining or restoring natural communities. The study objective specific to the vegetation sampling of the ground flora study is to determine the long-term, landscape-scale effects of different timber harvest regimes on the herbaceous natural community.

This project includes the following three treatments that represent the range of forest management practices on the forested land in Missouri:

1. No harvest treatment (control)
2. Even-aged management, including clear-cuts and intermediate thinning
3. Uneven-aged management, including single-tree selection and group openings

Group openings range in size and the prescription calls for 21 m in diameter on south-facing slopes, 32 m in diameter in level areas, and 43 m in diameter on north-facing slopes (Sheriff 2002). The average size of the clear-cut was 5 ha and that of the thinned areas was 4 ha (Kabrick et al. 2002).

There were nine total sites in this project, grouped into three replicated blocks. Each block was approximately 400 ha. Treatment type was randomly allocated to the sites. Within each site, there were 70–76 plots for a total of 648 plots across the entire study. Within each plot, data were collected from a total of 16 m^2 quadrats. Within each quadrat, each species <1 m tall was identified and the coverage of the species was estimated to the nearest 5%.

The first harvest occurred in 1996. Data were collected for 3 years before treatment. Vegetation data were collected again, 3 years after harvest, starting in 1999, 2000, and 2001. Vegetation sampling was conducted in a subsample of plots in 2002 and 2007. We completed a full sampling again in 2009 and 2010, with the next harvest scheduled for 2011.

9.3   Results and discussion

Invasive species accounted for a surprisingly minor component of the vegetation sampled. Only 0.1% of the records of the entire vegetation database from this expansive project was invasive plant species. In any given sampling year, there were no more than 15 nonnative species and most of them were rare. That alone is of relevance, since most areas of the temperate forest were likely to have a far greater number of species and abundance of invasive species. Over half the records of exotic plants were represented by five plant species; these species comprised the dominant nonnative species. Subsequent analysis consisted of these dominant nonnative species in the project: *Celastrus orbiculatus* (oriental bittersweet), *Elaeagnus umbellata* (autumn olive), *Lespedeza cuneata* (sericea lespedeza), *Rosa multiflora* (multiflora rose), and *Perilla frutescens* (beefsteak plant).

As of 2010, all plot types, irrespective of treatment, have roughly equivalent species richness (of all species) as they did in 1993. However, there were slight increases in all plots (including control) following the first harvest year, 1996 (Farrington, unpublished report). A total of 21 species that were present in 1993 were not evident in the data collection of 2010. Included in this figure were two common nonnative species: *Daucus carota* (Queen Anne's lace) and *Rumex acetosella* (sheep sorrel). *Lonicera japonica* (Japanese honeysuckle) was sampled but only in the control plots and only during the recent years, 2009 and 2010. Similarly, *C. orbiculatus* appeared only in 2009 and 2010.

The invasive shrub, *R. multiflora*, occurred more frequently than other invasive species; however, this species was present in just over 4% of all the plots (Table 9.1). *R. mutliflora* and *E. umbellata* (autumn olive) were the only species that occurred in the plots in each year. The latter notably increased in frequency of occurrence with successive years. Yates et al. (2004) suggest that *E. umbellate* can establish and continue development in both interior habitats and edge habitats and, therefore, may represent the most substantive ecological threat. These aggressive shrubs were planted by State and Federal agencies, as was *Lespedeza cuneata* (sericea lespedeza). *L. cuneata*, however, appeared to have declined in occurrence over time, not occurring in the sample since 2002. Bearing in mind that these data represent repeated measures, as the plots are permanent, *L. cuneata* may decline owing to increased overstory density.

Although both *R. muliflora* and *E. umbellate* were present each year, these were not found on all treatment types and their coverage varied (Table 9.2). Although *C. orbiculatus* appeared only in 2009 and 2010, it was evident in most treatment types. Contrary to expectations, the clear-cut treatment did not support the coverage of these nonnative species as well as other treatments. The control plots ("L" treatment) consistently supported

Table 9.1 The Frequency of Occurrence by Year of the Major Nonnative Plant Species on MOFEP Plots

Species	Year							
	1993	1994	1995	1999	2000	2001	2009	2010
Celastrus orbiculatus	–	–	–	–	–	–	2.16	2.93
Elaeagnus umbellata	0.31	0.46	1.70	1.23	1.39	1.85	2.16	2.47
Lespedeza cuneata	2.16	0.46	2.16	0.31	1.08	0.15	–	–
Lonicera japonica	–	–	–	–	–	–	0.15	0.15
Perilla frutescens	–	–	–	0.46	0.31	0.62	0.31	0.93
Rosa multiflora	3.55	2.62	2.16	4.17	2.16	3.55	2.47	4.01

Data reported are the percentage of plots in which the given species occurred in that year, across all treatment types. Only years in which a complete sample was conducted are reported.

Table 9.2 Mean Cover for Species Across All Treatment Types and Years

Year	Celastrus orbiculatus Treatment				Elaeagnus umbellata Treatment				Lespedeza cuneata Treatment					Perilla frutescens Treatment				Rosa multiflora Treatment				
	CC	I	L	UG	CC	I	L	U	CC	I	L	U	UG	CC	I	L	U	CC	I	L	U	UG
1993							2			1	<1	1							3	1	1	1
1994						10	3				1								1	2	2	1
1995						10	3	<1			<1									2	4	2
1999						4	3				1			3						4	9	
2000						9	5		2		<1	<1				1	3			6	8	
2001						4	4				<1				4	<1	<1			5	10	<1
2002						22							<1		2				3	2		
2007	1	12			1	3								2				2		8	7	
2009	2	9	12		<1	10	8	1								1	<1			4	48	25
2010	2	8	10	6	<1	6	7									<1	<1			6	27	8

In the years 2002 and 2007, a subsample across all treatment types was conducted. Treatment types: CC, clear-cut; I, intermediate, thinning; L, leave, control; U, uneven-aged management—single-tree selection; UG, uneven-aged management—group selection.

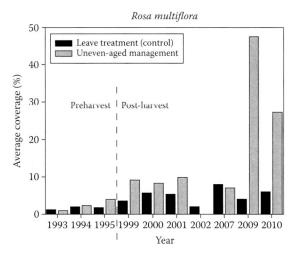

Figure 9.1 The mean percent plot coverage of *Rosa multiflora* at long-term study plots of the MOFEP project. Vegetation was not sampled in the years 1996–1998, 2003–2006, and 2008. Only a subsample was collected during 2002 and 2007. Uneven-aged management refers to small canopy openings or group openings.

the populations of *R. multiflora* and, except for the subsample years of 2002 and 2007, *E. umbellata*. Despite subsampling in 2002, *E. umbellata* reached its maximum coverage sampled in this study on an intermediate treatment site (even-aged management). The uneven-aged management, both single tree and group selection, promoted *R. multiflora* (Table 9.2).

When contrasting the control with uneven-aged management, it was obvious that disturbance and canopy opening increased the abundance of *R. multiflora* overtime (Figure 9.1). By 2009 and 2010, its coverage was significantly greater than in previous year and greater in plots that had been managed with uneven-aged approaches when compared with the controls. *E. umbellata* coverage was dissimilar on the sites before harvest, confounding the contrasts. However, control sites remained lower than the intermediate thinning treatment (Figure 9.2). *E. umbellata* increased in coverage on control sites and the abundance on intermediate thinning sites was comparable with preharvest levels.

Life history characteristics will dictate the success of invasive plant species (Myers and Bazely 2003). Many of the species identified in this project tend to be ruderal and under-story-intolerant. These prefer edge habitats possibly created by logging activity. However, perennial, long-lived, woody shrubs and lianas appear to be well established within this forested landscape of the Ozarks and also the invaders, *E. umbellata* and *R. multiflora*, among them. The conventional wisdom of assuming that edge habitat promoted invasive species must be reconsidered in the context of large block of forested landscape. Martin et al. (2009) reveal that 52% of North American invasive in forest was shade-tolerant. Although research has suggested that relatively undisturbed forests are resistant to invasion, 139 exotic plant species were known to have invaded deeply shaded forest understories that have not undergone substantial disturbance (Martin et al. 2009). As supporting evidence, the MOFEP data indicated that invasive species on control sites were neither more nor abundant nor more frequent than on harvested sites.

When posing the question of whether management activity, even small canopy opening, has promoted invasive species, our data were inconclusive. There were no trends in

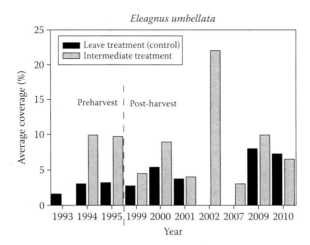

Figure 9.2 The mean percent plot coverage of *Elaeagnus umbellata* at long-term study plots of the MOFEP project. Vegetation was not sampled in the years 1996–1998, 2003–2006, and 2008. Only a subsample was collected during 2002 and 2007. The intermediate treatment represents thinning stands that will be eventually clear-cut under even-aged management approach.

the data; furthermore, the low incidence of nonnative plants challenged the ability to identify differences among treatments. Generally, single-tree selection might be considered the harvest type that caused the least disturbance. On these sites, however, *R. multiflora* appeared to do well. Burnham and Lee (2010) found that larger disturbance tend to promote the invasive *Frangula alnus*. An important finding was that small gaps were sinks. If *R. multiflora* or other species declines in uneven-aged management, we may be able to support that finding. However, more time is needed to evaluate the fate of specific populations of these species.

The findings of Belote et al. (2008) indicate that the relationship between native and nonnative plant species' richness responded to an experimentally applied disturbance gradient (from no disturbance up to clear-cut) in oak-dominated forests. The suggestion was that more species might invade species-rich communities than species-poor communities and that disturbance intensity increased both native and nonnative species richness. They found no evidence to suggest resistance of invasive by native species richness, rather disturbance and resource availability determined strength of invasion.

The MOFEP project does not include fire as a treatment and likely such a treatment will influence the abundance and potential invasion of a managed forest landscape. The response of the understory vegetation to fire contrasts sharply with response to clear-cutting and fire results in a more predictable assemblage of vegetation than harvesting (Sasseen and Muzika 2004).

A disproportionate amount of the research on invasive plant species focus on species that tend to interfere with productivity, negatively affect plant species composition and alter plant species diversity. For example, species such as *Alliaria petiolata* (garlic mustard), *Lonciera maackii* (Amur honeysuckle), and *Microstegium vimineum* (Nepalese browntop) tend to be common topics of research. These species can all invade shaded or semi-shaded habitat. Disturbance encourages *M. vimineum*, but growth of this species has been found to directly interfere with regeneration (Oswalt et al. 2007). Examination of species' competitive ability in the forest understory may be the net steps for understanding the long-term expansion of invasive species and the deleterious effects on favored woody species regeneration.

The Ozarks appear to be a moderately invaded landscape and is relatively free of invasive plant relative to other areas of the eastern deciduous forest. This area lacks an agricultural land use history and, although extensively logged in the late 19th and early 20th century, forest succession has occurred with very little additional, land conversion disturbances. Studying old-growth forest understory vegetation in Indiana, Weber and Gibson (2007) found that exotics accounted for <1% of total plant cover. There were only 25 species of nonnative plants that occurred in Missouri Forest Inventory and Analysis (FIA) data during 2005–2006 (Moser et al. 2008). *R. multiflora* rose was the most frequently found species, present on 36% of the plots.

The area has low invasion, and it is not clear why there are fewer species and lower abundance. The Ozarks represents a relatively large forested landscape, with low human population density. Consequently, a limited road network may account for the moderate number of invasive species, as distance from road was strongly associated with declining exotic species abundance (Joly et al. 2011, Flory and Clay 2006). Propagule limitation may account for the absence of invasive species, and the notable and most dominant species were those that were introduced by management in previous decades (Davis et al. 2000).

9.4   Conclusions

Continued evaluation of the vegetation community at MOFEP can serve both a monitoring goal and a role for understanding ecological characteristics that promote or hinder invasive species. Ecosystem and community-level responses to disturbance and management must be understood in the context of the long-term dynamics of forest communities. Although this manuscript provides a synopsis of 17 years of data, it represents only a brief perspective on the potential long-term effects of forest harvesting.

References

Belote, R.T., R.H. Jones, S.M. Hood, and B.W. Wender. 2008. Diversity-invasibility across an experimental disturbance gradient in Appalachian forests. *Ecology* 89:183–192.

Burnham, K.M., and T.D. Lee. 2010. Canopy gaps facilitate establishment, growth, and reproduction of invasive *Frangula alnus* in a *Tsuga canadensis* dominated forest. *Biological Invasions* 12:1509–1520.

Davis, M.A., J.P. Grime, and K. Thompson. 2000. Fluctuating resources in plant communities: A general theory of invasibility. *Journal of Ecology* 88:528–534.

Elton, C.S. 1958. *The Ecology of Invasions by Animals and Plants.* London, UK: Chapman and Hall.

Flory, S.L., and K. Clay. 2006. Invasive shrub distribution varies with distance to roads and stand age in eastern deciduous forests in Indiana, USA. *Plant Ecology* 184:131–141.

Joly, M., P. Bertrand, R.Y. Gbangou, M.C. White, J. Dubé, and C. Lavoie. 2011. Paving the way for invasive species: Road type and the spread of Common ragweed (*Ambrosia artemisiifolia*). *Environmental Management* 48:514–522.

Kabrick, J.M., R.G. Jensen, S.R. Shifley, and D.R. Larsen. 2002. Woody vegetation following even-aged, uneven-aged, and no-harvest treatments on the Missouri Ozark Forest Ecosystem Project Sites. In *Proceedings of the Second Missouri Ozark Forest Ecosystem Project Symposium: Post-Treatment Results of the Landscape Experiment*, eds. S.R. Shifley, and J.M. Kabrick, pp. 84–101. Gen. Tech. Rep. NC-227. St. Paul, MN: U.S. Department of Agriculture, Forest Service, North Central Forest Experiment Station.

Lozon, J.D., and H.J. MacIssac. 1997. Biological invasions: Are they dependent on disturbance? *Environmental Review* 5:131–141.

Martin, P.H., C.D. Canham, and P.L. Marks. 2009. Why forests appear resistant to exotic plant invasions: Intentional introductions, stand dynamics, and the role of shade tolerance. *Frontiers in Ecology and the Environment* 7:142–149.

Meekins, J.F. and B.C. McCarthy. 2001. Effect of environmental variation on the invasive success of a nonindigenous forest herb. *Ecological Applications* 11(5):1336–1348.

Moser, W.K., M.H. Hansen, and M.D. Nelson. 2008. The extent of selected non-native invasive plants on Missouri forestland. In *Proceedings of 16th Central Hardwood Forest Conference*, eds. D.F. Jacobs, and C.H. Michler, pp. 491–505. Gen. Tech. Rep. NRS-P-24. Newtown Square, PA: U.S. Department of Agriculture, Forest Service, Northern Research Station.

Myers, J.H., and D.R. Bazely. 2003. *Ecology and Control of Introduced Plants*. Cambridge, UK: Cambridge University Press.

Oswalt, C.M., S.N. Oswalt, and W.K. Clatterbuck. 2007. Effects of *Microstegium vimineum* (Trin.) A. Camus on native woody species density and diversity in a productive mixed-hardwood forest in Tennessee. *Forest Ecology and Management* 242:727–732.

Sasseen, A.N., and R.M. Muzika. 2004. Timber harvesting, prescribed fire and vegetation dynamics in the Missouri Ozarks. In *Proceedings of 14th Central Hardwood Forest Conference*, eds. D.A. Yaussy, D.M. Hix, R.P. Long, and P.C. Goebel, pp. 179–192. Gen. Tech. Rep. NE-316. Newtown Square, PA: U.S. Department of Agriculture, Forest Service, Northeastern Research Station.

Sheriff, S.L. 2002. Missouri Ozark forest ecosystem project: The experiment. In *Proceedings of the Second Missouri Ozark Forest Ecosystem Project symposium: Post-Treatment Results of the Landscape Experiment*, eds. S.R. Shifley, and J.M. Kabrick, pp. 1–25. Gen. Tech. Rep. NC-227. St. Paul, MN: U.S. Department of Agriculture, Forest Service, North Central Forest Experiment Station.

Weber, J.S., and K.D. Gibson. 2007. Exotic plant species in old-growth forest in Indiana. *Weed Science* 55:299–304.

Yates, E.D., D.F. Levia, and C.L. Williams. 2004. Recruitment of three non-native invasive plants into a fragmented forest in southern Illinois. *Forest Ecology and Management* 190:119–130.

chapter ten

Invasive plants and mutualistic interactions between fleshy fruits and frugivorous animals

M. C. Muñoz and J. D. Ackerman

Contents

10.1   Introduction

Plant–animal interactions are important for invasion success in many plant species, often determining patterns of recruitment and establishment in a new area. Frugivory and associated seed dispersal are among the most important of such interactions. In fact, most of the world's worst invasive plants are dispersed by animals, and some of the most aggressive weeds, *Miconia*, *Psidium*, and *Ardisia*, produce seeds in fleshy fruits that are dispersed by frugivores (Lowe et al. 2000). Frugivores not only disperse seeds, but the passage through the gut often has a positive effect on germination success, which can be crucial to the process of colonization. However, the efficacy of this relationship depends on both plant characteristics and animal behavior.

The relationship between frugivores and fleshy-fruited invasive plants is highly variable and deserves attention as such interactions can be the key to understanding how invasive plants spread so successfully. Invasion of fleshy-fruited plants may be facilitated by either native or invasive frugivores, and the consequences for native flora and fauna may vary from beneficial to devastating. In this chapter, we review the current literature on invasive plant species that bear fleshy fruits and the role of frugivorous animals in their establishment and spread in novel habitats. First and foremost, we will review the interactions between fleshy-fruit plants and frugivorous animals to elucidate the mechanisms involved under Section 10.2, "Frugivory in fleshy-fruit plants." Second, in Section 10.3, "Animals as seed dispersal vectors," we will examine the consequences for invasive plants after their fruit is consumed by frugivores and how animals transport seeds both internally and externally. Third, we will address why invasive plants are attractive to frugivores in Section 10.3, "Invasive plants as source of food." In Section 10.4, "Management of fleshy fruited invasive plants," we outline current strategies for control and management of the invasive plants that bear fleshy fruits.

10.2 Frugivory in fleshy-fruit plants

Seed dispersal in fleshy-fruited species begins with fruit maturation and display and ends with dispersal of seeds to appropriate sites for germination (Wang and Smith 2002, Schupp et al. 2010, Blendinger et al. 2011). A wide variety of vertebrates act as frugivores and seed dispersers, primarily mammals (e.g., monkeys, bats, ungulates) and birds (Clark et al. 2001), but occasionally lizards on islands (Valido and Nogales 1994) and fish in seasonally flooded tropical forests (Anderson et al. 2011). Plants signal frugivores with their fruit color (Lomascolo and Schaefer 2010) and sometimes odor (Sánchez et al. 2004), and the frugivores also make choices depending on fruit size and nutritional quality of the pulp, which may involve traits such as sugar, water, lipid, and protein content (Schaefer et al. 2003, Muñoz et al. 2007, Floerchinger et al. 2010). Considerable interspecific variation in fruit presentation strategies exists. Some plant species produce large fruit crops once a year and other species mature fewer fruits but with multiple fruiting peaks through the year (Genini et al. 2009). These fruiting phenology patterns influence frugivore diet choice, visitation rates, and foraging behavior (McCarty et al. 2002, Prasad and Sukumar 2010, Blendinger and Villegas 2011).

Frugivory is not always mutualistic; rather, it spans a continuum of possible plant–animal interactions. At one extreme is exploitation by animals, such as parrots, curassows, and deer mice that not only consume fruits but often destroy seeds as well (Santamaria and Franco 2000, Lobo et al. 2009, Villaseñor-Sánchez et al. 2010). At the other extreme is food deception, whereby fruits or seeds are mimetic, they appear nutritious but lack nutritional value, or are indigestible (Peres and Roosmalen 1996, Foster and Delay 1998, Galetti 2002, Herrera 2002, Cazetta et al. 2008).

When frugivores do not entirely destroy the seeds they ingest, their performance as dispersers may vary considerably. There are two components to effective seed dispersal. The first is quality, which includes dispersal distance to suitable places for seed germination, and the second is quantity, involving fruit removal rates and the number of seeds dispersed (Schupp et al. 2010). Effective seed dispersal not only involves plant traits but also frugivore characteristics such as body size, home range, seed passage time, seed passage treatment, or attachment to fur or feathers. All these are important determinants of seed movement across the landscape (Traveset et al. 2001, Luck and Daily 2003, Schupp et al. 2010, Blendinger et al. 2011, Holbrook 2011). Nevertheless, dispersal may not end after the initial transport from parent plant. Secondary dispersers such as ants, small rodents, and dung beetles are well known for their roles in moving seeds beyond frugivore deposition sites (Passos and Oliveira 2004, Vander Wall et al. 2005, Ponce-Santizo et al. 2006). Thus, seedling recruitment in fleshy-fruited plant populations is often closely tied to animal interactions: frugivores foraging fruits, removing pulp from their seeds, and effectively dispersing them. Throughout the next section of this chapter, we will look into how frugivores affect each stage of the dispersal cycle of invasive plants.

10.3   Animals as seed dispersal vectors

10.3.1   Gut treatment

The main consequence of frugivory is seed movement, a strategy upon which many plant species depend (Jordano 2000). When seeds are removed and transported from the parent tree, they have a better chance to find suitable places for germination and growth; new habitats are colonized, distributional area is expanded, and escape from pathogens and herbivores is enhanced (Howe and Smallwood 1982, Augspurger 1983, Wenny 2001, Wang and Smith 2002). Besides dispersal, frugivores can have effects on post-dispersal processes (Table 10.1). Seeds of many species have a higher percent germination or germination rate after passage through the gut of a frugivore (Traveset 1998, Heer et al. 2010, Reid and Armesto 2011), through scarification of the seed coat (Carpinelli et al. 2005, Bradford and Westcott 2010), or by removal of the inhibitory fruit or seed pulp (Yagihashi et al. 1998, Wenny 2000, Westcott et al. 2008). While others may not be affected by ingestion, there can be adverse consequences as some frugivores can destroy seeds in the gut or inhibit seed germination after defecation (Janzen et al. 1985, Traveset 1998).

Most dispersers of invasive plants are birds and mammals (Figure 10.1), and many of these animals have a positive effect on seed germination through scarification. Across a broad geographical sampling, invasive or exotic species have been shown to have higher percent seed germination after consumption by avian frugivores (Greenberg et al. 2001, Bartuszevige and Gorchov 2006, Linnebjerg et al. 2009), rats and rabbits (D'Antonio 1990, Bourgeois et al. 2005), jackrabbits (D'Antonio 1990), ungulates (Shiferaw et al. 2004, Vavra et al. 2007), and lizards (Padrón et al. 2011).

Sometimes the role of frugivores is indispensable to increase seed germination, although the effect on the seeds depends on the identity of the frugivores. Many plant species either cannot germinate or suffer much reduced germination when fruit pulp remains around the seeds. In Australia, seed germination of the invasive Brazilian pepper tree, *Schinus terebinthifolius*, is minimal without pulp removal, a task that is effectively accomplished by the native Silvereye, *Zosterops lateralis* (Panetta and McKee 1997). After fruit consumption, the seed germination increases to 87%–100%, reflecting how important this native bird is to this exotic plant. However, one cannot generalize from this. After gut passage through the native Cedar Waxwing, *Bombycilla cedrorum*, seed germination in the invasive plant *Lonicera maackii* is reduced to less than 50%, whereas the native American Robin, *Turdus migratorius*, had a positive effect on the same invasive species by improving germination to more than 80% (Bartuszevige and Gorchov 2006). Thus, the effect that frugivores have on seed germination can vary considerably.

Frequently, where there are invasive fleshy-fruited plants, invasive frugivores co-occur. Do invasive plants do better with native frugivores or are they most successful when paired with invasive frugivores? Although it is difficult to forecast future relationships among species, available data suggest that there are several possible components to consider. In the Canary Islands, frugivorous birds, mammals, and reptiles consume invasive cacti fruits (*Opuntia* spp.), but their performances varied. Only two of seven native frugivorous species enhanced germination: Western Canaries lizard (*Gallotia galloti*) and Stone Marten (*Martes foina*) (Padrón et al. 2011). The three exotics that also consumed the fruits (Hermann's tortoise, *Testudo hermanni*; Pine Marten, *Martes martes*; European rabbit, *Oryctolagus cuniculus*) had no effect on cactus seed germination. On the other hand, on the Canary Island of Fuerteventura, a comparison of frugivory by one native and two exotic frugivores found that the native, a lizard (*Gallotia atlantica*), was an effective disperser

Table 10.1 Invasive Plants and Consequences of Fruit Consumption by Frugivores

Invasive plant	Family	Locality	Seed dispersal process	Local dispersers	Exotic dispersers	Reference
Annona glabra	Annonaceae	Australia	Neutral effect after gut passage	1		Westcott et al. (2008)
Carpobrotus edulis	Aizoaceae	France	Positive effect after gut passage		4	Bourgeois et al. (2005)
Carpobrotus acinaciformis	Aizoaceae	France	Positive effect after gut passage		4	Bourgeois et al. (2005)
Carpobrotus edulis	Aizoaceae	U.S.A.	Positive effect after gut passage	2		Vila and D'Antonio (1998)
Carpobrotus edulis	Aizoaceae	U.S.A.	Positive effect after gut passage	4		D'Antonio and Vila (1990)
Celastrus orbiculatus	Celastraceae	U.S.A.	Positive effect after gut passage	1		Greenberg et al. (2001)
Clidemia hirta	Melastomataceae	Mauritius	Positive effect after gut passage		1	Linnebjerg et al. (2009)
Clidemia hirta	Melastomataceae	U.S.A.	Positive effect after gut passage	1	2	Smith (1992)
Lonicera maackii	Caprifoliaceae	U.S.A.	Neutral effect after gut passage	4	1	Bartuszevige and Gorchov (2006)
Ligustrum robustum	Oleaceae	Mauritius	Neutral effect after gut passage		1	Linnebjerg et al. (2009)
Maesopsis eminii	Rhamnaceae	Tanzania	Neutral effect after gut passage The greatest dispersion 3.97 km	3		Cordeiro et al. (2004)
Opuntia maxima	Cactaceae	Canary Islands	Positive effect after gut passage	1	2	Lopez-Diaz and Nogales (2008)
Opuntia spp.	Cactaceae	Canary Islands	Two species have a positive effect after gut passage/nine species have a negative effect	9	3	Padron et al. (2011)
Opuntia sp.	Cactaceae	South Africa	Positive effect after gut passage	1		Dean and Milton (2000)
Prunus serotina	Rosaceae	Belgium	Spatial clustering of perching birds	18		Deckers et al. (2008)
Schinus terebinthifolius	Anacardiaceae	Australia	Neutral effect after gut passage	1		Panneta and McKee (1997)

The potential effect was divided by three categories: neutral (any effect), positive (increase percent germination), and negative (seed damage or reduction on percent germination) effect.

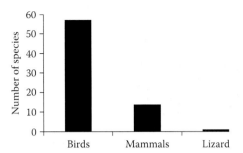

Figure 10.1 Frugivorous species reported as consumers of exotic plant species. We grouped the species per category based on a review of 17 articles published after 1996. Mammals include cats, hares, rabbits, deer, monkeys, mice, rabbits, and squirrels. Only one species of lizard was reported as a frugivore of exotic plants.

for three native species and one exotic plant on the island. Conversely, the exotic rabbit (*O. cuniculus*) preyed on all the seeds consumed including native and exotic plants, but the other exotic (Barbary ground squirrel, *Atlantoxerus getulus*) was an effective disperser of the exotic cactus, *Opuntia*. Seed germination was faster in one of the native plant species when the lizard ate the fruits than when the squirrel did (López-Darias and Nogales 2008). The former study suggests that exotic frugivores can have a negative effect on seed germination of the native plant species, and the exotic plant (*Opuntia*) has native and exotic frugivorous species facilitating its spread in the island.

One would not expect that all frugivores, native or exotic, would have the same effects on seed germination of a particular plant species. And by the same token, we would not expect the same frugivore to have the same effect on different invasive plant species. In Mauritius, the exotic frugivorous Red-whiskered Bulbul, *Pycnonotus jocosus*, had different mean gut passage times for seeds of two exotics *Clidemia hirta* (12.71 minutes) and *Ligustrum robustum* (15.49 minutes). Both exotic plants depend on frugivorous birds to remove pulp from the seeds, but the two species differed in their dependence on frugivores for seed germination. *C. hirta* seeds may not germinate without first passing through the gut of bulbuls, and they are more successful when they do. On the other hand, success of *L. robustum* is completely dependent on the bulbuls for germination (Linnebjerg et al. 2009).

Although our focus has been on frugivory and invasive plant species, exotic frugivores can have as wide ranging effects on native plant species as they do on exotic ones. Nogales et al. (2005) found that the endemic plant Tasiago (*Rubia fruticosa*) is well dispersed by native fauna such as lizards (*Gallotia atlantica*), warblers birds (*Sylbia* sp.), raven (*Corvus corax*), and the Yellow-legged Gull (*Larus cachinnans*), but the exotic rabbit (*O. cuniculus*) and the squirrel (*A. getulus*) damaged a great proportion of the seeds after fruit consumption and reduced seed germination compared to the positive effect of the native dispersers.

The consequences of differential effects of frugivores on native and exotic plant species can alter community structure and composition by facilitating invasion success while at the same time diminishing fitness of native species. In California, the ice plant, *Carpobrotus chilensis* (Aizoaceae) is becoming displaced by both the non-native *C. edulis* and the hybrid between the two. If the fruits are not consumed, seed germination of *C. chilensis* is almost 80%, whereas germination in the exotic *C. edulis* and their hybrids is less than 13%. However, the advantage is lost when native jackrabbits (*Lepus californicus*) and mule deer (*Odocoileus hemionus*) eat the fruits and disperse the seeds. Seed germination drops to 31% for the native *C. chilensis*, whereas germination improves to almost 40% for the exotic *C. edulis* and to 16% for the hybrid, which partially accounts for the invasive behavior of the two (Vilá and D'Antonio 1998). In Hawai'i, a recent study found that the Black Rat (*Rattus rattus*) consumes fruits of both native and invasive species, sometimes destroying seeds in the process. Although some species in both plant categories had high seed predation, 5 of 8 native species suffered, and the rat negatively affected only one of four invasive plants species (Shiels and Drake 2011). Such differential seed predation may facilitate invasion success of exotic plants. Considering the variation in performance of native and exotic frugivores on seed germination of either native or invasive plant species, it is clear that roles of frugivore are context dependent. Thus, a first step in the control of fleshy-fruited invasive plant species is to determine the role of each frugivore along the continuum of possible interactions.

10.3.2   Dispersion to habitats

Seed dispersal by frugivorous animals is usually directed, nonrandom. Frugivores often carry seeds to particular places such as nest sites, feeding roosts, and perch sites (Dean and Milton 2000, Corlett 2005, Deckers et al. 2008). They can move seeds among habitats such as ecotonal edges, wooded corridors, spurs, and open areas (Bartuszevige and Gorchov 2006; Linnebjerg et al. 2009). Seedling recruitment is also associated with disperser movements (Deckers et al. 2008) and the capacity of the seeds to germinate under tree canopies where the dispersers stay (Milton et al. 2007). When frugivorous animals forage from fleshy-fruited invasive plants, they may deposit seeds of many other species, even native species, and improve the general recruitment of seedlings in the forest (Tecco et al. 2006). Some birds with the ability to use human structures such as telegraph poles and wire fences have dispersed seeds along these structures and extended the distribution of invasive plants (Dean and Milton 2000). Other frugivorous species that feed on fruits of ornamental plants have been responsible for their escape from gardens, moving seeds to other anthropogenic areas as well as to native forests and savannas where the invasives may form dense stands (Lavergne et al. 1999, Koop 2004, Mack 2005, Milton et al. 2007, Foxcroft et al. 2008, Linnebjerg et al. 2009). Even though exotic frugivores can disperse high numbers of seeds of exotic plant species, in Hawai'i, they can also be important dispersers of native plants, helping in the restoration of the native forests (Foster and Robinson 2007). Thus, a first approach to predict the spatial patterns of invasive plant species with fleshy fruits is to identify which animals consume the fruits and where these animals travel.

10.3.3   Dispersal distances

Large animals usually have the capacity to move more seeds and over longer distances than small animals, which generally produce a smaller seed shadow. One of the largest frugivorous birds is the endangered Southern Cassowary, *Casuarius casuarius*, which can weigh 50 (male) to 70 kg (female). In Australia, these birds move seeds of an invasive species (Pond Apple; *Annona glabra*, L.) an average distance of 605 m, but 17% of the seeds were dispersed more than 1000 m, promoting the spread of the invasive plant (Westcott et al. 2008). Likewise, other large dispersers such as ungulates are very effective in moving seeds of invasive plants due to their wide home ranges. A comparative study of native elk (*Cervus elaphus*) and deer (*Odocoileus* sp.) found that these species can disperse both native and exotic species (Bartuszevige and Endress 2008). Over a 1 year period, the white-tailed deer (*Odocoileus virginianus*) can move seeds of approximately 70 exotic and native species. With a digestion time of 15–20 h and a range of 1–50 km depending on the season (Vellend 2002), these deer may be a significant dispersal vector of invasive species (Myers et al. 2004).

The estimate for seed dispersal distance by Mule deer, *Odocoileus hemionus*, is not quite significant, 10 to 800 m and some seeds to 1 km distance (D'Antonio 1990).

In contrast, small animals promote short seed dispersal distances. Bourgeois et al. (2005) estimated that on the French Mediterranean islands, rats move seeds an average of 40 m and rabbits may transport seeds for an average distance of 160 m. Brush rabbits (*Sylvilagus bachmani*) in the United States move seeds for even shorter distances, just an average of 10 m (D'Antonio 1990). In the case of small birds, the Silvereye retains seeds for 25 minutes after fruit consumption, but we have no estimate of the potential seed dispersal range for this species (Panetta and McKee 1997). Other small birds have a gut retention range for exotic fruits of 10–70 minutes and therefore their dispersal distance is short (Bartuszevige and Gorchov 2006). Additionally, McAlpine and Jasson (2008) reported that birds in New Zealand can move seeds of the invasive Darwin's barberry (*Berberis darwinii*) 100–450 m, but 65%–76% of the ingested seeds fell beneath the parental tree. On the other hand, migratory birds such as ducks (*Anas crecca*) may carry intact exotic seeds in their gut for distances up to 1000 km, helping to spread many exotic species across Europe (Brochet et al. 2009). So, much variation exists in dispersal distances of fleshy-fruited invasive species and this depends on which frugivores are involved.

10.3.4   Seed movement by epizoochory

Animals commonly disperse seeds through either frugivory (endozoochory) or epizoochory, which is when seeds or fruits are attached to hair, fur, feathers, or feet. Plants with this type of dispersal mechanism often have fruits or seeds with mucus, barbs, hooked hairs, or spines that attach the propagule to the animal's body. These can travel long distances and remain attached for a long time (Sorensen 1986, Manzano and Malo 2006, Couvreur et al. 2008). Epizoochory as a means of dispersal for invasive plants is not well known; however, some authors have suggested that exotic plant species have the tendency to adhere to a variety of mammal fur types, indicating flexibility for dispersal (Kulbaba et al. 2009). On Santa Catalina Island, off the coast of California, non-native species have spread across the island by becoming attached to the hair of another non-native, the American bison (*Bison bison*) (Constible et al. 2005). Likewise, horses (*Equus ferus caballus*) disperse seed of non-native species in their hooves (Gower 2008). Recently, migratory birds in Korea were observed dispersing propagules of the invasive weed, *Achyranthes japonica*. Choi et al. (2010) found that 3 of 3947 birds had seeds attached to the nape, back, or crural feathers. On one occasion, seeds were discovered on the lore and tail feathers. Although the authors did not know the origin of the seeds, they did note that these birds fly long distances. Thus, epizoochory is another means by which invasive species may be dispersed, though documentation is currently scant.

10.4   Invasive plants as source of food

10.4.1   Fruit crop attractiveness

The great advantage of fleshy-fruited invasive plants is that they can offer an alimentary resource for local fauna, contributing significantly to the portion of the total fruit available for frugivorous animals. Fruit abundance, fruit size, and pulpiness are used as cues by frugivores to select fruits (Sallabanks 1993). When invasive species have large fruit crops, they should be competitive, particularly where dispersal is frugivore-limited. Yet the spread of invasive species may simply occur through high fruit production, whether the plant–frugivore relationship is fruit- or frugivore-limited. Propagule pressure is expected to be greater for such plants. There are two key components of propagule pressure: (1) propagule size, which is the number of individuals in a propagule, and (2) propagule number, which is the rate at which propagules arrive per unit time (Simberloff 2009). Invasive plants generally have large fruit crops, and in many cases they produce numerous small fruits with bright and intense colors (Williams and Karl 1996, Shiferaw et al. 2004, Gosper and Vivian-Smith 2010). For example, in Australia, the most frequent color for fruits of invasive plants was purple to black, red, orange, and yellow (Gosper and Vivian-Smith 2010). And one of the most productive trees is *Miconia calvescens*, an aggressive invasive in Hawai'i, where these small trees average 45,000 fleshy berries per tree per year (Meyer 1998). Additionally, this tree can produce fruit for 6 months, with three major peaks, providing a steady resource for frugivores. *Chromolaena odorata* in South Africa can produce 2000 seeds/m^2/year (Witkowski and Wilson 2001). *Ardisia elliptica* in Puerto Rico produces an average of 2841 fruits per tree, with just one crop per year (Muñoz and Ackerman 2011). Fruit production by the same species in Florida varied across years and sites. However, the highest average number of seeds per tree was 929 (Pascarella 1998). *Carpobrotus* species (*C. edulis* and *C.* aff. *acinaciformis*), which is invading various localities in France, produces an estimated 25 fruits/m^2. The fruits are larger than the average for the region (6–10 mm vs. 19 mm) and have high energy and water content. Consequently, during drought and starvation periods, the fruits of these plants are highly desirable, attracting a number of different frugivores (Bourgeois et al. 2005). In California, Vilá and D'Antonio (1998) found that clones of the invasive *C. edulis* produce more fruits (average = 497) with a higher energy content than the native *C. chilensis*, which produces only 13 fruits per clone. As a consequence, fruits of the invasive species were preferred and removed more quickly by frugivores than those of the native *Carpobrotus*.

10.4.2   *Fruiting phenology*

Fruiting phenologies of local and exotic plants often differ (e.g., Gosper 2004, Tecco et al. 2006), which can facilitate the success of exotic plant species. In Hong Kong, fruit traits of native and exotic species are similar; however, there are differences between the two groups in fruiting phenologies. Fruits of native species are available only in February–April, whereas those of exotic species are available throughout the whole year, facilitating their establishment by native frugivores (Corlett 2005). In Australia, *Chrysanthemoides monilifera* produces fruit throughout the year, but with a peak in autumn to winter, when few native plants have fruits (Gosper 2004). This invasive species also has small fruits, which nearly all frugivorous birds in the region are able to manipulate and consume (Gosper et al. 2006). In Argentina, Tecco et al. (2006) found that native species are dispersed in the warmer months, whereas exotic species fruit in the cooler months. Likewise, in New Zealand, two exotic species are available in June and July, when no native fruits are available (Williams and Karl 1996). In the Canary Islands, local plants have their maximum fruit production in the winter and spring, but exotic species peak in autumn, thereby extending the period of fruit availability for frugivorous animals (López-Darias and Nogales 2008). Thus, exotic plants frequently fruit for longer periods than native species or fruit when native fruits are rare, providing reliable, long-term resources for both native and invasive frugivores.

Both native and invasive frugivores will incorporate exotic fruits in their diets (Williams and Karl 1996, Foster and Robinson 2007); when fleshy-fruited invasive plants occur with other fleshy fruited species in the same habitat, competition for frugivores may ensue. The presence of the invasive *C. monilifera* in Australia reduced fruit removal rates of three native plant species with fleshy fruits (Gosper et al. 2006). However, facilitation may occur if frugivores are in short supply, but respond positively to increased fruit availability. In the United States, where invasive *Lonicera maackii* and *L. morrowii* occur in high densities, fruit-removal rates for a native plant species increased (Gleditsch and Carlo 2011). In another case, Drummond (2005) conducted food choice experiments in Maine and found that birds ate similar proportions of the native *Viburnum opulus* and the invasive *Rosa multiflora* without discriminating between them. Similarly, field observations of the invasive *Lonicera tatarica* and the native *Cornus amomum* showed that they had the same fruit removal rates by local frugivores. Should populations of these invasive species increase, then seed dispersal of native species might be overwhelmed, negatively affecting mutualistic relationships (Drummond 2005), but this topic warrants more research. Additional evidence comes from the exotic *Celastrus orbiculatus* and the native *Ilex opaca*, two sympatric vines in the eastern United States whose fruit removal rates over a 7-month period were statistically similar (Greenberg et al. 2001). So, competition, facilitation, or no apparent effect on fruit removal may occur between fleshy-fruited invasive and native plant species in the first phase of seed dispersal. However, we need more comparative studies to detect whether there are indeed consequences of apparent neutrality to population growth of native and invasive frugivores and the native and invasive plant species that they disperse.

Invasive species are generally considered to be detrimental to native communities and ecosystems. However, under certain circumstances, invasive species may be beneficial, particularly in poorly managed, severely degraded ecosystems. Fleshy-fruited invasive species may become important resources for native frugivores whose original food resources have declined (Foster and Robinson 2007). Such substitution may actually benefit the depauperate populations of native plants if frugivores are limiting and they respond to increased food densities as discussed in Section 10.4.2. Another possibility is that invasive trees and shrubs may be able to penetrate abandoned agricultural lands thick with non-native grasses, setting the stage for native, animal-dispersed species to become established (Ewel and Putz 2004, Lugo 2004, Abelleira Martínez 2010). Thus, before devising eradication, control, or management strategies, one needs to know not only the relationship between fleshy-fruited invasive species and their frugivores, but also whether or not the invasive plants are actually detrimental to species richness, composition, or ecosystem function.

10.5 Management of fleshy-fruited invasive plants

There are three common types of management to control or eradicate invasive species, and the cost, efficacy, and collateral damage of each depends on particular situations. The first type of control is physical removal. This is done through felling, pulling, digging out stumps and saplings, and burning areas (Hobbs and Humphries 1995). Moreover, for fleshy-fruited plants, physical removal should be timed before their fruits mature (Cheplick 2010). The second method is chemical control, using toxic formulations or herbicides such as tripclopyr, glyphosate, picloram/2,2,4-D amine or Tordon (Hobbs and Humphries 1995, Simberloff 2003, Munniapan et al. 2005). The third one is biological control, which is one of the most controversial strategies because of negative impacts on the environment and non-target species, both plants and animals (Hobbs and Humphries 1995). Nevertheless, there are natural enemies such as beetles (Bruchidae, Carabidae, Chrysomelidae, Curculionidae), isopods (Oniscidea), and slugs and snails (Mollusca: Pulmonata) that consume seeds or seedlings, which may be manipulated to increase mortality in invasive plants. Such a strategy to control invasive species is in its infancy (Honek et al 2009, Maguire et al. 2011).

In addition to the three conventional mechanisms to eradicate invasive plants, additional measures are recommended for control of fleshy-fruited plants. Because naturalized fruit-bearing plants vary widely in their invasiveness, risk assessment is critical. Buckley et al. (2006) categorized fleshy-fruited species as at either high or low invasion risk based on fruit traits. High-risk species are those with small fruits, large crop size, long fruiting season, and gap-filling phenology. Moreover, the risk stays high if these fruits are dispersed by large, opportunistic, or migratory frugivores. Low-risk species have large fruits, a thick inedible peel, and are highly defended. Exceptions are when large frugivores are involved or when fruits are soft and multiseeded. Risk stays low when frugivores destructively process fruits during ingestion or digestion or when dispersers are confined to small habitat fragments. Natural resource managers should be able to effectively classify the type of risk simply with a sound knowledge of both plants and frugivorous animals.

Thus, the main objective in management of fleshy-fruited invasive species is to minimize the impact of seed dispersal. Gosper et al. (2005) suggested that managers should focus on four points: (1) reduce fruit production or fruit quality; (2) direct seed deposition; (3) identify and remove major sources of seed spread; and (4) provide alternative resources for frugivorous animals. Strategies to minimize seed dispersal of fleshy fruited invasive species may include the following: (1) create barriers around weed infestations; (2) after using an infested pasture, pen livestock for 48 hours before moving them; (3) deter or prevent animal use or movement through weed infestations when plants are fruiting; (4) do not feed livestock hay that is contaminated with seeds of invasive species in uninfected locations (Davies and Sheley 2007); (5) minimize habitat edges because they help frugivore movements; (6) create perches as managed seed sinks (Buckley et al. 2006); and (7) cultivate native plants that may compete with invasive fruit-bearing species or substitute as resources for frugivores where invasives are to be exterminated (Gosper and Vivian-Smith 2006).

Gosper and Vivian-Smith (2006) developed a methodology to determine which native species should substitute as replacements for invasive plants in restoration programs, garden settings, parks, and urban environments. All fruit-bearing species are assigned scores based on phenology, morphology, conspicuousness, and accessibility. Native species with scores similar to that of invasive species are targeted as potential substitutes. The main idea is to identify native species that might compete with invasives for seed dispersal services and seedling recruitment sites (Gosper and Vivian-Smith 2009).

For the success of any invasive species management program, one needs the support of those that live within or close to infected areas. Multifaceted educational strategies can be developed to raise public awareness of the consequences of a plant invasions and how everyone can help control the problem with simple measures, such as avoiding invasive plants as ornamentals or removing such species from gardens (Foxcroft et al. 2008).

10.6 Conclusions

Establishment and colonization of many invasive plant species is greatly facilitated by resident animals, whether they are native or invasive themselves. Many invasive plants with fleshy fruits have large, colorful fruit crops that are often available to frugivores at time periods different from those of local species, thus either avoiding competition or facilitating the maintenance of frugivore populations. Although recent attention has been given to the role of native and exotic frugivores, it appears that their performance is very much context-dependent, and so much remains unknown of the interaction, from differential effects on seed germination to seed dispersal distances. Despite gaps in our knowledge of the particulars of cases involving frugivores and invasive species, methods for invasive species risk assessments show some degree of success. These need to be implemented to target invasive plants for further study, and if they are determined to be detrimental to local communities, then strategies to control them can be developed.

References

Abelleira Martínez, O.J. 2010. Invasion by native tree species prevents biotic homogenization in novel forests of Puerto Rico. *Plant Ecology* 211:49–64.

Anderson, J.T., T. Nuttle, J.S. Saldaña Rojas, T.H. Pendergast, and A.S. Flecker. 2011. Extremely long-distance seed dispersal by an overfished Amazonian frugivore. *Proceedings of the Royal Society B* 278:3329–3335.

Augspurger, C.K. 1983. Seed dispersal of the tropical tree, *Platypodium elegans*, and the escape of its seedlings from fungal pathogens. *Journal of Ecology* 71:759–771.

Bartuszevige, A.M., and B.A. Endress. 2008. Do ungulates facilitate native and exotic plant spread? Seed dispersal by cattle, elk and deer in northeastern Oregon. *Journal of Arid Environments* 72:914–913.

Bartuszevige, A.M., and D.L. Gorchov. 2006. Avian seed dispersal of an invasive shrub *Biological Invasions* 8:1013–1022.

Blendinger, P.G., J.G. Blake, and B.A. Loiselle. 2011. Composition and clumping of seeds deposited by frugivorous birds varies between forest microsites. *Oikos* 120:463–471.

Blendinger, P.G., and M. Villegas. 2011. Crop size is more important than neighborhood fruit availability for fruit removal of *Eugenia uniflora* (Myrtaceae) by bird seed dispersers. *Plant Ecology* 212:889–899.

Bourgeois, K., M.C. Suehs, E. Vidal, and F. Médail. 2005. Invasional meltdown potential: facilitation between introduced plants and mammals on French Mediterranean islands. *Ecoscience* 12:248–256.

Bradford, M.G., and D.A. Westcott. 2010. Consequences of southern cassowary (*Casuarius casuarius*, L.) gut passage and deposition pattern on the germination of rainforest seeds. *Austral Ecology* 35:325–333.

Brochet, A.-L., M. Guillemain, H. Fritz, M. Gauthier-Clerc, and A.J. Green. 2009. The role of migratory ducks in the long-distance dispersal of native plants and the spread of exotic plants in Europe. *Ecography* 32:919–928.

Buckley, Y.M., S. Anderson, C.P. Catterall, R.T. Corlett, T.E. Carl, R. Gosper, R. Nathan, D.M. Richardson, M. Setter, O. Spiegel, G. Vivian-Smith, F.A. Voigt, J.E.S. Weir, and D.A. Westcott. 2006. Management of plant invasions mediated by frugivore interactions. *Journal of Applied Ecology* 43:848–857.

Carpinelli, M.E., C.S. Schauer, D.W. Bohnert, S.P. Hardegree, S.J. Falck, and T.J. Svejcar. 2005. Effect of ruminal incubation on perennial pepperweed germination. *Rangeland Ecology and Management* 58:632–636.

Cazetta, E., L.S. Zumstein, T.A. Melo-Júnior, and M. Galetti. 2008. Frugivory on *Margaritaria nobilis* L.f. (Euphorbiaceae): Poor investment and mimetism. *Brazilian Journal of Botany* 31:303–308.

Cheplick, G.P. 2010. Limits to local spatial spread in a highly invasive annual grass (*Microstegium vimineum*). *Biological Invasions* 12:1759–1771.

Choi, C.-Y., H.-Y. Nam, and H.-Y. Chae. 2010. Exotic seeds on the feathers of migratory birds on a stopover island in Korea. *Journal of Ecology and Field Biology* 33:19–22.

Clark,C.J., J.R. Poulsen, and V.T. Parker. 2001. The role of arboreal seed dispersal groups on the seed rain of a lowland tropical forest. *Biotropica* 33:606–620.

Constible, J.M., R.A. Sweitzer, D.H. VanVuren, P.T. Schuyler, and D.A. Knapp. 2005. Dispersal of non-native plants by introduced bison in an island ecosystem. *Biological Invasions* 7:699–709.

Cordeiro, N.J., Patrick, D.A.G., Munisi, B. and V. Gupta. 2004. Role of dispersal in the invasion of an exotic tree in an East African submontane forest. *Journal of Tropical Ecology* 20:449–457.

Corlett, R.T. 2005. Interactions between birds, fruit bats and exotic plants in urban Hong Kong, South China. *Urban Ecosystems* 8:275–283.

Couvreur, M., K. Verheyen, M. Vellend, I. Lamoot, E. Cosyns, M. Hoffmann, and M. Hermy. 2008. Epizoochory by large herbivores: Merging data with models. *Basic and Applied Ecology* 9:204–212.

D'Antonio, M.C. 1990. Seed production and dispersal in the non-native, invasive succulent *Carpobrotus edulis* (Aizoaceae) in coastal strand communities of central California. *Journal of Applied Ecology* 27:693–702.

Davies, K.W., and Sheley R.L. 2007. A conceptual framework for preventing the spatial dispersal of invasive plants. *Weed Science* 55:178–184.

Dean, W.R.J., and S.J. Milton. 2000. Directed dispersal of *Opuntia* species in the Karoo, South Africa: Are crows the responsible agents? *Journal of Arid Environments* 45:305–314.

Deckers, B., K. Verheyen, M. Vanhellemont, E. Maddens, B. Muys, and M. Hermy. 2008. Impact of avian frugivores on dispersal and recruitment of the invasive Prunus serotina in an agricultural landscape. *Biological Invasions* 10:717–727.

Drummond, B.A. 2005. The selection of native and invasive plants by frugivorous birds in Maine. *Northeastern Naturalist* 12:33–44.

Ewel, J.J., and F.E. Putz. 2004. A place for alien species in ecosystem restoration. *Frontiers in Ecology and the Environment* 2:354–360.

Floerchinger, M., J. Braun, K. Boehning-Gaese, and H.M. Schaefer. 2010. Fruit size, crop mass, and plant height explain differential fruit choice of primates and birds. *Oecologia* 164:151–161.

Foster, J.T., and S.K. Robinson. 2007. Introduced birds and the fate of Hawaiian rainforests. *Conservation Biology* 21:1248–1257.

Foster, M.S., and L.S. Delay. 1998. Dispersal of mimetic seeds of three species of *Ormosia* (Leguminosae). *Journal of Tropical Ecology* 14:389–411.

Foxcroft, L.C., D.M. Richardson, and J.R.U. Wilson. 2008. Ornamental plants as invasive aliens: Problems and solutions in Kruger National Park, South Africa. *Environmental Management* 41:32–51.

Galetti, M. 2002. Seed dispersal of mimetic seeds: Parasitism, mutualism, aposematism or exaptation? In *Seed Dispersal and Frugivory: Ecology, Evolution and Conservation*, eds. D. Levey, W.R. Silva, and M. Galetti, pp. 177–192. Oxon, United Kingdom: CABI Publishing.

Genini, J., M. Galetti, L. Patricia, and C. Morellato. 2009. Fruiting phenology of palms and trees in an Atlantic rainforest land-bridge island. *Flora* 204:131–145.

Gleditsch, J.M., and T.A. Carlo. 2011. Fruit quantity of invasive shrubs predicts the abundance of common native avian frugivores in central Pennsylvania. *Diversity and Distributions* 17:244–253.

Gosper, C.R. 2004. Fruit characteristics of invasive bitou bush, *Chrysanthemoides monilifera* (Asteraceae), and a comparison with co-occurring native plant species. *Australian Journal of Botany* 52:223–230.

Gosper, C.R., C.D. Stansbury, and G. Vivian-Smith. 2005. Seed dispersal of fleshy-fruited invasive plants by birds: Contributing factors and management options. *Diversity and Distributions* 11:549–558.

Gosper C.R., and G. Vivian-Smith. 2006. Selecting replacements for invasive plants to support frugivores in highly modified sites: A case study focusing on *Lantana camara*. *Ecological Management & Restoration* 7:197. 203.

Gosper, C.R., and Vivian-Smith G. 2009. The role of fruit traits of bird-dispersed plants in invasiveness and weed risk assessment. *Diversity and Distribution* 15:1037. 1046.

Gosper, C.R., and G. Vivian-Smith. 2010. Fruit traits of vertebrate-dispersed alien plants: Smaller seeds and more pulp sugar than indigenous species. *Biological Invasions* 12:2153–2163.

Gosper, C.R., R.J. Whelan, and K. French. 2006. The effect of invasive plant management on the rate of removal of vertebrate-dispersed fruits. *Plant Ecology* 184:351–363.

Gower, S.T. 2008. Are horses responsible for introducing non-native plants along forest trails in the eastern United States? *Forest Ecology and Management* 256:997–1003.

Greenberg, C.H., L.M. Smith, and D.J. Levey. 2001. Fruit fate, seed germination and growth of an invasive vine -an experimental test of 'sit and wait' strategy. *Biological Invasions* 3:363–372.

Heer, K., L. Albrecht, and E.K.V. Kalko. 2010. Effects of ingestion by neotropical bats on germination parameters of native free-standing and strangler figs (*Ficus* sp., Moraceae). *Oecologia* 163:425–435.

Herrera, C.M. 2002. Seed dispersal by vertebrates. In *Plant Animal Interactions: An Evolutionary Approach*, eds. C.M. Herrera, and O. Pellmyr, pp. 705–727. Oxford, United Kingdom: Blackwell Publishing.

Hobbs, R.J., and S.E. Humphries. 1995. An integrated approach to the ecology and management of plant invasions. *Conservation Biology* 9:761–770.

Holbrook, K.M. 2011. Home range and movement patterns of Toucans: Implications for seed dispersal. *Biotropica* 43:357–364.

Honek, A., Z. Martinkova, P. Saska, and S. Hoprdova. 2009. Role of post-dispersal seed and seedling predation in establishment of dandelion (*Taraxacum agg.*) plants. *Agriculture, Ecosystems and Environment* 134:126–135.

Howe, H.F., and J. Smallwood. 1982. Ecology of seed dispersal. *Annual Review of Ecology and Systematics* 13:201–228.

Janzen, D.H., M.W. Demment, and J.B. Robertson. 1985. How fast and why do germinating Guanacaste seeds (*Enterolobium cyclocarpum*) die inside cows and horses? *Biotropica* 17:322–325.

Jordano, P. 2000. Fruits and frugivory. In *Seeds: The Ecology of Regeneration in Natural Plant Communities*, ed. M. Fenner, pp. 125–166. Wallingford, UK: CABI Publishing.

Koop, A.L. 2004. Differential seed mortality among habitats limits the distribution of the invasive non-native shrub *Ardisia elliptica*. *Plant Ecology* 172:237–249.

Kulbaba, M.W., J.C. Tardif, and R.J. Staniforth. 2009. Morphological and ecological relationships between burrs and furs. *American Midland Naturalist* 161:380–391.

Lavergne, C., J.C. Rameau, and J. Figier. 1999. The invasive woody weed *Ligustrum robustum* subsp. *walkeri* threatens native forests on La R´eunion. *Biological Invasions* 1:377–392.

Linnebjerg, J.F., D.M. Hansen, and J.M. Olesen. 2009. Gut passage effect of the introduced red-whiskered bulbul (*Pycnonotus jocosus*) on germination of invasive plant species in Mauritius. *Austral Ecology* 34:272–277.

Lobo, N., M. Duong, and J.S. Millar. 2009. Conifer-seed preferences of small mammals. *Canadian Journal of Zoology* 87:773–780.

Lomascolo, S.B., and H.M. Schaefer. 2010. Signal convergence in fruits: A result of selection by frugivores? *Journal of Evolutionary Biology* 23:614–624.

López-Darias, M., and M. Nogales. 2008. Effects of the invasive barbary ground squirrel (*Atlantoxerus getulus*) on seed dispersal systems of insular xeric environments. *Journal of Arid Environments* 72:926–939.

Lowe, S., M. Browne, S. Boudjelas, and M. De Poorter. 2000. *100 of the World's Worst Invasive Alien Species A selection from the Global Invasive Species Database*. Published by The Invasive Species Specialist Group (ISSG) a specialist group of the Species Survival Commission (SSC) of the World Conservation Union (IUCN), 12pp.

Luck, G.W., and G.C. Daily. 2003. Tropical countryside bird assemblages: Richness, composition, and foraging differ by landscape context. *Ecological Applications* 13:235–247.

Lugo, A.E. 2004. The outcome of alien tree invasions in Puerto Rico. *Frontiers in Ecology and the Environment* 2:265–273.

Mack, R.N. 2005. Predicting the identity of plant invaders: Future contributions from horticulture. *HortScience* 40:1168–1174.

Maguire, D., R. Sforza, and S.M. Smith. 2011. Impact of herbivory on performance of *Vincetoxicum* spp., invasive weeds in North America. *Biological Invasions* 13:1229–1240.

Manzano, P., and J.E. Malo. 2006. Extreme long-distance seed dispersal via sheep. *Frontiers in Ecology and Environment* 4:244–248.

McAlpine, K.G., and L.A. Jesson. 2008. Linking seed dispersal, germination and seedling recruitment in the invasive species *Berberis darwinii* (Darwin's barberry). *Plant Ecology* 197:119–129.

McCarty, J.P., D.J. Levey, C.H. Greenberg, and S. Sargent. 2002. Spatial and temporal variation in fruit use by wild life in a forested landscape. *Forest Ecology and Management* 164:277–291.

Meyer, J.Y. 1998. Observations on the reproductive biology of *Miconia calvescens* DC (Melastomataceae), an alien invasive tree on the island of Tahiti (South Pacific Ocean). *Biotropica* 30:609–624.

Milton, S.J., J.R.U. Wilson, D.M. Richardson, C.L. Seymour, W.R.J Dean, D.M. Iponga, and S. Proche. 2007. Invasive alien plants infiltrate bird-mediated shrub nucleation processes in arid savanna. *Journal of Ecology* 95:648–661.

Munniapan, R., G.V.P. Reddy, and P.-Y. Lai. 2005. Distribution and biological control of *Chromolaena odorata*. In *Invasive Plants: Ecological and Agricultural Aspects*, ed. Inderjit, pp. 223–233. Switzerland: Springer.

Muñoz, M.C., and J.D. Ackerman. 2011. Spatial distribution and performance of native and invasive *Ardisia* (Myrsinaceae) species in Puerto Rico: The anatomy of an invasion. *Biological Invasions* 13:1543–1558.

Muñoz, M.C., G.A. Londoño, M.M. Rios, and G.H. Kattan. 2007. Diet of the Cauca Guan: Exploitation of a novel food source in times of scarcity. *Condor* 109:841–851.

Myers, J.A., M. Vellend, S. Gardescu, and P.L. Marks. 2004. Seed dispersal by white-tailed deer: Implications for long-distance dispersal, invasion, and migration of plants in eastern North America. *Oecologia* 139:35–44.

Nogales, M., C. Nieves, J.C. Illera, D.P. Padilla, and A. Traveset. 2005. Effect of native and alien vertebrate frugivores on seed viability and germination patterns of *Rubia fruticosa* (Rubiaceae) in the eastern Canary Islands. *Functional Ecology* 19:429–436.

Padrón, B., M. Nogales, A. Traveset, M. Vilá, A. Martínez-Abraín, D.P. Padilla, and P. Marrero. 2011. Integration of invasive *Opuntia* spp. by native and alien seed dispersers in the Mediterranean area and the Canary Islands. *Biological Invasions* 13:781–792.

Panetta, F.D., and J. McKee. 1997. Recruitment of the invasive ornamental, *Schinus terebinthifolius*, is dependent upon frugivore. *Australian Journal of Ecology* 22:432–438.

Pascarella, J.B. 1998. Hurricane disturbance, plant-animal interactions, and the reproductive success of a tropical shrub. *Biotropica* 30:416–424.

Passos, L., and P.S. Oliveira. 2004. Interaction between ants and fruits of *Guapira opposita* (Nyctaginaceae) in a Brazilian sandy plain rainforest: Ant effects on seeds and seedlings. *Oecologia* 139:376–382.

Peres, C.A., and M.G.M. Roosmalen. 1996. Avian dispersal of mimetic seeds of *Ormosia lignivalis* by terrestrial granivores: Deception or mutualism? *Oikos* 75:249–258.

Ponce-Santizo, G., E. Andresen, E. Cano, and A.D. Cuaron. 2006. Primary dispersal of seeds by primates and secondary dispersal by dung beetles in Tikal, Guatemala. *Biotropica* 38:390–397.

Prasad, S., and R. Sukumar. 2010. Context-dependency of a complex fruit-frugivore mutualism: Temporal variation in crop size and neighborhood effects. *Oikos* 119:514–523.

Reid, S., and J.J. Armesto. 2011. Avian gut-passage effects on seed germination of shrubland species in Mediterranean central Chile. *Plant Ecology* 212:1–10.

Sallabanks, R. 1993. Hierarchical mechanisms of fruit selection by an avian frugivore. *Ecology* 74:1326–1336.

Sánchez, F., C. Korine, B. Pinshow, and R. Dudley. 2004. The possible roles of ethanol in the relationship between plants and frugivores: First experiments with Egyptian fruit bats. *Integrative and Comparative Biology* 44:290–294.

Santamaria, M., and A.M. Franco. 2000. Frugivory of Salvin's Curassow in a rainforest of the Colombian Amazon. *Wilson Bulletin* 112:473–481.

Schaefer, H.M., V. Schmidt, and F. Bairlein. 2003. Discrimination abilities for nutrients: Which difference matters for choosy birds and why? *Animal Behaviour* 65:531–541.

Schupp, E.W., P. Jordano, and J.M. Gómez. 2010. Seed dispersal effectiveness revisited: A conceptual review. *New Phytologist* 188:333–353.

Shiels, A.B., and D.R. Drake. 2011. Are introduced rats (*Rattus rattus*) both seed predators and dispersers in Hawaii? *Biological Invasions* 13:883–894.

Shiferaw, H., D. Teketay, S. Nemomissa, and F. Assefa. 2004. Some biological characteristics that foster the invasion of *Prosopis juliflora* (Sw.) DC. At Middle Awash Rift Valley Area, north-eastern Ethiopia. *Journal of Arid Environments* 58:135–154.

Simberloff, D. 2003. How much information on population biology is needed to manage introduced species? *Conservation Biology* 17:83–92.

Simberloff, D. 2009. The role of propagule pressure in biological invasions. Annual *Review of Ecology, Evolution and Systematics* 40:81–102.

Smith, C.W. 1992. Distribution, status, phenology, rate of spread, and management of *Clidemia* in Hawai'i. In *Alien Plant Invasions in Native Ecosystems of Hawaii*, ed. C.P. Stone, C.W. Smith, and J.T. Tunison, pp. 241–253. Honolulu, Hawaii: University of Hawaii Cooperative National Park Resources Studies Unit.

Sorensen, A.E. 1986. Seed dispersal by adhesion. *Annual Review of Ecology and Systematics* 17:443–263.

Tecco, P.A., D.E. Gurvich, S. Díaz, N. Perez-Harguindeguy, and M. Cabido. 2006. Positive interaction between invasive plants: The influence of *Pyracantha angustifolia* on the recruitment of native and exotic woody species. *Austral Ecology* 31:293–300.

Traveset, A. 1998. Effect of seed passage through vertebrate frugivores' guts on germination: A review. *Perspectives in Plant Ecology, Evolution and Systematics* 1:151–190.

Traveset, A., N. Riera, and R.E. Mas. 2001. Passage through bird guts causes interspecific differences in seed germination characteristics. *Functional Ecology* 15:669–675.

Valido, A., and M. Nogales. 1994. Frugivory and seed dispersal by the lizard *Gallotia galloti* (Lacertidae) in a xeric habitat of the Canary Islands. *Oikos* 70:403–411.

Vander Wall, S.B., K.M. Kuhn, and J.R. Gworek. 2005. Two-phase seed dispersal: Linking the effects of frugivorous birds and seed-caching rodents. *Oecologia* 145:282–287.

Vavra, M., C.G. Parks, and M.J. Wisdom. 2007. Biodiversity, exotic plant species, and herbivory: The good, the bad, and the ungulate. *Forest Ecology and Management* 246:66–72.

Vellend, M. 2002. A pest and an invader: White-tailed deer (*Odocoileus virginianus zimm*) as a seed dispersal agent for Honeysuckle Shrubs (*Lonicera L*). *Natural Areas Journal* 22:230–234.

Vilá, M., and C. D'Antonio. 1998. Fruit choice and seed dispersal of invasive vs. noninvasive *Carpobrotus* (Aizoaceae) in coastal California. *Ecology* 79:1053–1060.

Villaseñor-Sánchez, E.I., R. Dirzo, and K. Renton. 2010. Importance of the lilac-crowned parrot in pre-dispersal seed predation of *Astronium graveolens* in a Mexican tropical dry forest. *Journal of Tropical Ecology* 26:227–236.

Wang, B.C., and T.B. Smith. 2002. Closing the seed dispersal loop. *Trends in Ecology and Evolution* 17:379–385.

Wenny, D.G. 2000. Seed dispersal, seed predation, and seedling recruitment of a neotropical montane tree. *Ecological Monographs* 70:331–351.

Wenny, D.G. 2001. Advantages of seed dispersal: A re-evaluation of directed dispersal. *Evolutionary Ecology Research* 3:51–74.

Westcott, D.A., M. Setter, M.G. Bradford, A. McKeown, and S. Setter. 2008. Cassowary dispersal of the invasive pond apple in a tropical rainforest: The contribution of subordinate dispersal modes in invasion. *Diversity and Distributions* 14:432–439.

Williams, P.A., and B.J. Karl. 1996. Fleshy fruits of indigenous and adventive plants in the diet of birds in forest remnants, Nelson, New Zealand. *New Zealand Journal of Ecology* 20:127–145.

Witkowski, E.T.F., and M.Wilson. 2001. Changes in density, biomass, seed production and soil seed banks of the non-native invasive plant, *Chromolaena odorata*, along a 15 year chronosequence. *Plant Ecology* 152:13–27.

Yagihashi, T., M. Hayashida, and T. Miyamoto. 1998. Effects of bird ingestion on seed germination of *Sorbus commixta*. *Oecologia* 114:209–212.

chapter eleven

Alien plant invasion and its ecological implications

An Indian perspective with particular reference to biodiversity-rich regions

R.S. Tripathi

Contents

11.1 Introduction

John Harper (1967), in his presidential address to the British Ecological Society, emphasized the importance of studies on population biology of invading plants. Mack (1985, p. 127), while discussing the potential contribution of invading plants to population biology, mentioned that "the importance of plant invasions to population dynamics and the transformation of landscapes did not escape the attention of Victorian biologists including Darwin and Wallace." Darwin (1872) pointed to the rapid spread of alien cardoon (*Cynara cardunculus*) and a tall thistle (*Silybum marianum*) in Argentina. Wallace (1905) reported that in New Zealand, alien *Rumex acetosella* covered hundreds of acres with a sheet of red. In recent times, the problem of invasion by alien plant species has become a matter of great concern all over the world. The migration of plant species from one geographical region to another across natural barriers such as high mountains, seas, and oceans has been taking place since time immemorial, but the movement of plant species through natural dispersal agents has been rather slow. However, with globalization there has been a phenomenal increase in trade, tourism, travel, and other human activities, and this has caused both intentional and unintentional introduction of species from one country to another at a pace that was never witnessed before (Tripathi 2009). If a plant species arrives in a territory where the habitat conditions are similar to those of its native place, it germinates, survives, grows, reproduces, and produces self-sustaining populations in natural and seminatural ecosystems in the course of time. Many exotic species may grow luxuriantly in the new environment, expand their range of distribution at a fast rate, and even pose a serious threat to the native species of the invaded area. These alien invasive species are characterized by rapid growth, high reproductive capacity, efficient dispersal mechanism, strong competitive ability, and ability to adapt physiologically to new environmental conditions. Thus, they are able to cope successfully with the biological and physical conditions of the invaded territory. It may be noted that all alien/exotic species may not be invasive.

The problem of biological invasion has been recognized by the Scientific Committee on the Problems of the Environment (SCOPE) as a central problem in the conservation of biological communities. Invasive alien plants have serious ecological implications for the conservation of native biodiversity, maintenance of plant community structure, plant succession, and ecosystem processes in the areas invaded by them. The problem of plant invasion has engaged the attention of ecologists, conservationists, and environmentalists all over the world over the past 3 or 4 decades, especially after the launch of the Global Invasive Species Programme (GISP) by SCOPE. However, in India, so far the problem of plant invasion has not been addressed as adequately as it should have been, although in India too, several exotic plants have invaded the high-value biodiversity areas and have adversely affected the natural and seminatural ecosystems.

Biological invasions and climate change are among the hottest topics in ecology these days. The extent of distribution, rate of spread, and persistence of invasive alien species directly influence the native biodiversity of an invaded region; therefore, the trends in invasion by alien species have been identified as an important indicator of the loss of biodiversity. The Convention on Biological Diversity's (CBD) framework for monitoring progress toward its "2010 Target," which commits CBD to achieve by 2010 a significant reduction in the current rate of biodiversity loss, regards biological invasion as one of the major threats to biodiversity, next only to habitat destruction. It has been reported by a number of researchers that the invasion of plant species in new environments is triggered by human-induced habitat fragmentation, land degradation, forest degradation, land-use and land-cover changes, fire regime changes, and other kinds of anthropogenic stresses that impact natural ecosystems.

11.2   Role of population dynamics in plant invasion

Studies on population dynamics and growth of a number of exotic species such as *Chromolaena odorata, Ageratina adenophora, Ageratina riparia*, and *Imperata cylindrica* in relation to burning, age of "jhum" fallows, associated vegetation, and varied density, light, and soil conditions indicate that these weeds are particularly successful in disturbed habitats (Tripathi 1985). The facilitative effect of these drivers on plant invasion may presumably be mediated through the reduction in various kinds of biotic and physical resistances that would have been offered by the undisturbed host plant community. The quantification of the extent of influence exercised by different kinds of environmental resistances to an invading plant species in a host community could be a very challenging area of ecological study. Apart from causing depletion of native biodiversity, invasive alien species alter species composition; affect physical, chemical, and biological properties of soil; and affect community development and ecosystem processes adversely; but no reliable quantitative data is available on any of these aspects. The effects of invasive alien species on the distribution, abundance, and population dynamics of native plant species in natural ecosystems, hydrology, soil biology, and ecosystem processes need to be studied in detail. In India, a good number of high-value biodiversity sites have been invaded by invasive alien plants; but unfortunately, studies on the biology of plant invasion are rather scanty. The biology and population dynamics of a number of exotic weeds were studied by the author and his collaborators in the department of botany at North-Eastern Hill University in Shillong, Meghalaya, from the 1980s onwards. Besides, research on weed biology has been conducted at several other universities and research organizations in India, but the plant invasion perspective is missing in most of these studies.

Although the problem related to the invasion of alien plant species has engaged the attention of ecologists, foresters, agricultural scientists, and government agencies all over the world, nothing tangible has been done to eradicate and manage the ever-increasing populations of invading plant species that have already established themselves in the invaded areas and are still extending their range of distribution due to human-induced habitat fragmentation and several other kinds of stresses. The economic and ecological costs associated with the invasion of these alien plant species are indeed staggering. Some of the interesting aspects, and exciting points, emerging as a result of analysis and synthesis of the scientific information gathered on relevant aspects of plant invasion (Tripathi 2009) are presented as follows:

- Genetic changes are likely to occur in a species subsequent to invasion in a new region, and these changes may hold the key to its success in the invaded land. An invading species that colonizes a novel environment has to face a genetic challenge, because it has not experienced the selective pressures presented by the new environment. Despite this, most alien species become successful invaders, although they have to face challenges from the already well-adapted native species. Biologists need to find out the underlying mechanisms and processes that make the invading species so successful in their new environments.

- Invasive alien species (e.g., *Eupatorium odoratum*) are intrinsically better competitors, and so they offer strong competition and pose a serious threat to native species in the invaded region (Yadav and Tripathi 1981). The native species of an area show a decline in resource use, and the invaders can increase their distribution and abundance at the expense of the resident species. This may cause a drastic reduction in the population size of several native species; some may even be eliminated from their natural habitats.

- Most invasive plant species possess high phenotypic plasticity (Rai and Tripathi 1983) coupled with hybridization capacity and highly efficient reproductive strategies. Rai and Tripathi (1983) have reported that although the reproductive effort of *Galinsoga ciliata* and *Galinsoga parviflora* showed considerable plasticity, it is maintained at fairly high level even under the stressed ecological conditions. Their populations are characterized by the presence of at least three seedling cohorts, which emerge at different times, and these seedling cohorts differ in their half-life, survivorship, and seed output (Rai and Tripathi 1984a). This adds to their level of plasticity and contributes to their adaptability, ecological success, and ability to invade new areas.

- Many invasive plant species, for example, *Ageratina riparia* or *Eupatorium riparium* (Rai and Tripathi 1984b), *Parthenium hysterophorus*, *Chromolaena odorata*, and *Ageratina adenophora* (Tripathi et al. 2012a), release chemical compounds into the environment, which are not generally harmful to them, but those chemicals suppress the growth of plants of other species growing in the close proximity of such invasive species. This negative effect (often referred to as an allelopathic effect) of invaders on the native species confers a tremendous competitive advantage on the former. The "chemical release" hypothesis offers a plausible explanation for the spectacular success of invasive plant species in the areas that they invade.

- The herbivores and parasites or pathogens, the natural enemies of invasive species, that were regulating the population growth of invaders in their native place are absent in an invaded region. Invading species generally arrive in new environments without their coevolved natural enemies from their natural habitats. This may provide invaders opportunities for luxuriant growth and more prolific reproduction, which allow them to outcompete native species and expand their range of distribution. This is the basis of the so-called "enemy release" or "escape" hypothesis, which is employed to explain the spectacular success of invasive alien plants in new environments.

The aforementioned hypotheses or approaches explain why and how alien species become more successful in the invaded land compared to their native place. It may be mentioned that the majority of studies on invasive alien species have been conducted in the invaded territory; surprisingly, we do not have any quantification regarding their abundance, competitive success, aggressiveness, and response to natural enemies in their native land. The soundness of these "invasion hypotheses" can be tested only when we apply a comparative biogeographical approach toward the problem of biological invasion and have sufficient relevant data from the native as well as invaded regions (Tripathi 2009). In one of his earlier articles, Tripathi (1985) emphasized that a comparative study of population behavior, individual fitness, and reproductive strategies of invasive alien species in their countries of origin and newly invaded host countries could be quite revealing and rewarding.

For any species to become successful in a new environment, it is essential that the species genetically adapts itself to its new environment. It may also be mentioned here that preserving genetic diversity is absolutely necessary for a species to continually adapt genetically to a changing environment. Therefore, investigating the genetic adaptability of invasive alien plant species in new environments should also be an issue of focus for population ecologists and conservation biologists.

Besides impacting native biodiversity through direct competitive suppression, invasive alien species have strong allelopathic potential and influence plant diversity by converting a complex plant community into a much simpler one, which is characterized by the dominance of only a few species. In extreme situations where the impact is severe, the plant community could even show single-species dominance. This kind of effect of invasive species has serious implications for food chain length and food web complexity, which are prerequisites for the stability and smooth functioning of ecosystems. Unfortunately, research studies related to the impact of invasive plant species on ecosystem processes are scarce. Ecological impacts of plant invasion are of grave concern to all of us. As the ecological consequences of plant invasion are manifold, we need to have zero tolerance toward invasive alien species.

Of late, the problem of invasion by alien plant species has assumed alarming proportions in several parts of India, and many biodiversity-rich areas in the plains and hills are being adversely impacted due to the luxuriant growth of invasive exotic plants. India is particularly vulnerable to biological invasions due to its wide range of climatic conditions, soil types, and other factors that promote invasion. Therefore, plant and animal species native to various countries with different climatic conditions can find suitable niches in some parts of India or the other. The literature on population dynamics of invasive alien plant species shows that they grow quite fast and are capable of invading and colonizing new areas within a short span of time if ecological conditions are favorable. They have a strong competitive ability and exert strong allelopathic effects on native plants, and this contributes to their invasive success.

11.3   *Invasive plants in the northeastern region of India*

Earlier, the northeastern region of India comprised the states of Assam, Arunachal Pradesh, Manipur, Meghalaya, Mizoram, Nagaland, and Tripura; subsequently, the state of Sikkim was also included in this region. Large variations in altitude, topography, soil types, rainfall patterns, temperature, and other climatic factors have caused tremendous niche diversity in this region, and diverse forest types ranging from tropical humid forests to subalpine forests and alpine vegetation are found there. The interactive influence of various ecological factors has been instrumental in promoting successful colonization and perpetuation of a large variety of plants and animals. The edapho-climatic conditions of the area favor the growth and multiplication of a variety of plant species, making northeast India a very rich floristic region. In fact, this region together with Eastern Himalaya, which is a global hot spot of biodiversity, forms the richest botanical zone of India. The slash-and-burn agricultural practices (locally called jhum) prevalent in the region, shortening of the jhum cycle (the intervening fallow period after which the vegetated land is again cultivated) due to increased human population, and deforestation have created habitats suitable for successful colonization by arriving exotics. Some of these exotic species, especially those having invasive attributes, have successfully colonized the newly created habitats and disturbed areas, and thus they have undergone tremendous range expansion in the northeastern region. Their ever-increasing population and luxuriant growth show that they are highly aggressive and exert strong competitive influence on the native species. Among the exotic weeds growing in the northeastern region, three species of *Eupatorium*, that is, *Eupatorium odoratum* L. (syn. *Chromolaena odorata* (L.) King & Robinson), *Eupatorium adenophorum* Spreng. (syn. *Ageratina adenophora* (Spreng.) King & Robinson), and *Eupatorium riparium* Regel. (syn. *Ageratina riparia* (Regel.) King & Robinson); *Ageratum conyzoides; Galinsoga ciliata; Galinsoga parviflora; Lantana camara*; and *Mikania micrantha* are of special interest as they have become dominant components of the weed communities in the region. Our studies on population dynamics of three invasive species of *Eupatorium* in relation to varied ecological conditions indicate that they are particularly successful on the disturbed and exposed habitats of northeast India (Kushwaha et al. 1981, Yadav and Tripathi 1981, Tripathi 1985, Tripathi and Yadav 2012, Tripathi et al. 2012a). Being aggressive, they have been spreading very fast and are posing a serious threat to many useful elements of the native flora and influencing the structure and function of various natural ecosystems of the region. The shortening of the jhum (slash-and-burn agriculture) cycle from 20–30 years to 4–5 years due to increased population pressure has also resulted in the spread of these invasive alien species of *Eupatorium* to large areas of the northeastern region of India. They have also invaded vast areas in the Western Ghats, Uttarakhand, Sikkim, and Himachal Pradesh. *Parthenium hysterophorus*, whose population was negligible in the region about 30 years ago, has now become an invasive species in the plains of this region. Several alien invasive species choke the waterways, small rivers, ponds, and lakes in northeastern India and take a heavy toll on the wetlands. Many invasive alien species, such as *Chromolaena odorata, Ageratina adenophora, Ageratina riparia* (Tripathi and Yadav 2012, Tripathi et al. 2012a), and *Mikania micrantha* (Tripathi et al. 2012b), are now naturalized and have become serious pests of plantation crops and agriculture in this region. According to Swamy and Ramakrishnan (1987) and Tripathi et al. (2012b), *Mikania micrantha* has been spreading very fast in the plains of this region. These perennial species belonging to the family Asteraceae produce enormous numbers of seeds annually. For example, a single plant of *Chromolaena odorata* can produce 153,710 seeds (Tripathi and Yadav 1981), which are dispersed over long distances by wind. The seed population density of the three *Eupatorium* spp. in soil is very high and a large proportion of the viable seeds acquire enforced and induced dormancy during burial (Yadav and Tripathi 1982). The population dynamics of *Galinsoga quadriradiata* (*Galinsoga ciliata*) and *Galinsoga parviflora* have been studied in detail (Rai and Tripathi 1983, 1984a, 1984b), and it has been shown that these alien species have become naturalized in the northeastern region of India and are suppressing the seedling populations of several indigenous herbaceous species. Sudhakar Swamy and Ramakrishnan (1987) and Tripathi et al. (2012b) have shown that *Mikania micrantha*, a climbing perennial weed belonging to the family Asteraceae, has been spreading very fast and has now become a serious weed in tea gardens and forest plantations in the region. Tripathi et al. (2012b) have reviewed the literature on the biology of this weed and have mentioned that it is widely distributed in the states of Assam and Arunachal Pradesh, particularly at low altitudes.

The economic and ecological costs associated with the invasion of these exotic weeds are enormous. Many alien invasive plants adversely affect seed germination and seedling growth of native species in invaded plant communities. Their allelopathic effects appear to be one of the most important means by which invasive alien species gain dominance over native species. The ability of *Chromolaena odorata*, *Ageratina adenophora*, and *Ageratina riparia* to outcompete and suppress the growth of other neighboring plant species through allelopathic influence has been reported by Tripathi and his collaborators (Tripathi, Singh and Rai 1981; Rai and Tripathi 1982). These fast-spreading and luxuriantly growing plant invaders have serious ecological implications for the diversity, distribution, and abundance of native species and ecosystem functioning in most parts of the northeastern region of India.

The status of alien plant species in the region has been discussed by Uma Shankar et al. (2012), who have listed some of the most frequently recorded invasive plants growing in different ecosystems of northeast India. Besides those invasive alien plants whose population dynamics have been studied in detail by Tripathi and his collaborators, a good number of other alien species grow luxuriantly in this region and pose a serious threat to native species, for example, *Ageratum conyzoides*, *Ageratum houstonianum*, *Artemisia nilagirica*, *Bidens pilosa*, *Mikania micrantha*, *Parthenium hysterophorus*, *Urena lobata*, *Mimosa invisa*, *Oxalis corniculata*, *Saccharum spontaneum*, *Rumex nepalensis*, *Lantana camara*, *Ligustrum robustum*, *Eichhornia crassipes*, *Borreria hispida*, and *Solanum sisymbrifolium* (Uma Shankar et al. 2012). The invasion of subtropical forests of Meghalaya, which are dominated by *Pinus kesya* (Khasi pine), by the invasive alien shrubs *Artemisia nilagirica*, *Lantana camara*, and *Ligustrum robustum* has now become a common phenomenon. Rao and Sagar (2012a) stated that exotic invasive plants have started gradually upsetting the original floristic elements in this region. They have also listed a number of wild plant species of northeast India whose populations have been adversely affected by the spread of invasive weeds.

11.4   Invasive alien plants in the Western Ghats

Invasion by alien plants has brought about a conspicuous change in biodiversity, natural landscape, plant community structure, and ecosystem processes not only in the northeastern region of India but also in another biodiversity hot spot of India, that is, the Western Ghats. It is a part of the triangular Deccan Plateau, which comprises the greater part of southern and central India. On both its western and eastern edges, the Deccan Plateau rises to the ranges of mountains called the Ghats, which meet near Kanyakumari at the southern tip of the peninsula. The Western Ghats is a narrow stretch of land extending about 1500 km as far south as Kanyakumari and as far north as the hills south of the river Rapti along the west coast of India. The wide altitudinal variation ranging from below 500 m in the Deccan plains in the east to about 2900 m in the west causes a large gradient of rainfall and temperature regime. Owing to the varied climatic conditions in different parts of the Ghats, a large variety of plant and animal species are found here, making this region one of the richest floristic regions in the world. The Western Ghats have several types of forests, that is, evergreen, semievergreen, moist deciduous, and dry deciduous forests; dense tropical rain forests in its southern part; and "sholas" (southern montane wet temperate forests) in the Nilgiri and Anamalai regions. The Western Ghats covers only 5% of India's total geographical area, but it contains over 30% of the plant species found in India: about 12,000 species, which includes both flowering plants and plants belonging to lower groups (Rao and Sagar 2012b). The area is also characterized by high endemism.

However, increasing population pressure, unsustainable utilization of bioresources, raising of plantation crops, forest degradation and deforestation, fire, movement of humans and material, and several other kinds of anthropogenic stresses have caused tremendous changes in the structure and function of various ecosystems in the Western Ghats. All these factors have contributed to invasions by a large number of alien species in the region. Some of the alien species have spread to vast areas in the region, becoming naturalized and now posing a serious threat to the native species, of which many are endemic to the Western Ghats. Rao and Sagar (2012b) have enumerated the invasive alien species in this region and have listed their description and nativity. Their list comprising the 12 worst invasive alien plant species in the Western Ghats is given in Table 11.1.

Besides the 12 worst invasive alien plants listed in Table 11.1, there are many other species that are quite abundant in disturbed habitats, such as degraded sparse forests with large gaps and openings, waste places, roadsides, along railways, and forest fringes. Rao and Sagar (2012b) have described a number of them giving their nativity and distribution in the Western Ghats. Some of them are *Bidens biturnata* (a native of tropical America), *Conyza bonariensis* (a native of Argentina), *Galinsoga quadriradiata* or *Galinsoga ciliata* (a native of Mexico), *Galinsoga parviflora* (from South America), *Cassia occidentalis* (a native of West Indies distributed in waste places, meadows, and foothills throughout the Western Ghats), *Cassia tora* (a native of South America growing abundantly in the central and northern parts of Karnataka on wastelands, along forest margins, along roadsides, and along

Table 11.1 Twelve Worst Invasive Alien Plant Species in the Western Ghats

Invasive plant species	Family
Ageratina adenophora	Asteraceae
Alternanthera paronychioides	Amaranthaceae
Cassia odorata	Caesalpiniaceae
Cassia uniflora	Caesalpiniaceae
Chromolaena odorata	Asteraceae
Eichhornia crassipes	Pontederiaceae
Hyptis suaveolens	Lamiaceae
Ipomoea fistula	Convolvulaceae
Lantana camara	Verbenaceae
Mikania micrantha	Asteraceae
Parthenium hysterophorus	Asteraceae
Prosopis chilensis	Mimosaceae
Ricinus communis	Euphorbiaceae

Source: Rao, R.R., and K. Sagar, Invasive alien weeds of the Western Ghats: Taxonomy and distribution, In *Invasive Alien Plants: An Ecological Appraisal for the Indian Subcontinent*, CAB International, Oxford, United Kingdom, 139–161, 2012.

railway lines); *Croton bonplandianum* (a native of South America) and *Mimosa invisa* (a native of South America) grow abundantly in Karnataka, Kerala, and Tamil Nadu in open, moist areas. *Salvinia molesta* (a native of South America) commonly occurs in the wetlands of Kerala and Karnataka (Rao and Sagar 2012b). It has been mentioned by Rao and Sagar (2012b, p. 140) that "*Mikania micrantha, Puereria phaseoloides*, and others have eliminated much of the indigenous flora in the Western Ghats." According to them, the diversity of about 50 wild species in India has been affected by invasive alien species of the family Asteraceae. Perhaps this conclusion is based only on visual observations; I am of the view that extensive field studies should be undertaken and credible quantitative data be gathered to substantiate this claim.

There is every likelihood that invasive alien species would adversely affect vegetation pattern and processes by progressively replacing existing indigenous flora in not only high-value biodiversity areas such as the northeastern region of India and the Western Ghats but also in other parts of the country. There is, however, a strong need for undertaking long-term research programs covering different aspects of alien plant invasion in a network mode covering the entire length and breadth of India, as the problem of invasive alien plants is already quite severe in this country and it is going to worsen a few decades from now.

It has been reported by a number of researchers that the invasion of plant species is triggered by human-induced habitat fragmentation, land degradation, degradation of forests, depletion of forest cover, changes in land-use pattern, and several other kinds of anthropogenic stresses that impact natural ecosystems. However, the exact mechanism underlying the facilitatory influence of these drivers on plant invasion is yet to be clearly and fully understood. It is often claimed that plant invasion causes depletion of native biodiversity; alters species composition; affects physical, chemical, and biological properties of soil; and adversely affects community development and ecosystem processes; but no reliable quantitative data is available on any of these aspects vis-à-vis plant invasion. The effects of invasive alien species on the distribution, abundance, and population dynamics of native plant species in natural ecosystems, hydrology, soil biology, and ecosystem processes need to be studied in detail. Biology and population dynamics of a number of exotic weeds were studied intensively by the author and his collaborators at North-Eastern Hill University, but the plant invasion perspective was not adequately addressed in those studies. However, in view of the enormity of the plant invasion problem in this country, research related to plant invasion must be accorded high priority. There is a need to launch a national website on biological invasion, set up regional biological invasion centers across the country, and create a national biological invasion authority that covers the entire gamut of problems associated with biological invasion.

11.5   *Conclusions and the way forward*

The problem of alien plant invasion needs to be addressed very seriously, like climate change. Indeed, the problem has been engaging the attention of ecologists all over the globe. However, in India, so far the problem of alien plant invasion has not been adequately addressed. Some of the points related to alien plant invasion in India that need to be addressed without further delay are as follows:

- The status of alien plant invasion in India
- Identification of the worst alien species that have invaded natural ecosystems in different biogeographic regions of India and their distribution, rates of invasion, and population dynamics

- Invasive alien species and their impact on native biodiversity, plant community composition, and ecosystem processes
- Attributes and ecological strategies of the worst invasive alien species and their pathways of invasion
- Identification of the habitats and ecosystems most vulnerable to plant invasion
- Causes of the spectacular success of invasive alien plants in the natural and semi-natural ecosystems invaded by them
- Plant invasion in relation to various kinds of anthropogenic disturbances, fire regimes, species richness and composition, and habitat characteristics of host plant communities
- Plant invasion in relation to elevated concentrations of carbon dioxide, climate change, and other global changes
- Genetic adaptability of invasive exotic species in new environments
- Impact of invasive alien plants on physical, chemical, and biological properties of the soil of areas invaded by them
- Effects of invasive alien plants on ecosystem processes
- Strategies and action plans for the effective management of some of the worst invasive plant species at the local, ecoregion, and national levels

There is a need to launch a coordinated/networked research program at the national level in India encompassing the aforementioned points. We need to develop a policy framework for tackling the problem of plant invasion in India. Special emphasis must be laid on some of the worst invasive alien weeds that have been spreading very fast and have become established and naturalized in several parts of India. Notable among such alien species are *Ageratum conyzoides, Eichhornia crassipes, Eupatorium odoratum* L. (syn. *Chromolaena odorata* (L.) King & Robinson), *Eupatorium adenophorum* Spreng. (syn. *Ageratina adenophora* (Spreng.) King & Robinson), *Eupatorium riparium* Regel. (*Ageratina riparia* (Regel.) King & Robinson), *Galinsoga ciliata, Lantana camara, Imperata cylindrica, Mikania micrantha, Parthenium hysterophorus*, and *Prosopis juliflora*.

There is a strong need to launch a national website and create a national authority that covers the entire gamut of problems associated with biological invasion. The proposed authority could monitor the invasion and spread of alien species and educate people about the adverse impact of invasive alien species on native flora, the integrity of natural ecosystems, and human and animal health. The authority could also formulate strategies and action plans for preventing alien invasion and for controlling and eradicating invasive exotic species, and suggest suitable mitigation measures where the preventive and control measures fail to yield the desired results.

11.6   Acknowledgments

Funding support from the Indian National Science Academy (INSA), New Delhi, in the form of INSA Senior Scientist and INSA Honorary Scientist positions awarded to me is gratefully acknowledged. I thank the successive directors (Drs. P. Pushpangadan, R. Tuli, and C.S. Nautiyal) of the CSIR-National Botanical Research Institute (NBRI) in Lucknow, Uttar Pradesh for providing infrastructural facilities. The article on alien plant invasion that I wrote in 2009 in *Environews*, which is published by the International Society of Environmental Botanists, NBRI Campus, Lucknow, has provided the basic framework for this chapter. I sincerely thank my former PhD and postdoctoral students for generating huge amounts of data and information on the biology and population dynamics of exotic weeds in the northeastern region of India.

References

Darwin, C. 1872. *On the origin of species*. 6th ed. London, UK: John Murray.

Harper, J.L. 1967. A Darwinian approach to plant ecology. *Journal of Ecology* 55:247–270.

Kushwaha, S.P.S., P.S. Ramakrishnan, and R. S. Tripathi. 1981. Population dynamics of *Eupatorium odoratum* in successional environments following slash and burn agriculture. *Journal of Applied Ecology* 18:529–535.

Mack, R.N. 1985. Invading plants: Their potential contribution to population ecology. In *Studies on plant demography: A Festschrift for John L. Harper*, ed. James White, pp. 127–142. London, UK: Academic Press.

Rai, J.P.N., and R.S. Tripathi. 1982. Allelopathy as a factor contributing to dominance of Eupatorium riparium Regel. *Indian Journal of Ecology* 9:14–20.

Rai, J.P.N., and R.S. Tripathi. 1983. Population regulation of *Galinsoga ciliata* and G. *parviflora*: Effect of sowing pattern, population density and soil moisture and texture. *Weed Research* 23:151–1163.

Rai, J.P.N., and R.S. Tripathi. 1984a. Population dynamics of different seedling cohorts of two co-existing annual weeds, *Galinsoga ciliata* and *G. parviflora*, on two contrasting sites. *Acta Oecologica: Oecologia Plantarum* 5:357–368.

Rai, J.P.N., and R.S. Tripathi. 1984b. Allelopathic effect of *Eupatorium riparium* on population regulation of two species of *Galinsoga* and soil microbes. *Plant and Soil* 23:105–117.

Rao, R.R., and K. Sagar. 2012a. Invasive alien weeds in the tropics: The changing pattern in the herbaceous flora of Meghalaya in north-east India. In *Invasive alien plants: An ecological appraisal for the Indian subcontinent*, eds. J.R. Bhatt, J.S. Singh, S.P. Singh, R.S. Tripathi, and R.K. Kohli, pp. 189–198. Oxford, UK: CAB International.

Rao, R.R., and K. Sagar. 2012b. Invasive alien weeds of the Western Ghats: Taxonomy and distribution. In *Invasive alien plants: An ecological appraisal for the Indian subcontinent*, eds. J.R. Bhatt, J.S. Singh, S.P. Singh, R.S. Tripathi, and R.K. Kohli, pp. 139–161. Oxford, UK: CAB International.

Shankar, U., A.S. Yadav, J.P.N. Rai, and R.S. Tripathi. 2012. Status of alien plant invasion in the northeastern region of India. In *Invasive alien plants: An ecological appraisal for the Indian subcontinent*, eds. J. R. Bhatt, J.S. Singh, S.P. Singh, R.S. Tripathi, and R.K. Kohli, pp. 174–188. Oxford, UK: CAB International.

Swamy, P.S., and P.S. Ramakrishnan. 1987. Effect of fire on population dynamics of *Mikania micrantha* H.B.K. during early succession after slash and burn agriculture (jhum) in northeast India. *Weed Research* 27:397–403.

Tripathi, R.S. 1985. Population dynamics of a few exotic weeds in north-east India. In *Studies on plant demography: A Festschrift for John L. Harper*, ed. James White, pp. 157–170. London, UK: Academic Press.

Tripathi, R.S. 2009. Alien plant invasion: A hot ecological issue. *Environews* 15:7–10.

Tripathi, R.S., M. L. Khan, and A. S. Yadav. 2012b. Biology of *Mikania micrantha* H.B.K.: A review. In *Invasive alien plants: An ecological appraisal for the Indian subcontinent*, eds. J.R. Bhatt, J.S. Singh, S. P. Singh, R.S. Tripathi, and R. K. Kohli, pp. 99–107. Oxford, UK: CAB International.

Tripathi, R.S., and A.S. Yadav. 2012. Population dynamics of invasive alien species of *Eupatorium*. In *Invasive alien plants: An ecological appraisal for the Indian subcontinent*, eds. J.R. Bhatt, J.S. Singh, S. P. Singh, R.S. Tripathi, and R.K. Kohli, pp. 257–270. Oxford, UK: CAB International.

Tripathi, R.S., A.S. Yadav, and S.P.S. Kushwaha. 2012a. Biology of *Chromolaena odorata, Ageratina adenophora* and *Ageratina riparia*: A review. In *Invasive alien plants: An ecological appraisal for the Indian subcontinent*, eds. J.R. Bhatt, J.S. Singh, S.P. Singh, R.S. Tripathi, and R.K. Kohli, pp. 43–56. Oxford, UK: CAB International.

Tripathi, R.S., R.S. Singh, and J.P.N. Rai. 1981. Allelopathic potential of Eupatorium adenophorum—a dominant ruderal weed of Meghalaya. *Proccedings of the Indian National Science Academy* 7 (B):458–465.

Wallace, A.R. 1905. *Darwinism: An exposition of the theory of natural selection, with some of its applications*. 3rd ed. New York, US: Macmillan.

Yadav, A.S., and R.S. Tripathi. 1981. Population dynamics of the ruderal weed *Eupatorium odoratum* and its natural regulation in nature. *Oikos* 36:355–361.

Yadav, A.S., and R.S. Tripathi. 1982. A study on seed population dynamics of three weedy species of *Eupatorium*. *Weed Research* 22:69–76.

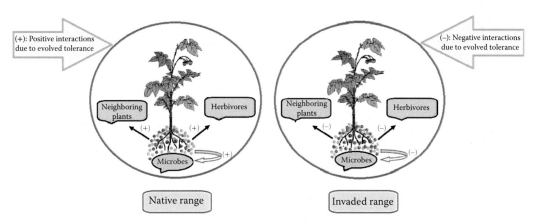

Figure 3.1 Pictorial representation of the novel weapon hypothesis: In the native range, chemicals released by the plants exhibit positive interactions toward neighboring plants, herbivores, and microbes possibly due to tolerance acquired over the years. In the invaded range, on the other hand, chemicals released by the exotic plants serve as novel weapons to the other naive neighboring plants, herbivores (generalists), and microbes that were not earlier exposed to such chemicals. Thus, there is no evolved tolerance in them toward novel chemicals; this results in negative interactions that provide selective advantage to the exotic plant (donor of novel chemicals).

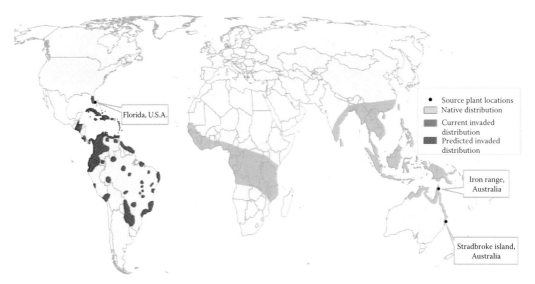

Figure 4.1 Native, introduced, and predicted distributions of *Lygodium microphyllum*: Spore–source populations of plants used in the two studies include the introduced Florida population, the native Iron Range population (reputed source of the plants originally introduced to Florida), and the native Stradbroke Island population. The study conducted in Florida used only plants originating from the Florida source population, whereas the study conducted in southeast Queensland included plants from all the three source populations. (Adapted from Pemberton, R., and A. Ferriter. *Am. Fern J.*, 88(4), 165–175, 1998; J. A. Goolsby. *Nat. Areas J.*, 24, 351–353, 2004; Volin et al. *Diversity and Distributions*, 10, 439–446, 2004; Volin et al. *Plant Ecol.*, 208(2), 223–234, 2010.)

Figure 4.2 Old World climbing fern (*Lygodium microphyllum*) smothering understory and overstory native plant species in a bald cypress (*Taxodium distichum*) swamp in southern Florida. (Courtesy of Peggy Greb, United States Department of Agriculture/Agricultural Research Service.)

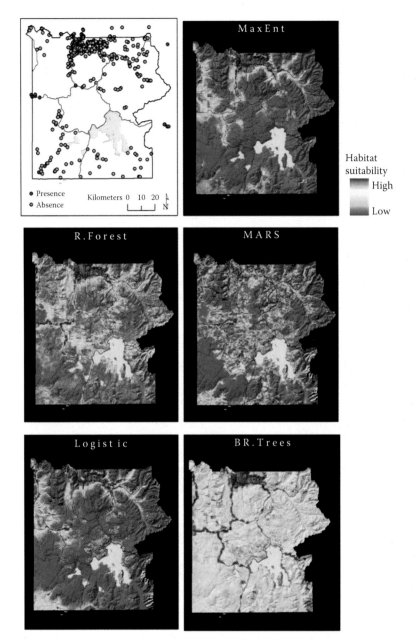

Figure 14.2 Point distribution of the invasive plant *Linaria dalmatica* (Dalmatian toadflax) in Yellowstone National Park, Wyoming, United States, and habitat suitability maps generated using five common species distribution models: Maxent, random forest (R. Forest), multivariate adaptive regression splines (MARS), logistic regression (Logistic), and boosted regression trees (BR. Trees). All models predicted suitable habitat in the north-central part of the park, which overlapped the majority of reported presence locations, otherwise the models accentuated different locations, for example, the boosted regression model highlighted road corridors compared to the MARS model, which emphasized habitat suitability in the park's interior. (Reprinted from Stohlgren, T.J. et al., Ensemble habitat mapping of invasive plant species. *Risk Analysis.* 2010. 30:224–235. Copyright Wiley. Reproduced with permission.)

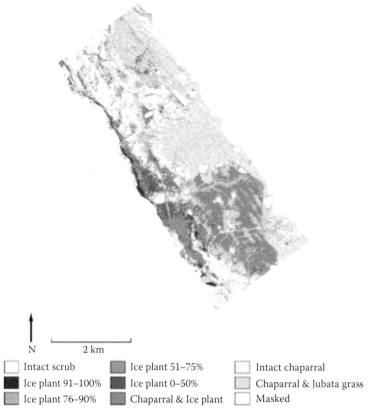

Intact scrub | Ice plant 51–75% | Intact chaparral
Ice plant 91–100% | Ice plant 0–50% | Chaparral & Jubata grass
Ice plant 76–90% | Chaparral & Ice plant | Masked

Figure 14.4 A supervised classification of an AVIRIS flight line in coastal California using bands generated from a minimum noise fraction procedure. Invasive ice plant (*Carpobrotus edulis*) was detected based on its water-absorption features in the spectral profile (around 0.9 μm) and different depths of this feature permitted four classes of ice plant density to be mapped with a high level of accuracy.

(b)

Figure 14.5 (b) Photograph of perennial pepperweed (*Lepidium latifolium*) highlighting the white flowers. (Photo courtesy of M. Andrew.)

Figure 16.1 Fire flows (solid lines) along ridges and jumps (dotted lines) across gullies to contribute to the fragmented forest location pattern within fynbos shrubland near Swellendam (left) and within grassland in the Catherdral Peak area, KwaZulu-Natal Drakensberg (middle and right), South Africa. (Adapted from Geldenhuys, C.J., *Journal of Biogeography*, 21, 49–62, 1994.)

(a)   (b)   (c)

Figure 16.4 Natural pioneer species and invasive alien species become established in the natural forest environment with the protection of the landscape against fire: (a) Natural regeneration of *Juniperus procera* on the Nyika Plateau, Malawi, inside the firebreak around a mixed evergreen natural forest with a few large trees of this species; (b) Invasion of grassland by the native legume pioneer tree *Acacia karroo* after the private nature reserve near East London, South Africa, was protected against fire; (c) Invasion of fynbos shrubland vegetation above the natural forest edge by *Pinus radiata* trees when the total area was protected against fire to safeguard the plantation forestry estate behind the photographer.

chapter twelve

Impact of invasive alien plant species on aboveground and belowground species diversity in the Kashmir Himalaya, India

Zafar A. Reshi, Nazima Rasool, Pervaiz A. Dar,
Waheeda Rehman, and Manzoor A. Shah

Contents

12.1 Introduction

A key driver of global environmental change is the invasion of ecosystems by alien species, many of which attain sufficiently high abundance to alter ecosystem structure and function (D'Antonio and Vitousek 1992, Ogle et al. 2003, Meffin et al. 2010). Biological invasions affect virtually all ecosystems on earth, but the extent of invasion of different regions and biomes varies greatly (Foxcroft et al. 2010). The invasions by alien species are also known to impact ecosystem services (Charles and Dukes 2007) and human well-being (Pejchar and Mooney 2009, Vilà et al. 2011). Invasive alien plants, because of their ability to alter ecological processes, such as carbon and nitrogen cycling (Liao et al. 2008, Ehrenfeld 2010), hydrological cycles (Calder and Dye 2001), frequency and/or intensity of fire (Brooks et al. 2004), and alteration of the normal disturbance regimes in the native communities (D'Antonio and Meyerson 2002, Werner et al. 2010), have transformed many ecosystems by outcompeting native species (Lankau 2010) and, thus, are rightly regarded as one of the most substantial threats to biodiversity on Earth (Cronk and Fuller 1995, Chapin et al. 2000, Kowarik 2003, Werner et al. 2010). Recent global meta-analysis of 199 research studies dealing with 1041 field studies involving 135 alien plant taxa revealed that abundance and diversity of resident species decreased in invaded sites, whereas primary production and several ecosystem processes were enhanced (Vilà et al. 2011) due to invasion by different plant species.

Most of the studies exploring the effects of plant invasions have focused on aboveground flora and fauna (Levine et al. 2003), although soil organisms play important roles in regulating ecosystem-level processes (Wardle et al. 2004), and soils contain much of the biodiversity of terrestrial ecosystems (Torsvik et al. 1990, Vandenkoornhuyse et al. 2002). This is mainly because aboveground communities are relatively easy to observe and quantify (Belnap and Phillips 2001) and also because there are methodological limitations in studying belowground diversity.

As a result, few studies to date have considered the effects of invasive organisms on the abundance, composition, and activity of the soil biota. However, with the advent of tools and techniques that exploit the presence of signature biomolecules, such as phospholipid fatty acid (PLFA) profiling, denaturing gradient gel electrophoresis (DGGE), terminal restriction fragment length plymorphism (T-RFLP), and so on, have been used to monitor changes in microbial communities in many plant invasion studies (Meyer 1994, Kourtev et al. 2002a, 2003, Angeloni et al. 2006, Batten et al. 2006, Li et al. 2006, Kulmatiski and Beard 2008, Zhang et al. 2010). These studies have revealed that invasive alien plants may suppress some harmful rhizospheric soil microbes (Bais et al. 2004, Lorenzo et al. 2010) and enrich the beneficial ones, thereby establishing positive feedback, which could contribute to their proliferation (Klironomos 2002, Batten et al. 2006, Kulmatiski and Beard 2008, Sanon et al. 2011), to the detriment of native biodiversity (Callaway et al. 2004, Lorenzo et al. 2010). On the other hand, several studies have also revealed the negative effects of invasive plants on soil fungi due to invasion by *Bromus tectorum* (Belnap et al. 2005), arbuscular mycorrhizal fungi in response to dominance of nonmycorrhizal *Alliaria petiolata* in North American forests (Roberts and Anderson 2001, Wolfe and Klironomos 2005, Stinson et al. 2006, Callaway et al. 2008, Wolfe et al. 2008, Pringle et al. 2009, Vogelsang and Bever 2009), microbial biomass, C and ratio of fungi to bacteria due to *Falcataria moluccana* (Allison et al. 2006), and both soil fungi and bacteria due to *Acacia dealbata* invasion (Lorenzo et al. 2010).

Invasive alien plants also significantly influence catabolic diversity of the soil microbial communities through their impact on the activities of soil enzymes, which represent a link between litter decomposition, microbial activity, and nutrient availability (Sinsabaugh et al. 2000, Elk 2010). The influence of exotic plants on the activity of soil enzymes has been reported by several workers (Kourtev et al. 2002b, Allison et al. 2006, Chapuis-Lardy et al. 2006, Li et al. 2006, Fan et al. 2010). Thus, it is clear from the growing number of studies that invasive alien species can alter ecosystem processes through a wide variety of mechanisms over a variety of spatial and temporal scales (Ehrenfeld 2010). Indeed, multiple mechanisms have been indentified that interact and reinforce each other in bringing about ecosystem change. Thus, it is necessary to search for mechanisms of impact of invasive alien species through documentation of interacting mechanisms, rather than to focus on single causative pathways as the "holy grail" of universal explanation (Simberloff 2010).

In the Kashmir Himalaya, significant progress has been made in documenting the alien plant species in different ecosystems and habitats. Khuroo et al. (2007) reported the occurrence of 571 alien plant species harbored by different ecosystems, while 223 alien plant species were recorded in freshwater habitats of India (Shah and Reshi 2012). Operational characterization of terrestrial alien plant species by Khuroo et al. (2008) has revealed that out of a total of 436 alien plant species (excluding cultivated aliens) recorded in this region, the number of species belonging to invasion stages II, III, IVa, IVb, and V was 119, 107, 56, 77, and 77, respectively. Though some recent studies have shown that invasions by alien plants alter the structure and functioning of the natural ecosystems in the region, such as the grasslands (Reshi et al. 2008a) and wetlands (Reshi et al. 2008b), detailed investigations examining the impact of different invasive species on various attributes of species diversity are of vital importance for policy makers, nature conservation authorities, and land managers for developing effective management programs. Hence, the present chapter provides an overview of the series of ongoing studies on the impact of alien plant invasions in the Kashmir Himalayan region on aboveground and belowground biodiversity.

12.2   Impact on aboveground diversity

12.2.1   Species richness

Comparing species richness of invaded and uninvaded plots has provided some crucial insights into the community-level impacts of alien species (Gordon 1998, Manchester and Bullock 2000, Hejda et al. 2009), and species diversity, in particular, has been shown to vary because of the influence of invasive alien species. For example, the grassland not invaded by *Centaurea iberica* in the Kashmir Himalaya supported 21 plant species, with *Themeda anathera, Lespedeza elegans*, and *Trifolium pratense* being the most frequent species (Table 12.1). In comparison, vegetation in the invaded area of the grassland comprised only seven species; *Centurea iberica* was the most frequent species, which incidentally is also a non-native invasive species (Reshi et al. 2008a).

Likewise, Khuroo et al. (2010) also studied the impact of invasion by *Leucanthemum vulgare* on species richness of the forest understory in the Kashmir Himalaya. The study also reported decrease in species richness in the uninvaded, moderately invaded, and highly invaded sites wherein the number of species was 20, 11, and 7, respectively (Table 12.2).

Table 12.1 Frequency of Plant Species in the Grasslands Invaded and Uninvaded (Control) by *Centaurea iberica* in the Kashmir Himalaya

Name of taxa	Family	Intact site	Invaded site	Frequency (%)
Androsace rotundifolia Hardw.	Primulaceae	+	−	42
Artemesia annua L.	Asteraceae	+	−	66
Asparagus filicinus Ham.	Liliaceae	+	−	20
Bothriochloa ischaemum Keng	Poaceae	−	+	70
Bupleurum hoffmiestri L.	Apiaceae	+	−	50
Carex nubigena D.Don.	Cyperaceae	+	−	28
Centaurea iberica Trev. Ex Spreng	Asteraceae	−	+	100
Cichorium intybus L.	Asteraceae	−	+	10
Clinopodium vulgare L.	Lamiaceae	+	−	56
Cynodon dactylon Pers.	Poaceae	−	+	24
Dactylis glomerata L.	Poaceae	+	−	32
Erigeron multicaulis Wall. Ex DC.	Asteraceae	+	−	50
Euphorbia wallichi Hook.f.	Euphorbiaceae	+	−	26
Hieracium umbellatum L.	Asteraceae	+	−	18
Hypericum perforatum L.	Hypericaceae	+	−	56
Indigofera heterantha Wall. Ex Brand	Fabaceae	+	−	72

(Continued)

Table 12.1 (continued) Frequency of Plant Species in the Grasslands Invaded and Uninvaded (Control) by *Centaurea iberica* in the Kashmir Himalaya

Name of taxa	Family	Intact site	Invaded site	Frequency (%)
Leonurus cardiaca L.	Lamiaceae	+	−	22
Lespedeza cuneata G.Don	Fabaceae	−	+	28
Lespedeza elegans Camb.	Fabaceae	+	−	58
Lespedeza tomentosa Sieb. Ex Mexim.	Fabaceae	+	−	52
Lithospermum arvense L.	Boraginaceae	+	−	42
Origanum vulgare L.	Lamiaceae	+	−	56
Plantago lanceolata L.	Plantaginaceae	−	+	10
Silene vulgaris Garcke	Caryophyllaceae	+	−	42
Themeda anathera Hack.	Poaceae	+	−	100
Thymus serpyllum L.	Lamiaceae	−	+	36
Trifolium pratense L.	Fabaceae	+	−	64

+, present; −, absent.

Table 12.2 Comparison of Species Diversity in the Sites Highly Invaded, Moderately Invaded, and Uninvaded by *Leucanthemum vulgare* in the Kashmir Himalaya

Species composition	Highly invaded site	Moderately invaded site	Uninvaded site
Leucanthemum vulgare Lam.	34.2	28.6	−
Trifolium repens L.	3.0	6.2	1.2
Fragaria nubicola Lindel.	0.8	−	4.4
Cynodon dactylon Pers.	1.3	102.0	−
Prunella vulgaris L.	0.6	2.2	0.8
Sambucus wightiana Wall. ex Wight & Arn.	1.2	−	−
Trifolium pratense L.	1.2	4.2	3.2
Achillea millefolium L.	−	6.8	−
Bothriochloa ischaemum Keng.	−	3.2	−
Rumex nepalensis Spreng.	−	1.8	2.0
Rumex hastatus D. Don	−	0.2	1.2
Viola odorata L.	−	0.2	1.6
Geranium nepalense Sweet	−	1.8	2.4
Taraxacum officinale Weber	−	−	1.4
Polygonum amplexicaule D. Don	−	−	1.4
Plantago lanceolata L.	−	−	1.8
Plantago major L.	−	−	0.8
Chenopodium foliosum Aschers	−	−	1.6
Cichorium intybus L.	−	−	2.0
Oxalis acetosella L.	−	−	0.8
Potentilla reptans L.	−	−	2.6
Oxytropis mollis Royle ex Bth.	−	−	1.2
Poa bulbosa L.	−	−	2.0
Bromus japonicus Thumb.	−	−	1.8
Stellaria media Cyr.			1.4

Table 12.3 Impact of Three Alien Species on Species Richness in the Kashmir Himalaya

Species richness	Anthemis cotula		Conyza canadensis		Sisymbrium loeselii	
	Invaded	Uninvaded	Invaded	Uninvaded	Invaded	Uninvaded
Total	55	44	61	49	71	69
Native	15	9	16	15	24	25
Alien	40	35	45	34	47	44
Invasive	28	19	27	19	24	22

Contrary to the above results, a detailed investigation of the impact of three invasive plant species, namely *Anthemis cotula, Conyza canadensis,* and *Sisymbrium loeselii,* on the richness of native and alien species in the Kashmir Himalaya (Rehman, unpublished) revealed that species richness was higher in invaded sites compared to uninvaded (control) sites (Table 12.3). This higher richness was mainly due to the increased numbers of alien species in the invaded plots. The number of alien species in *Anthemis, Conyza,* and *Sisymbrium* invaded plots was 40, 45, and 47, respectively, while the number of alien species in uninvaded plots was 35, 34, and 44, respectively. The data suggest that invasions facilitate further invasions, hence aggravating the impact of alien species on resident communities, a process described as *invasional meltdown* by Simberloff and Von Holle (1999).

12.2.2   Species diversity

Despite being an important descriptor of communities, very few studies have been conducted to explore the impact of invasive alien species on species diversity (Hejda et al. 2009). In the Kashmir Himalaya, the impact of *Centaurea iberica* on species diversity and dominance was compared in invaded and uninvaded (control) sites (Table 12.4). It was observed that the Shannon–Weiner index was lower in invaded sites compared to uninvaded sites. Similarly, H_{max} was only 1.95 for the invaded site, while it was 3.05 for the uninvaded site. Dominance, whether computed using Simpson's or the Berger–Parker index, was higher for the invaded site (Reshi et al. 2008a).

Khuroo et al. (2010) also studied the impact of *Leucanthemum vulgare* on the species diversity and evenness of the forest understory vegetation in the Kashmir Himalaya. Perusal of the data in Table 12.5 reveals that species diversity ranged from 2.82, 1.21, and 0.83 in the uninvaded, moderately invaded, and highly invaded sites, respectively. Compared to uninvaded sites, both the moderately and highly invaded sites showed significantly lower values for species evenness.

Table 12.4 Comparison of Species Diversity and Dominance in Sites Invaded and Uninvaded by *Centaurea iberica*

Index	Uninvaded sites	Invaded site
Shannon *H*	2.95	1.66
Shannon H_{max}	3.05	1.95
Simpson diversity	0.05	0.22
Simpson diversity ($1/d$)	18.37	4.46
Beger–Parker dominance (*D*)	0.10	0.36
Berger–Parker dominance (1/D)	9.80	2.78

Table 12.5 Measures of Species Diversity and Species Evenness in the Sites Highly Invaded, Moderately Invaded, and Uninvaded by *Leucanthemum* in Kashmir Valley, India

Sampling sites	Species diversity (H)	Species evenness (J)
Highly invaded	0.83	0.12
Moderately invaded	1.21	0.11
Uninvaded	2.82	0.94

12.2.3   Community assembly

Community assembly provides a conceptual foundation for understanding the processes that determine which and how many species can occur in a particular community (Chase 2003, Santoro et al. 2011) and it is directly related to community function. It is becoming increasingly abundant that invasion by alien species could disrupt community assembly and alter both structure and function of an ecosystem. Null models have been used to determine whether the invaded communities differ from randomly created null communities with respect to species co-occurrence. Reshi et al. (2008a) used four co-occurrence indices (C score; number of checkerboard units; number of species combinations, and V ratio) in the Kashmir Himalaya to examine the impact of invasion by *Centaurea iberica* on the community assembly. Perusal of the data in Tables 12.6 and 12.7 reveal that the co-occurrence of the plant species in the intact grassland was determined by competitive interactions between the species (Table 12.6). Disassembly of the community structure due to invasion by *C. iberica* was evident as none of the indices was significant with any of the used null model algorithm in the *Centaurea*-invaded grassland. Comparison between intact and invaded communities (Table 12.8) revealed that C-score, Checker, and Combo were larger for the intact grassland than for the invaded grassland, pointing toward differences in the community structure of the two grasslands.

12.2.4   Biotic homogenization

Large-scale effects of plant invasions include the homogenization of floras because of which originally distinct phytogeographical units become largely similar (Kühn and Klotz 2006, Qian and Ricklefs 2006, Schwartz et al. 2006). Biotic homogenization is related with decrease in beta (β) diversity, which is a fundamental component of biodiversity. Since habitat type and traits of alien species are important predictors of the extent of invasion, the degree of biotic homogenization could also differ depending upon the habitats being homogenized and the traits of invaders. To determine which habitats are more homogenized and which groups of alien plant species promote homogenization, Dar (2011) used two measures of beta diversity, namely overall beta diversity (β_O) and beta diversity for aliens (β_A), for assessing the role of alien plant invasions in biotic homogenization of different ecosystems in the Kashmir Himalaya. The study revealed that the beta diversity for aliens, β_A, was lower than the overall beta diversity, β_O (Table 12.9), which indicated that alien species decrease beta diversity and as such increase the similarity. This decrease in beta diversity differed depending upon habitat type, invasion status, life span, and habit of alien species. Among stages of invasion, the average of overall beta diversity and the average of beta diversity for aliens were lowest for invasives followed by naturalized and casual species, which indicated

Table 12.6 Results of Null Model Analyses of the Co-Occurrence Matrix of the Intact Site

Index	Row constraint	Column constraint	Observed index	Simulated index (SS)	Simulated index (IS)
C-score	Fixed	Fixed	112.12	111.49*	111.48*
	Fixed	Equiprobable	112.12	110.89	110.95
	Fixed	Proportional	112.12	106.33*	106.36*
Checker	Fixed	Fixed	2.00	0.67	0.68
	Fixed	Equiprobable	2.00	0.69	0.69
	Fixed	Proportional	2.00	0.60	0.59
Combo	Fixed	Fixed	49.00	49.97*	49.99*
	Fixed	Equiprobable	49.00	49.99*	49.99*
	Fixed	Proportional	49.00	49.99*	49.99*
V-ratio	Fixed	Equipropable	0.96	1.00	0.99
	Fixed	Proportional	0.96	1.35*	1.35*

SS, expected score generated using sequential swap algorithm; IS, expected score generated using the independent swap algorithm.

*Observed score is significantly ($p < .05$) different from the expected score.

Table 12.7 Results of Null Model Analyses of the Co-Occurrence Matrix of the Invaded Site

Index	Row constraint	Column constraint	Observed index	Simulated index (SS)	Simulated index (IS)
C-score	Fixed	Fixed	45.29	43.93	43.92
	Fixed	Equiprobable	45.29	44.62	44.62
	Fixed	Proportional	45.29	41.16	41.02
Checker	Fixed	Fixed	0.00	0.63	0.60
	Fixed	Equiprobable	0.00	1.17	1.17
	Fixed	Proportional	0.00	1.01	1.03
Combo	Fixed	Fixed	18.00	18.82	18.81
	Fixed	Equiprobable	18.00	19.09	19.11
	Fixed	Proportional	18.00	19.07	19.09
V-ratio	Fixed	Equipropable	1.08	0.99	0.99
	Fixed	Proportional	1.08	1.12	1.13

SS, expected score generated using sequential swap algorithm; IS, expected score generated using the independent swap algorithm.

Table 12.8 Co-Occurrence Structure in Intact versus Invaded Sites

Index	Observed value		Mean simulated value	
	Intact site	Invaded site	Intact site	Invaded site
C-score	112.12	45.29	67.79	65.02
Checker	2.00	0.00	155.24	154.87
Combo	49.00	18.00	37.29	36.44
V-ratio	0.96	1.08	2.96	2.94

Table 12.9 Overall Beta Diversity (β_O) and Beta Diversity (β_A) for Each Sub-Category of Habitat and Plant Groups

Category	Sub-category	Average β_O	Average β_A
Habitat	Grasslands	0.7429	0.7355
	Forests	0.7289	0.7219
	Roadsides	0.6491	0.6167
Stage of invasion	Casual	0.9523	0.9523
	Naturalized	0.7923	0.7923
	Invasive	0.5993	0.5993
Life span	Annual/Biennial	0.7544	0.7301
	Perennial	0.715	0.6596
Habit	Herbs	0.7259	0.6918
	Shrubs/Sub-shrubs/Lianas	0.9573	0.9402
	Trees	0.979	1

that invasive species promote homogenization greater than casual and naturalized species. Similarly, perennial species (average β_A = 0.6596) promoted homogenization more than annual or biennial species (average β_A = 0.7309), and herbaceous species (average β_A = 0.6918) promoted homogenization more than shrubs (average β_A = 0.9402) and trees (average β_A = 1). β_A among habitats was lowest for roadsides followed by grasslands and forests, while β_O was lowest for roadsides followed by forests and grasslands, thereby indicating that roadsides were more homogenous than forests and grasslands. Thus, the extent of biotic homogenization was related to habitat type and the traits (invasion status, life span, and growth form) of alien species.

12.3   Impact on belowground diversity

12.3.1   Species richness

In view of the importance of belowground diversity in the structural organization and functional integrity of ecosystems, few studies have been carried out in the Kashmir Himalaya to study the impact of invasive alien plant species on microbial species richness. For example, Rasool (2012), using a conventional plate count technique, studied the impact of *Conyza canadensis, Sambucus wightiana*, and *Anthemis cotula* on soil microbes in the Kashmir Himalaya, and it was observed that bacterial numbers were stimulated due to invasion by all the three invasive species, while the numbers of soil fungi were more or less decreased (Figure 12.1). However, using DGGE, it was found that soil bacterial diversity increased in *Conyza canadensis*–invaded patches. Though the number of species in *Sambucus wightiana*–invaded and uninvaded patches remained unchanged, the nature of species changed completely in the invaded patches. The total number of species of Ascomycete fungi increased in *Conyza*-invaded patches, whereas the same in *Sambucus*-invaded patches decreased. The Basidiomycete diversity increased in *Sambucus*-invaded patches (Figure 12.2). Thus, these alterations in the belowground richness and diversity point towards the fact that the alien invasive species impact the belowground diversity, which has implications for other ecosystem functions.

Shah et al. (2010) examined the impact of two invasive alien plant species, namely *Anthemis cotula* and *Conyza canadensis*, on arbuscular mycorrhizal fungi (AMF) at different

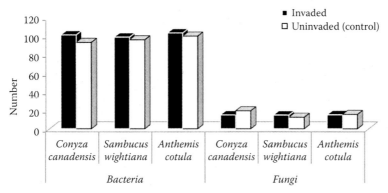

Figure 12.1 Impact of invasive plant species on belowground diversity using conventional pour plate technique.

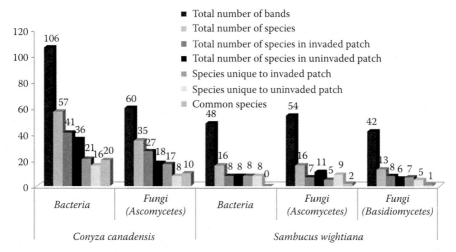

Figure 12.2 Impact of invasive plant species on belowground diversity using DGGE technique.

invaded and uninvaded sites in the Kashmir Himalaya (Table 12.10). Significant differences in AMF associated with the two invasive plant species, *Anthemis cotula* and *Conyza canadensis*, were observed (Table 12.10). Nine AMF species were recovered from the rhizosphere of *A. cotula* and seven from that of *C. canadensis*. In contrast, the uninvaded (control) sites yielded 17 and 11 AMF species, respectively (Table 12.10). Shah et al. (2010) reported that invasion by alien species in the Kashmir Himalaya brings about a shift in the Glomalean spore diversity. Spore density of AMF, unlike their species richness, was higher in the invaded rhizospheric soils than in the nearby uninvaded soils. The number of AMF spores per gram of soil in the rhizosphere of *A. cotula* and *C. canadensis* also varied across sites.

12.3.2   Soil microbial and enzyme activity

Invasive alien plants also significantly influence catabolic diversity of the soil microbial communities through their impact on the activity of soil enzymes, which represent a link between litter decomposition, microbial activity, and nutrient availability (Sinsabaugh

Table 12.10 Arbuscular Mycorrhizal Species Composition in the Rhizospheric Soils Invaded (I) by *Anthemis cotula* and *Conyza canadensis* in Comparison to Nearby Uninvaded Habitats (UI)

	A. cotula		C. canadensis	
AMF taxa	UI	I	UI	I
Glomaceae				
Glomus caledonium (Nicol. & Gerd.) Trappe & Gerd.	P	A	P	A
G. lamellosum Dalpe, Koske & Tews	P	A	A	A
G. etunicatum Becker & Gerdemann	P	A	A	P
G. intraradices Schenck & Smith	A	A	P	A
G. mosseae (Nicol. & Gerd.) Gerd. & Trappe	P	P	A	A
G. claroideum Schenck & Smith	P	A	A	A
G. dimorphicum Boyetchko & Tewari	A	P	A	P
G. fasciculatum (Thaxter) Walker & Koske	A	A	P	P
G. diaphanum Morton & Walker	A	P	A	A
G. luteum	A	P	A	P
Gigasporaceae				
Gigaspora decipiens Hall & Abbott	P	A	A	A
G. margarita Becker & Hall	A	P	P	A
Scutellospora erythropa (Koske & Walker) Walker & Sanders	P	A	A	A
S. verrucosa (Koske & Walker) Walker & Sanders	A	P	A	A
S. heterogama (Nicol. & Gerdemann) Walker & Sanders	P	P	A	A
S. dipurpurascens Morton & Koske	P	A	A	A
S. calospora (Nicol. & Gerd.) Walker & Sanders	P	A	A	A
S. pellucida (Nicol. & Schenck) Walker & Sanders	P	A	A	A
Scutellospora sp.	A	A	P	A
Scutellospora sp.	A	A	P	A
Acaulosporaceae				
Acaulospora spinosa Walker & Trappe	P	P	A	P
A. laevis Gerdemann & Trappe	P	P	A	A
A. foveata Trappe & Janos	A	A	A	P
A. paulineae Blaszkowski	P	A	A	A
Acaulospora sp.	P	A	P	A
Entrophospora infrequens (Hall) Ames & Schneider	P	A	A	A
Archaeosporaceae				
Archaeospora trappei Morton & Redecker	A	A	P	A
Paraglomaceae				
Paraglomus occultum (Walker) Morton & Redecker	P	A	A	P

et al. 2000, Elk 2010). The influence of exotic plants on the activity of soil enzymes has been reported by several workers (Kourtev et al. 2002b, Allison et al. 2006, Chapuis-Lardy et al. 2006, Li et al. 2006, Fan et al. 2010). Rasool (2012) studied the effect of three invasive plant species, namely *Conyza canadensis*, *Sambucus wightiana*, and *Anthemis cotula*, on soil microbial activity (measured as FDase activity), soil microbial biomass, and activity of several soil enzymes in the Kashmir Himalaya (Figure 12.3). The study revealed that invasion by

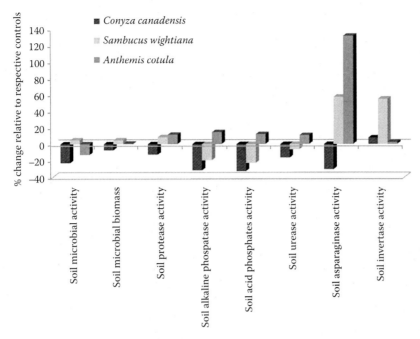

Figure 12.3 Effect of certain invasive alien species on the microbial activity, microbial biomass, and activity of various soil enzymes.

C. canadensis had a depressing effect on most of the soil attributes studied, but soil invertase activity was stimulated. Similarly, invasion by *S. wightiana* influenced soil attributes, soil microbial biomass, soil microbial activity, and activities of asparaginase and invertase, positively, whereas activities of protease, alkaline and acid phosphatase, and urease were negatively influenced. *A. cotula* increased protease, urease, asparaginase, and invertase activities; however, it decreased soil microbial activity, soil microbial biomass, and alkaline and acid phosphatase activities (Figure 12.3).

12.4   Conclusion

Based on the foregoing discussion, it can be safely concluded that the studies carried out so far in the Kashmir Himalaya reveal that invasive alien species not only affect aboveground diversity but also affect belowground component of the biological diversity, which is often ignored. Since the belowground and aboveground components of biodiversity are closely linked together, a change in any of these components because of invasive alien species could trigger a cascade of changes, which will have an impact on both structural and functional aspects of ecosystems.

12.5   Future challenges

Keeping in view the multidimensional impacts of alien species, it becomes necessary to document the impact of alien species on all possible aspects of ecosystems and at all possible levels or scales. Given limited financial resources, it is necessary to quantify the impact of as many invasive alien species as possible so that the species are ranked according to their overall impact for setting management priorities. This will also help in developing generalizations about alien plant invasion, which could prove very useful in conservation practices.

References

Allison, S.D., C. Nielsen, and R. Hughes. 2006. Elevated enzyme activities in soils under the invasive nitrogen-fixing tree *Falcataria moluccana*. *Soil Biology & Biochemistry* 38:1537–1544.

Angeloni, N.L., K.J. Jankowski, N.C. Tuchman, and J.J. Kelly. 2006. Effects of an invasive cattail species (*Typha × glauca*) on sediment nitrogen and microbial community composition in a freshwater wetland. *FEMS Microbiology Letters* 263:86–92.

Bais, H.P., S. Park, T.L. Weir, R.M. Callaway, and J.M. Vivanco. 2004. How plants communicate using the underground information superhighway. *Trends in Plant Science* 9:26–32.

Batten, K.M., K.M. Scow, K.F. Davies, and S.P. Harrison. 2006. Two invasive plants alter soil microbial community composition in serpentine grasslands. *Biological Invasions* 8:217–230.

Belnap, J., and S.L. Phillips. 2001. Soil biota in an ungrazed grassland: Response to annual grass (*Bromus tectorum*) invasion. *Ecological Applications* 11:1261–1275.

Belnap, J., S.L. Phillips, S.K. Sherrod, and A. Moldenke. 2005. Soil biota can change after exotic plant invasion: Does this affect ecosystem processes? *Ecology* 86:3007–3017.

Brooks, M.L., C.M. D'Antonio, D.M. Richardson, J.B. Grace, J.E. Keeley, J.M. DiTomaso, R.J. Hobbs, M. Pellant, and D. Pyke. 2004. Effects of invasive alien plants on fire regimes. *Bioscience* 54:677–688.

Calder, I.R., and P. Dye. 2001. Hydrological impacts of invasive alien plants. *Land Use and Water Resources Research* 1:1–12.

Callaway, R.M., D. Cipollini, K. Barto, G.C. Thelen, S.G. Hallet, D. Prati, K. Stinson, and J. Klironomos. 2008. Novel weapons: Invasive plant suppresses fungal mutualists in America but not it's native Europe. *Ecology* 89:1043–1055.

Callaway, R.M., G.C. Thelen, A. Rodriguez, and W.E. Holben. 2004. Soil biota and exotic plant invasion. *Nature* 427:731–733.

Chapin, F.S., E.S. Zavaleta, V.T. Eviner, R.L. Naylor, P.M. Vitousek, H.L. Reynolds, D.U. Hooper, et al. 2000. Consequence of changing biodiversity. *Nature* 405:232–242.

Chapuis-Lardy, L., S. Vanderhoeven, N. Dassonville, L.S. Koutika, and P. Meerts. 2006. Effect of the exotic invasive plant *Solidago gigantea* on soil phosphorus status. *Biology and Fertility of Soils* 42:481–489.

Charles, H., and J.S. Dukes. 2007. Impacts of invasive species on ecosystem services. In *Biological Invasions*, ed. W. Nentwig, pp. 217–237. Berlin: Springer.

Chase, J.M. 2003. Community assembly: When should history matter? *Oecologia* 136:489–498.

Cronk, Q.C.B., and J.L. Fuller. 1995. *Plant Invaders: The Threat to Natural Ecosystems*. London: Chapman & Hall.

D'Antonio, C.M., and L.A. Meyerson. 2002. Exotic plant species as problems and solutions in ecological restoration: A synthesis. *Restoration Ecology* 10:703–713.

D'Antonio, C.M., and P.M. Vitousek. 1992. Biological invasions by exotic grasses, the grass/fire cycle, and global change. *Annual Review of Ecology and Systematics* 23:63–87.

Dar, P.A. 2011. *Plant invasions in relation to biotic homogenization in the Kashmir Himalaya, India*. MPhil Dissertation (unpublished). University of Kashmir, India.

Ehrenfeld, J.G. 2010. Ecosystem consequences of biological invasions. *Annual Review of Ecology and Systematics* 41:59–80.

Elk, M.R. 2010. *A survey of plant root extracellular enzyme activity in native and invasive exotic plants of oak openings*. PhD Thesis. The University of Toledo, USA.

Fan, P., L. Terriera, A. Haya, A. Marstona, and K. Hostettmann. 2010. Antioxidant and enzyme inhibition activities and chemical profiles of *Polygonum sachalinensis* F. Schmidt ex Maxim (Polygonaceae). *Fitoterapia* 81:124–131.

Foxcroft, L.C., D.M. Richardson, M. Rejmanek, and P. Pysek. 2010. Alien plant invasions in tropical and sub-tropical savannas: Patterns, processes and prospects. *Biological Invasions* 12:3913–3933.

Gordon, D.R. 1998. Effects of invasive, non-indigenous plant species on ecosystem processes: Lessons from Florida. *Ecological Applications* 8:975–989.

Hejda, M., P. Pyšek, and V. Jarošík. 2009. Impact of invasive plants on the species richness, diversity and composition of invaded communities. *Journal of Ecology* 97:393–403.

Khuroo, A.A., A.H. Malik, Z.A. Reshi, and G.H. Dar. 2010. From ornamental to detrimental: Plant invasion of *Leucanthemum vulgare* Lam. (Ox-eye Daisy) in Kashmir Valley, India. *Current Science* 98:600–602.

Khuroo, A.A., I. Rashid, Z. Reshi, G.H. Dar, and B.A. Wafai. 2007. The alien flora of Kashmir Himalaya. *Biological Invasions* 9:269–292.

Khuroo, A.A., Z. Reshi, I. Rashid, G.H. Dar, and Z.S. Khan. 2008. Operational characterization of alien invasive flora and its management applications. *Biodiversity and Conservation* 17:3181–3194.

Klironomos, J.N. 2002. Feedback with soil biota contributes to plant rarity and invasiveness in communities. *Nature* 417:67–70.

Kourtev, P.S., J.G. Ehrenfeld, and M. Haggblom. 2002a. Exotic plant species alter the microbial community structure and function in the soil. *Ecology* 83:3152–3166.

Kourtev, P.S., J.G. Ehrenfeld, and M. Haggblom. 2002b. Enzyme activities during litter decomposition of two exotic and two native plant species in hardwood forests of New Jersey. *Soil Biology and Biochemistry* 34:1207–1218.

Kourtev, P.S., J.G. Ehrenfeld, and M. Haggblom. 2003. Experimental analysis of the effect of exotic and native plant species on the structure and function of soil microbial communities. *Soil Biology & Biochemistry* 35:895–905.

Kowarik, I. 2003. *Biological Invasions: Neophytes and Invasive Species in Central Europe*. Stuttgart, Germany: Ulmer.

Kühn, I., and S. Klotz. 2006. Urbanization and homogenization: Comparing the floras of urban and rural areas in Germany. *Biological Conservation* 127:292–300.

Kulmatiski, A., and K.H. Beard. 2008. Decoupling plant-growth from land-use legacies in soil microbial communities. *Soil Biology & Biochemistry* 40:1059–1068.

Lankau, R. 2010. Soil microbial communities alter allelopathic competition between *Alliaria petiolata* and a native species. *Biological Invasions* 12:2059–2068.

Levine, J.M., M. Vila, C.M. D'Antonio, J.S. Dukes, K. Grigulis, and S. Lavorel. 2003. Mechanisms underlying the impacts of exotic plant invasions. *Proceedings of Royal Society London Series B-Biological Sciences* 270:775–781.

Li, W.H., C.B. Zhang, H.B. Jiang, G.R. Xin, and Z.Y. Yang. 2006. Changes in soil microbial community associated with invasion of the exotic weed, *Mikania micrantha* HBK. *Plant and Soil* 281:309–324.

Liao, C., R. Peng, Y. Luo, X. Zhou, X. Wu, C. Fang, J. Chen, and B. Li. 2008. Altered ecosystem carbon and nitrogen cycles by plant invasion: A meta-analysis. *New Phytologist* 177:706–714.

Lorenzo, P., S. Rodrıguez-Echeverrıa, L. Gonzalez, and H. Freitas. 2010. Effect of invasive *Acacia dealbata* Link. On soil microorganisms as determined by PCR-DGGE. *Applied Soil Ecology* 44:245–251.

Manchester, S.J., and J.M. Bullock. 2000. The impacts of non-native species on UK biodiversity and the effectiveness of control. *Journal of Applied Ecology*: 37:845–864.

Meffin, R., A.L. Miller, P.E. Hulme, and R.P. Duncan. 2010. Experimental introduction of the alien plant *Hieracium lepidulum* reveals no significant impact on montane plant communities in New Zealand. *Diversity and Distributions* 16:804–815.

Meyer, J.Y. 1994. *Mechanisms of Invasion "Miconia calvescens Dc." in French Polynesia*. PhD Thesis. Université Montpellier II Sciences et Techniques du Languedoc, Montpellier.

Ogle, S.M., W.A. Reiners, and K.G. Gerow. 2003. Impacts of exotic annual brome grasses (*Bromus* spp.) on ecosystem properties of northern mixed grass prairie. *American Midland Naturalist* 149:46–58.

Pejchar, L., and H.A. Mooney. 2009. Invasive species, ecosystem services and human well-being. *Trends in Ecology and Evolution* 24:497–504.

Pringle, A., R.I. Adams, H.B. Cross, and T.D. Bruns. 2009. The ectomycorrhizal fungus *Amanita phalloides* was introduced and is expanding its range on the West Coast of North America. *Molecular Ecology* 18:817–833.

Qian, H., and R.E. Ricklefs. 2006. The role of exotic species in homogenizing the North American flora. *Ecology Letters* 9:1293–1298.

Rasool, N. 2012. *Effect of some alien invasive plant species on soil microbial structure and function.* PhD Thesis. University of Kashmir, India.

Reshi, Z., I. Rashid, A.A. Khuroo, and B.A. Wafai. 2008a. Effect of invasion by *Centaurea iberica* on community assembly of a mountain grassland of Kashmir Himalaya, India. *Tropical Ecology* 49:147–156.

Reshi, Z., M.A. Shah, I. Rashid, and A.A. Khuroo. 2008b. Alien plant invasions: Threat to wetland health and public wealth. *Sarovar Saurabh* 4:2–5.

Roberts, K.J., and R.C. Anderson. 2001. Effect of garlic mustard [*Alliaria Petiolata* (Beib. Cavara & Grande)] extracts on plants and arbuscular mycorrhizal (AM) fungi. *American Midland Naturalist* 146:146–152.

Sanon, A., T. Beguiristain, A. Cébron, J. Berthelin, and S.N. Sylla. 2011. Differences in nutrient availability and mycorrhizal infectivity in soils invaded by an exotic plant negatively influence the development of indigenous *Acacia* species. *Journal of Environmental Management* 95:S275–S279.

Santoro, R., T. Jucker, M. Carboni, and A.T.R. Acosta. 2011. Patterns of plant community assembly in invaded and non-invaded communities along a natural environmental gradients. *Journal of Vegetation Science* DOI: 10.1111/j.1654-1103.2011.01372.x.

Schwartz, M.W., J.H. Thorne, and J.H. Viers. 2006. Biotic homogenization of the California flora in urban and urbanizing regions. *Biological Conservation* 127:282–291.

Shah, M.A., Z.A. Reshi, and N. Rasool. 2010. Plant invasion induces shift in Glomalean spore diversity. *Tropical Ecology* 51(2S):317–323.

Shah, M.A. and Z.A. Reshi. 2012. Invasion by alien macrophytes in freshwater ecosystems of India. In *Invasive Alien Plants: An Ecological Appraisal for the Indian Subcontinent.* eds. J. R. Bhatt, J. S. Singh, R.S. Tripathi, S. P. Singh and R.K. Kohli, pp. 199-216. United Kingdom:CAB International.

Simberloff, D. 2010. Invasions of plant communities—more of the same, something very different, or both? *American Midland Naturalist* 163:220–233.

Simberloff, D., and B. Von Holle. 1999. Positive interactions of nonindigenous species: Invasional meltdown? *Biological Invasions* 1:21–32.

Sinsabaugh, R.L., H. Reynolds, and T.M. Long. 2000. Rapid assay for amidohydrolase (urease) activity in environmental samples. *Soil Biology & Biochemistry* 32:2095–2097.

Stinson, K.A., S.A. Campbell, J.R. Powell, B.E. Wolfe, R.M. Callaway, G.C. Thelen, S.G. Hallett, D. Prati, and J.N. Klironomos. 2006. Invasive plant suppresses the growth of native tree seedlings by disrupting belowground mutualisms. *PLoS Biology* 4:e140.

Torsvik, V., J. Goksoyr, and F.L. Daae. 1990. High diversity of DNA in soil bacteria. *Applied and Environmental Microbiology* 56:783–787.

Vandenkoornhuyse, P., R. Husband, T.I. Daniell, J.M. Duck, A.H. Fitter, and J.P.W. Young. 2002. Arbuscular mycorrhizal community composition associated with two plant species in a grassland ecosystem. *Molecular Ecology* 11:1555–1564.

Vilà, M., J.L. Espinar, M. Hejda, P.E. Hulme, V. Jarošík, J.L. Maron, J. Pergl, U. Schaffner, Y. Sun, and P. Pyšek. 2011. Ecological impacts of invasive alien plants: A meta-analysis of their effects on species, communities and ecosystems. *Ecology Letters* 14:702–708.

Vogelsang, K.M., and J.D. Bever. 2009. Mycorrhizal densities decline in association with nonnative plants and contribute to plant invasion. *Ecology* 90:399–407.

Wardle, D.A., R.D. Bardgett, J.N. Klironomos, H. Setala, W.H. van der Putten, and D.H. Wall. 2004. Ecological linkages between aboveground and belowground biota. *Science* 304:1629–1633.

Werner, C., U. Zumkier, W. Beyschlag, and C. Maguas. 2010. High competitiveness of a resource demanding invasive acacia under low resource supply. *Plant Ecology* 206:83–96.

Wolfe, B.E., and J.N. Klironomos. 2005. Breaking new ground: Soil communities and exotic plant invasion. *BioScience* 55:477–487.

Wolfe, B.E., V.L. Rodgers, K.A. Stinson, and A. Pringle. 2008. The invasive plant *Alliaria petiolata* (garlic mustard) inhibits ectomycorrhizal fungi in its introduced range. *Journal of Ecology* 96:777–783.

Zhang, Q., R. Yang, J. Tang, H. Yang, S. Hu, and X. Chen. 2010. Positive feedback between mycorrhizal fungi and plants influences plant invasion success and resistance to invasion. *PLoS ONE* 5:e12380.

chapter thirteen

Ecology and management of invasive plants in Africa

Roland A.Y. Holou, E.G. Achigan-Dako, and Brice Sinsin

Contents

13.1 Introduction

Africa's biodiversity is rich and diverse with, for instance, more than 40,000 plant species found in forests, savannahs, and other ecosystems. Most of these plants are basic resources for many Africans who use them as multipurpose species. Many of these plants are grown to feed the ever-growing population on the continent. Unfortunately, these productions are hampered by the infestation of unwanted plants generally referred to as weeds. The consequences of invasive plants infestation in Africa are multiple and rise from the damage caused to agriculture to the ecological effects on the biodiversity. For instance, weeds can reduce corn yield by 90% (Gianessi 2009), tremendously reducing farmers' income and the already low yield of most African crops (Holou and Dakpogan 2003). This situation calls for adequate technologies to control weeds so as to increase the production. Unfortunately, technologies used to control invasive plants in Africa are still rudimentary in most cases. For example, small holder farmers in Africa still hold onto hoeing to control invasive plants, although this practice is very tedious and labor consuming. This is a result of poverty and a lack of education and appropriate means to afford advanced technologies. The time allotted for weed control depends on the weed species and the crop. As an illustration, farmers spend more than 370 h/ha to hand-weed fields planted with groundnut (*Arachis hypogaea*) as compared to 150 hours for sorghum (*Sorghum bicolor*) (Akobundu 1987). Effective weed control can therefore provide African farmers with more time that can be used for other production activities. Moreover, the weeding of one hectare in Africa is equivalent to a 10-km walk in the "stooped position" (Gianessi 2009). This is why repeated weeding causes back pain for many African farmers. The fight against weeds can consequently improve the health of many African farmers. Not only are specific herbicides to appropriately treat each crop lacking in Africa, but also the use of herbicides is not always efficient. For instance, in most African countries, there is no efficient herbicide to control weeds in rice (Rodenburg and Demont 2009), and yet that crop is one of the main consumed staple crops on the continent. The damages caused by invasive plants to rangelands and forests are also alarming. The improvement of weed control is, therefore, of paramount importance to increase crops yield and to alleviate poverty in Africa. And yet, very little research is done to fully understand invasive plant ecology, distribution, and management in Africa and to reduce the pain of smallholders in their fight against them. However, it is necessary to pinpoint that invasive plants are not always perceived as harmful components to all production systems in Africa. In fact, many invasive plants are used in traditional medicine, diets, and other utilization forms like rope, basket, and net making (Geldenhuys 2007, PROTA 2010).

This chapter reviews some of the main invasive plants encountered in African agroecosystems, gives some insight into their utilization by local communities, their economic and environmental impact on agricultural production systems, and the perception of African communities of weed management, and finally addresses the need to improve invasive weed science researches in Africa through an innovated and interdisciplinary approach of collaboration.

13.2   *Diversity of invasive plants and their impacts on the African agroecosystems*

A plant is considered a weed when it grows in a place where others are wanted. Weeds can grow and flourish in various ecological landscapes such as aquatic areas, forests, savannahs, industrial areas, roadsides, railways, and airfields (Rao 2000). Based on their life span, weeds can be divided into two groups: annuals and perennials.

To date, the list of the invasive plants in Africa is incomplete. However, some species have been mentioned as invasive, a major constraint in agriculture across the continent. These invasive plants displace native plants and disturb the environment. These weeds can be classified by their chorology and ecology. The geographical and ecological distribution of weeds can give some insight into their history and that of the ecosystem where they are found and the effort needed to fight them. Indeed, certain species become invasive after the natural ecosystem in place is disturbed, opening space for other opportunist plants acting as weeds. Another factor that makes weeds specific to a station is the climate and that explains why the same weed species could be found in tropical Africa, Asia, Oceania, and South America. According to their ecology and life form, invasive plants can be classified into six categories (Table 13.1). These categories include rhizomatous weeds (sprout just after hoeing from the underground rhizome) and prolific seed producers (Asteraceae and opportunist grass species, e.g., *Digitaria*, *Brachiaria*, and *Dactyloctenium*). Other species are characteristic of the soil condition (e.g., nitrophilous species such as *Sida*, *Amaranthus*, *Acanthospermum*, *Solanum*, *Physalis*, and *Ageratum*), others are hygrophilous (Cyperaceae species, *Paspalum*, and *Portulaca*), others are aquatic (e.g., *Eichhornia crassipes*, *Pistia*, *Echinochloa*, and Cyperaceae species), and others are parasitic (e.g., *Striga hermonthica* and *Rhamphicarpa fistulosa*).

Rhizomatous species have underground stems; one of the most devastating is *Imperata cylindrica*. Farmers usually dig up the rhizomatous weeds in an attempt to increase the chance of killing them. Unfortunately, because the depth of the hoeing is not deep enough

Table 13.1 Chorological and Ecological Classification of Some African Weeds

Chorological and Ecological Group	Weed Species
Rhizomatous	*Imperata cylindrica, Cynodon dactylon, Cyperus rotundus, Cyperus* spp., and *Commelina benghalensis*
Prolific (high seed producers)	*Chromolaena odorata, Dactyloctenium aegyptium, Bidens pilosa, Commelina benghalensis, Eleusine indica, Solanum americanum, Acanthospermum glabratum, Amaranthus retroflexus, Aspilia africana, Brachiaria deflexa, Brachiaria mutica, Brachiaria lata, Cleome monophylla, Crotalaria zanzibarica, Crotalaria retusa, Digitaria velutina, Digitaria horizontalis, Euphorbia hirta, Vernonia ambigua,* and *Euphorbia heterophylla*
Nitrophilous	*Acanthospermum hispidum, Amaranthus hybridus, Eragrostis aspera, Physalis angulata, Physalis peruviana, Ageratum conyzoides, Phyllanthus tenellus, Physalis philadelphica, Sida acuta, Sida rhombifolia, Sida cordifolia, Trianthema portulacastrum,* and *Solanum* spp.
Hygrophilous	*Cyperus esculentus, Cyperus difformis, Cyperus iria, Oryza longistaminata, Oryza barthii, Portulaca oleracea, Paspalum commersonii,* and *Paspalum scrobiculatum*
Aquatic	*Cyperus rotundus, Cyperus esculentus, C. difformis, C. iria, O. longistaminata, O. barthii, Echinochloa colona, Eichhornia crassipes, Salvinia molesta, Pistia stratiotes, Azolla africana,* and *Rhamphicarpa* spp.
Parasite	*Striga hermonthica, Striga asiatica,* and *Rhamphicarpa fistulosa*

to dig up all their rhizomes, the rhizomatous species can sprout easily and quickly. This makes the rhizomatous the most harmful invasive plants and also the most difficult to control in Africa. The time spent in digging a rhizomatous weed can be more than thrice that needed to just hoe it. The nitrophytes grow mostly in places that are rich in nitrogen. The hygrophilous weeds colonize wet ecosystems in contrast to the aquatic weeds that reside in water. Finally, parasites live at the expense of their host. Parasitic weeds that prefer to attack the roots of their hosts are referred to as epirrhizous (e.g. *S. hermonthica*), whereas those that develop on the aerial part of their hosts are referred to as epiphytes (e.g., *Tapinanthus* spp.) (Hoffmann et al. 1997).

These invasive plants affect the yield of major crops such as corn (*Zea mays*), cotton (*Gossypium hirsutum*), rice (*Oryza sativa*), sorghum, and potatoes (*Solanum tuberosum*), to list a few. Some of the most frequent weeds found in rice fields in Africa include *Brachiaria lata, Cleome rutidosperma,* and *Commelina benghalensis* (Dzomeku et al. 2007). However, the scope of the invasiveness varies according to the habitat and the rice cultivars. Invasive plants found in the upland rice field in the sub-Saharan region include *Cyperus rotundus, Imperata cylindrica, Chromolaena odorata, Euphorbia heterophylla, Digitaria horizontalis,* and *Striga* spp. (Rodenburg and Johnson 2009). In contrast, the invasive plants encountered in the lowland rice fields in Africa include *Cyperus* spp., *Oryza longistaminata, Sphenoclea zeylanica, Echinochloa* spp., *Fimbristylis littoralis,* and *Ischaemum rugosum* (Rodenburg and Johnson 2009). *Lantana camara,* which originated from Central and South America, is an invasive plant that is harming forests, pastures, and farmlands in Africa (Ambika et al. 2003; Williams and Madire 2008). In South Africa, *L. camara* infested more than 10 crops (Ambika et al. 2003). This weed generally secretes allelopathic compounds through its roots or leaves in order to inhibit the normal growth of the surrounding crops (Ambika et al. 2003). Because of its toxicity, *Lantana* can become very dangerous for livestock. Sugarcane

(*Saccharum* spp.) is one of the main crops in South Africa. Unfortunately, in that region, *Cynodon dactylon* is reducing its yield (Campbell 2008).

Many invasive plants in West and Central Africa are found in the genus of *Tephrosia* and species of the Convolvulaceae family (Perret et al. 1997, Carrara et al. 1998). *Parthenium hysterophorus*, which originated from Central America and which is considered the worst weed in the world, is also abundant in Africa (Javaid 2010).

Besides the terrestrial ecosystems, weeds are also invading diverse aquatic ecosystems in Africa. These aquatic weeds block water transportation, change the water's ecology, and alter the habitat of the aquatic organisms (Mitchell 1985). Especially, *E. crassipes* is among the top aquatic invasive plant in the world. The other dominant invasive plants encountered in African agricultural ecosystems include *Argemone mexicana, Boerhavia diffusa, Centella asiatica, Chenopodium schraderianum, Convolvulus arvensis, Corchorus trilocularis, Digitaria velutina, Guizotia scabra, Lolium multiflorum, Lolium temulentum, Malva verticillata, Melochia corchorifolia, Nicandra physalodes, Oxalis latifolia, Oxygonum sinuatum, Pueraria phaseoloides, Rottboellia cochinchinensis, Melinis repens, Rhynchosia malacophylla, Setaria verticillata, Ipomoea cordofana, Ipomoea kourankoensis, Xanthium brasilicum, Galinsoga parviflora, Kikuyuochloa clandestina, Fallopia convolvulus,* and *Tagetes minuta.*

Some of the invasive plants that are considered refusals (nonpalatable) for livestocks in West Africa's rangelands are *C. odorata, Combretum paniculatum, Waltheria indica, Macrosphyra longistyla, Cnestis ferruginea, Clerodendrum capitatum, Croton lobatus, Agelaea obliqua, Indigofera garckeana, Hippocratea indica, Zanthoxylum zanthoxyloides, Dichapetalum guineense, Deinbollia pinnata, Triclisia subcordata, Triumfetta rhomboidea,* and *Cassia hirsuta* (Holou and Sinsin 2002). In addition, more than 12 cacti are damaging pastures in Africa (Zimmermann and Moran 1982). If they contain a poison, invasive plants can be a real threat for animals during grazing and can intoxicate livestock. These refusals significantly reduce the productivity of pastures dominated by *Andropogon gayanus, Panicum maximum, C. dactylon, Brachiaria ruziziensis, Paspalum scrobiculatum,* and so on, which are some of the best forages grazed by cattle in Africa (Holou 1998, Holou and Sinsin 2002). The success of weed infestation in pastures depended on "the life form and the morphology of the dominant species, its density, the age of the pasture, the influence of human activity and the selective grazing action of herds" (Holou and Sinsin 2002, p. 42). *C. odorata* is among the main invasive plants that are dominating degraded African grasslands (Holou and Sinsin 2001, 2002). Since its introduction in the humid zone of West Africa in the seventies, its geographical distribution did not cease increasing and this trend is similar in many countries across Africa. The recovery of *C. odorata* can exceed 50% according to the degradation level. In fields where *C. odorata* had high recovery percentage, no other species can occur. In old fallows, cattle cannot move easily in search of grazing areas while *C. odorata* regrowth increases. Because of its high seed productivity (45,151 seeds per plant), *C. odorata* spread quickly through farmlands (Aboh et al. 2008). Besides *C. odorata, Hyptis suaveolens* is another disturbing invasive plant found in the farming systems in central and southern Benin (Aboh et al. 2008, Oumourou et al. 2010). The degree of spreading of *C. odorata* is often higher than that of *H. suaveolens* (Aboh et al. 2008).

In general, no clear estimate exists on the economical impacts of invasive plants in Africa. However, weeds generally reduce crop yields up to 25% (Rao 2000). In Uganda, when *Bidens pilosa's* population increased beyond 8 plants/m², the yield of cowpea (*Vigna unguiculata*) was significantly reduced (Sebuliba-Mutumba 1995). Uncontrolled weeds in Kenya diminished the yield of onion bulbs (*Allium cepa*) by 75%–88% compared to the hand-weeded controls (Mburu et al. 1997). Weeds increased stem etiolating and reduced the leaf area index, dry matter accumulation, and yield of sorghum (Bisikwa 1997). *Striga* alone damaged more than 2.4 million hectares of farmland in Africa (Woomer et al. 2008). The annual yield loss caused by *Striga* is more than 1.5 million tons in Africa (Woomer et al. 2008). This loss is equivalent to more than $380 million. In rice production, the yield loss is even more striking. Commonly, weeds can reduce rice's yield up to 72% (Dzomeku et al. 2007). The impairment of rice production by weeds in sub-Saharan Africa is estimated to more than 2 million tons loss per year (Rodenburg and Johnson 2009). This loss is equivalent to about $1.5 billion.

Invasive plants control is one of the most expensive and time- and energy-consuming tasks in Africa. Usually, to control weed infestation, farmers have no choice other than hoeing or hand-weeding. Because of the pain associated with weed control, weeds are viewed as a symbol of misery by many African farmers.

13.3 Utilization of invasive plants in Africa

Weeds interfere with the utilization of land, water, nutrient resources, and consequently, they negatively affect human welfare (Rao 2000, Monaco et al. 2002). In general, the perception of weeds in Africa highly depends on the region, cultural practices, and human activities. Weeds can be beneficial. In reality, the attribute of the weed is related to a lack of knowledge on the usefulness of plant resources. This is much truer in Africa where many of the species ordinarily referred to as weeds exhibit interesting properties and are used as commodities by the African communities. Local communities in Benin, Togo, and Kenya, for instance, use many invasive plants as vegetables or as ingredients in herbal medicine; hence, these plants are spared during land clearance. This is the case of *Amaranthus spinosus* and *B. pilosa*, which are used as green vegetables in the diet of many Africans (Maundu et al. 2009, Achigan-Dako et al. 2010).

Despite its invasiveness along roadsides and farms in Africa, *P. maximum* is a highly appreciated fodder and is therefore very much appreciated by households who rear livestock. Long known as one of the worst invasive plants in Africa, *C. odorata* (Siam weed) is however used in the cosmetics industry, livestock production, soil fertilization, and horticulture (Holou and Sinsin 2001, Aro et al. 2009). The harsh weed *Commelina diffusa* is used by African farmers to feed animals because its leaves are rich in proteins (Holou and Dakpogan 2003, Lanyasunya et al. 2006). In areas dominated by livestock raising such as Niger, most weeds are a tremendous source of crude protein during dry seasons (Lamers et al. 1996). By using weeds to feed animals, farmers can significantly increase their income.

Weeds can be used for environmental protection as well. For instance, in Africa, livestock can seriously damage grazed ecosystems by overencroaching and eroding equipment, such as pastoral hydraulic dams. In Benin, a biological approach was taken to control this kind of environmental effect of cattle encroachment by using *Mezoneuron benthamianum*, a highly refused weed. Indeed, *M. benthamianum* is a short and thorny shrub that the Laboratory of Applied Ecology in Benin Republic successfully used to protect dams from cattle trampling in Benin. This was done by planting hose refusals on the walking tracks of livestock. Because of the thorns present on *M. benthamianum*, and the wounds that walking through the shrubs can cause, cattle herds avoid the places where those weeds were planted. By doing so, the erosion of the dam was reduced and its life span was significantly increased, proving the use of thorny weeds as a biological fence in Africa.

In rangelands, weeds decrease the productivity and quality of forages (Munyasi and Nichols 2007). Generally, invasive plants reduce the pasture's carrying capacity. However, because of the value they can add to a farmer's income, most woody weeds are not considered as weeds in some African countries such as Kenya (Munyasi and Nichols 2007). Instead, they are usually integrated into the agroforestry management systems where value is added to them. Because most of the alien weeds in Africa were introduced as ornamental plants, they still provide an esthetic comfort to some people. A database was built to describe the multiple uses of plant species (including invasive plants) in Africa and this can be viewed at www.prota.org.

13.4   Management of invasive plants in Africa

The management of invasive plants in Africa has consisted in their suppression regardless of the benefit they can provide to the ecosystems or to the communities. The methods used (Table 13.2) can be divided into cultivation practices, mechanical control, chemical control, and biological control (Mwaja and Masiuna 1997). The mechanical control is usually done by hand, whereas the chemical control is by herbicide spraying. For effective management of invasive plants, integrated systems that combine the methods listed above would be better. However, due to financial problems and a lack of advanced technologies and education, this systemic approach to weed management is less frequently used in Africa.

Table 13.2 Synthesis of the Methods of Invasive Plants Control in Africa

	Method of Control	Crop Protected
Cultivation practices	Crop rotation and spacing	Most crops
	Mulching with species such as *Colophospermum mopane*	Most cereals
	Cultivar selection	Most crops
	Use of fire	Most pastures
Mechanical control	Hand-hoeing and hand-weeding	Peanut
	Burial (e.g., *Commelina* spp.)	Most cereals
	Tillage	Most crops
Chemical control	Pendimethalin (stomp)	Cowpea
	Atrazine, primagarm, alachlor, and gesaprim	Corn
	Imazethapyr, pendimethalin, oxyfluorfen, and metribuzin (sencor)	Soybean
	Fenoxapropo-P-ethyl (Puma Super®)	Bread wheat
	Imazapyr (e.g., to control *Cynodon dactylon*)	Sugarcane and pasture
	Glyphosate	Pastures invaded by *Chromolaena odorata*
Biological control	Introduction of *Neochetina eichhorniae*, *Neochetina bruchi*, *Niphograpta albiguttalis*, *Orthogalumna terebrantis*, and *Eccritotarsus catarinensis* to control *Eichhornia crassipes*	Protection of aquatic ecosystems
	Use of insects, e.g., *Dactylopius austrinus*, *Dactylopius opuntiae*, and *Cactoblastis cactorum* to control cacti invasion	Forests and pastures
	Beetle, *Uroplata girardi* (e.g., to control *Lantana camara*)	Pastures and crops
	Cover crops (e.g., *Mucuna pruriens, Citrullus lanatus, Lagenaria siceraria, Cucurbita* spp., and *Sesamum indicum*)	Most cereals

13.4.1   Cultivation practices

These practices include crop spacing and rotation, mulching, cultivar selection, and fire management. Traditionally, in Africa invasive plants are removed from the cropping areas before and during the life cycle of the desired crop. Crop rotation and intercropping are primarily used to reduce the competition between weeds and crops. In Malawi, a 3-year field experiment conducted to evaluate the effect of a continuous cropping and a rotational cropping of maize indicated that crop rotation reduced the weed flora and density (Mloza-Banda 1997). According to Mloza-Banda, continuous cropping stabilized the seed complex, particularly under no weeding where density-dependent factors may have contributed to a lower diversity index.

Mulching is an efficient tool to suppress weeds in Africa. This is because mulching prevents weeds from covering the soil and deprives their seeds from finding enough light to germinate. In Malawi, mulches from three indigenous tree species (*Colophospermum mopane*, *Acacia natalitia*, and *Acacia nilotica*) reduced the incidence of *Striga asiatica* and delayed its emergence and flowering (Chanyowedza and Chivinge 1999). The success of mulching in controlling weeds depended on the species. Mulch from *C. mopane* was the most effective in suppressing *Striga* between the 4th and the 5th weeks after its application (Chanyowedza and Chivinge 1999). For a long time, most African stakeholders believed that *Striga* was widespread only in areas of low soil fertility. However, a collaborative study testing the effectiveness of eight *Striga*'s management technologies in Kenya and in Uganda found no direct or simple relationship between *Striga* seed counts and soil fertility parameters at the beginning of the experiment (Okalebo et al. 2005).

Another way of combating invasive plants is by cultivar improvement. An off-farm experiment conducted in Zimbabwe indicated that certain cowpea cultivars reduced *S. asiatica* emergence by more than 40%, with IT82D-849 exhibiting the highest percentage reduction (Kasembe et al. 2001). Additionally, some leguminous plants can stimulate the germination of *Striga* in the absence of a host plant, therefore causing the weed to die prematurely (Woomer et al. 2008). An advanced investigation of the mechanism of action of these leguminous plants can provide a great output about the control of *Striga*. The use of corn cultivars that are resistant to herbicide eased the fight against *Striga* and improved corn yield by more than 1 ton/ha (Woomer et al. 2008). Efforts are being made to improve maize resistance to *Striga* in Africa (Rich and Ejeta 2008). In Nigeria, a field study conducted to assess the effect of weed competition on the performance of upland rice varieties revealed that the grain yield of NERICA 4 was significantly higher than that of the other varieties when two hoe-weedings were done (Ekeleme et al. 2007).

The optimization of crops density and spacing improved weeds control and crop yield in Africa (Mashingaidze et al. 2009). For instance, weed control was improved in Zimbabwe by planting corn at a spacing of 60 cm instead of 75 cm or more (Mashingaidze et al. 2009). However, when rows are too narrow, the yield can be obviously reduced because of the increased competition between crops. In contrast, when crops are spaced too much, weeds can colonize the field more easily. The development of cultivars that can do well even in high density, therefore, seems to be a means to improve weeds control in Africa. In Uganda, rice fields planted in rows were less colonized by weeds than their pairs where the seeds were broadcasted (Okot 1995). In Nigeria, intra-row spacing increased the efficiency of weed control (Smith and Ojo 2007).

Weed invasion in African savannahs is generally affected by grazing management, pasture improvement, fire regimes, plant ecology, anthropogenic activity, and the morphological and physiological characteristics of the dominant species (Holou 1998, Holou and

Sinsin 2002, Foxcroft et al. 2010). Fire was used in Benin to improve the yield of savannahs invaded by weeds (Houinato et al. 2001, Holou 2002a, Houehounha et al. 2010). Without the grazing of weeds by some mammalians and the seasonal fire, alien weeds would have already devastated African savannahs (Holou 2002b, Foxcroft et al. 2010). However, few studies were devoted to weed control using fire in Africa. In Benin Republic, the Laboratory of Applied Ecology has been extensively monitoring the effects of different types of fire (early burning at the end, wet season fire, and late fire in dry season) on the management of rangeland ecosystems.

13.4.2 Mechanical control

Hand-hoeing and hand-weeding are still the most common and easily available methods used by the smallholder farmers across Africa. And yet, they are neither the cheapest nor the best ways to control weeds. Hoeing is occasionally combined with hand-weeding during the growth (production) cycle, particularly around the flowering stage (period). In fields planted at a high density such as those of peanuts (*A. hypogaea*) or Bambara groundnuts (*Vigna subterranea*), which bury their pods underground, hand-weeding is preferred to hoeing. Farmers can hoe their fields two to three times per season depending on the weed infestation and the available labor. Specific tasks include uprooting and burning (Sebuliba-Mutumba 1995, Wanjala 1997, Okalebo et al. 2005). Sometimes, to increase the success of the mechanical control, invasive plants are buried. This is the case of *Commelina* spp., which, because of their high water content, do not quickly die, even when their roots are shaken to free the soil and then exposed to the sun. Different tillage methods can be used to exhaust invasive plants' seed bank in the soil. A continued tillage in an experiment in Kenya decreased weed emergence from 79% in the first season to 2% in the fifth season (Muthamia et al. 1997, Ngesa et al. 1997). A number of engines are used to facilitate tillage for invasive plants suppression in African agriculture. However, the extent of utilization is limited because smallholder farmers cannot afford to own or use most of them. The commonly used engines include rotary hoe, flex-time harrow, and row cultivators. Depending on the type of engine used, the weeding process consists of uprooting, burying, or mutilating weeds. Weeders can be manual-push, ox-drawn, or motor-led. Not all push weeders are efficient. A manual-push weeder designed in Sudan proved to be promising (Alaeldin 2008).

13.4.3 Chemical control

Because of their cost, herbicides are less adopted in Africa. However, herbicides are by far the most used tool in large-scale production where man power is a rare commodity. More than 400 herbicides have been developed and used in the last 70 years (Rao 2000). Herbicides can be divided into two categories depending on the time of application: preemergence herbicides (generally soil-applied) and postemergence herbicides (foliage-applied). Preemergence herbicides used or tested so far in African agriculture include pendimethalin, linuron, and alachlor tested on onions (Mburu et al. 1997); metolachlor (dual), metolachlor/metobromuron (galex), prometryn/metolachlor (codal), and pendimethalin (stomp) for control of weed in cowpeas; atrazine, primagarm, alachlor, and gesaprim in maize cultivation (Zewdu 1985, Rupende et al. 1995); imazethapyr in mixture with pendimethalin or oxyfluorfen, metribuzin (sencor), in soybeans (*Glycine max*) (Kanyomeka and Schmid 1995, Wanjala 1997); and actrial DS [a formulated mixture of loxynic (4-hydrox-3, 5-diiodobenzonitrile) and 2,4-D (2,4-(dichlorophenoxy) acetic acid)], glyphosate (Roundup).

Postemergence herbicides are not commonly used in African agriculture. The relatively few used include dimethametryn (4-1, 2-dimethyl-propylamino)-ethylamino-6 methylthio-s-triazine. Other herbicides used in Africa comprise ronstar, stam, kamata (Mon 8751), and gramoxone in tea; sulfentrazone, chlormesulome, prosulfuron, and laddock in maize (Kamidi and Mulati 1997); and fenoxaprop-P-ethyl (Puma super) against annual grassweed in bread wheat in Tanzania showed efficacy (Mkunga 1991). Finally, imazapyr is efficient at controlling *Cynodon* (Campbell 2008), whereas glyphosate was also effectively used against *C. odorata*.

Because of the increasing environmental concerns regarding chemical pollution, mechanical erosion of soils, and the destruction of beneficial insects, the use of herbicides is under strict regulations in most countries in Africa.

13.4.4   Biological control

Biological agents have been successfully used to control certain invasive plants in Africa. The principle is the use of one or more organisms to manage the population of another harmful organism without detrimentally impacting the surrounding ecosystem. Various approaches have been used on case-by-case basis. A well-known example is the biological control of *E. crassipes* on water surfaces with various agents. For instance, to control water hyacinth in South Africa, the following agents have been introduced: *Neochetina eichhorniae, Neochetina bruchi, Niphograpta albiguttalis, Orthogalumna terebrantis,* and *Eccritotarsus catarinensis* (Coetzee and Hill 2008). Unfortunately, these insects were not efficient because of incompatibility and "eutrophication of water hyacinth impoundments" (Coetzee and Hill 2008). A biological approach has been applied to many other weeds in Africa. For example, the insects *Dactylopius austrinus, Dactylopius opuntiae,* and *Cactoblastis cactorum* were able to control cacti invasion in Africa (Zimmermann and Moran 1982). *L. camara* was controlled by the use of the beetle *Uroplata girardi* in South Africa (Cilliers 1991).

The use of cover crops increases the efficiency of the control of weeds (Van Gils et al. 2004). This is the case of the well-known cover crop *Mucuna pruriens* (Antwi-Boasiako 2000), which not only kills weeds by covering them but also improves soil fertility. Indeed, 3 months after sowing, *Mucuna* can cover about 90% of the cultivation plot, hence its effective use and adoption by African farmers (Marfo-Ahenkorah et al. 2001). Many species of the Cucurbitaceae family have been used in rotation with other crops in West Africa for their creeping capacity and ability to control weeds and to improve soil fertility (Achigan-Dako et al. 2008). These species include *Citrullus lanatus, Lagenaria siceraria,* and *Cucurbita* spp. The ability of the sweet potato to play a similar role has been recognized in Ghana (Orkwor et al. 1994). Other species are used as trap crops for weeds in Africa. This is the case of *Sesamum indicum*, which was used as a trap crop for the control of *S. hermonthica* in maize fields in Kenya (Omuony 2006). Although the use of this trap crop did not affect maize yield, however, it significantly reduced *Striga* seed production.

13.5   Conclusion and implications for capacity building in Africa

Like tropical biodiversity, weeds are well diversified in species and life forms in Africa. Weeds also represent an important constraint to agricultural development and ecosystem management in Africa. They hinder the production by decreasing yields. However, weeds are also useful in communities as much utilization can be made out of them. Usually, weed control methods in Africa barely consider endogenous knowledge and usage of these species. This partially explains the lack of integration and link between the beneficial aspects of weeds and their management in Africa. Unfortunately, instead of adding value to the local practices, the majority of the few advanced methods used to control weeds are imported. In most cases, there is no much real biochemical study applied to weeds in Africa. However, an increasing interest exists to train African scientists in the field of agriculture. As agriculture is the foundation of the economy in most African countries, the development of the continent irrevocably depends on the improvement of the farming systems. In addition, if nothing is done to improve control weeds in Africa, the economical and the ecological consequences will be more devastating. This is because weeds are reducing farmers' income and the biodiversity in some ecosystems, and the pesticides used to control them are amplifying the damage. Therefore, efforts must be made to better understand weeds in Africa and implement that knowledge into actions to help develop that continent. For instance, the understanding and the application of the indigenous knowledge of weeds in Africa will lead to the discovery of interesting photochemical or pharmaceutical products. To sustainably achieve this goal, multidisciplinary and transdisciplinary investigations need to be carried on weeds in Africa. Advanced investigations must explore areas of research related to agriculture, ecology, ethnobotany, biology, management, biochemistry, and economy of weeds in Africa in accordance with the African's cultural realities. A deep investigation of the indigenous knowledge of African farmers on the various usages of weeds will surely yield outstanding discoveries for a sustainable use of African plant diversity and for the betterment of contemporary societies. Moreover, a thorough interdisciplinary study of the knowledge of Africans on weeds may even lead to breakthroughs of active principles that can treat certain recurring diseases. A better understanding of the local knowledge of weeds in Africa can advance weed control and usage. To sustain this effort, Africa needs to train more weed scientists. Extension agents and farmers are key stakeholders who should also be educated as well on how to sustainably manage weeds. This is because capacity development is a critical step toward sustainable management of human and natural resources in Africa. Collaboration among African weed scientists must be improved in order to advance communication and exchange of knowledge. To do so, African scientists need to create a society that can address weed issues specific to African contexts. This organization may be called the "Weed Science Society of Africa." The Weed Science Society of Eastern Africa, the Weed Science Society of South Africa, and the African Crop Science Society are already a starting point that African scientists can build upon. African scientists must come together to join their efforts and enlarge their vision for the whole Africa in order to integrate the diversity of Africa's ecosystems into the fight against weeds in that continent. Besides the solutions already known (Holou 2010), it is crucial to keep in mind that the development of Africa will also depend on the improvement of agriculture, and to reach that point, particular attention needs to be paid to factors that are devastating yields, such as weeds. How can biotechnology be applied to advance the fight against weeds in a continent such as Africa, where molecular biology and related technologies are still poorly understood and supported by many governments? Therefore, from now on, weed sciences must be a key component of the capacity-building programs for agriculture development in Africa.

Reasons in favor of capacity building are multiple and include irrelevant fear of biotechnology, impact of bad utilization of pesticides, outcome from pharmaceutical business, underexploitation of local resources, and high motivation of many institutions and donors to improve Africa's higher education in agriculture (World Bank, United States Agency for International Development, European Union, African Union, China, Economic Community of West African States, etc.).

References

Aboh, B.A., M. Houinato, M. Oumorou, and B. Sinsin. 2008. Capacités envahissantes de deux especes exotiques, *Chromalaena odorata* (Asteraceae) et *Hyptis suaveolens* (Lamiaceae), en relation avec l'exploitation des terres de la region de Betecoucou (Benin). *Belgian Journal of Botany* 141:113–128.

Achigan-Dako, E.G., M.W. Pasquini, F.K. Assogba, S. N'Danikou, A. Dansi, and B. Ambrose-Oji. 2010. *Traditional Vegetables in Benin.* Imprimeries du CENAP, Cotonou: Institut National des Recherches Agricoles du Bénin.

Achigan-Dako, G.E., R. Fagbemissi, T.H. Avohou, S.R. Vodouhe, O. Coulibaly, and A. Ahanchede. 2008. Importance and practices of egusi crops (*Citrullus lanatus, Cucumeropsis mannii* and *Lagenaria siceraria* cv. 'Aklamkpa') in socio-linguistic areas in Benin. *(Biotechnology, Agronomy, Society, and Environment)* 12:393–403.

Akobundu, O. 1987. *Weed Science in the Tropics. Principles and Practices.* New York, NY: John Wiley & Sons.

Alaeldin, M.E. 2008. Design performance evaluation of a manual push weeder for weed control between ridges in faba bean in River Nile state. *Sudan Journal of Agricultural Research* 11:109–116.

Ambika, S.R., S. Poornima, R. Palaniraj, S.C. Sati, and S.S. Narwal. 2003. Allelopathic plants. 10. *Lantana camara* L. *Allelopathy Journal* 12:147–161.

Antwi-Boasiako, F. 2000. *Comparing the efficiency of some traditional legumes: Cowpea, soybean, groundnut and fallow with mucuna in restoration of soil fertility and weed control in dry mulch cropping systems.* B.Sc. Agric. (Hons) degree thesis, Department of Crop Sciences, School of Agriculture, University of Cape coast, Cape Coast, Ghana.

Aro, S.O., I.B. Osho, V.A. Aletor, and O.O. Tewe. 2009. *Chromolaena odorata* in livestock nutrition. *Journal of Medicinal Plant Research* 3:1253–1257.

Bisikwa, J. 1997. *Effect of weed competition on the growth and yield of sorghum [Sorghum bicolor (L.) Moench].* M.Sc. dissertation, Faculty of Agriculture, Makere University, Kampala, Uganda.

Campbell, P.L. 2008. Efficacy of giyphosate, alternative post-emergence herbicides and tillage for control of *Cynodon dactylon. South African Journal of Plant and Soil* 25:220–228.

Carrara, A., N. Viarouge, T. Le Bourgeois, and P. Marnotte. 1998. Practical determination of certain species of the *Tephrosia* genus, weeds found in West and Central Africa. *Agriculture et Développement (Agriculture and Development)* 17:51–59.

Chanyowedza, R.M., and O.A. Chivinge. 1999. Effect of *Acacia nilotica, Acacia karroo* & *Colophospermum mopane* leaf mulches on the incidence of with weed [(*Striga asiatica* (L) Kuntze)]. In *Proceedings of the 17th Biennial Weed Science Society for Eastern Africa,* eds. O.A. Chivinge, G. Tusiime, P. Nampala, and E. Adipala, pp. 123–130. Harare, Zimbabwe: Weed Science Society for Eastern Africa.

Cilliers, C.J. 1991. Biological control of water hyacinth, *Eichhornia crassipes* (Pontederiaceae), in South Africa. *Agriculture, Ecosystems & Environment* 37:207–217.

Coetzee, J.A., and M.P. Hill. 2008. Biological control of water hyacinth – The South African experience. *EPPO Bulletin* 38:458–463.

Dzomeku, I.K., W. Dogbe, and E.T. Agawu. 2007. Responses of NERICA rice varieties to weed interference in the Guinea savannah uplands. *Journal of Agronomy* 6:262–269.

Ekeleme, F., A.Y. Kamara, S.O. Olkeh, D. Chikoye, and L.O. Omolgui. 2007. Effect of weed competition on upland rice production in north-eastern Nigeria. In *Proceedings of African Crop Science Conference,* 27–31 October, eds. Z.H. Kasem, M.M. Abdel Hakim, and SI., pp. 61–65. Minia, Egypt: African Crop Science Society.

Foxcroft, L.C., D.M. Richardson, M. Rejmánek, and P. Pyšek. 2010. Alien plant invasions in tropical and sub-tropical savannas: Patterns, processes and prospects. *Biological Invasions* 12:3913–3933.

Geldenhuys, C.J. 2007. Weeds or useful medicinal plants in the rural home garden? *Food and Nutrition Bulletin* 28:392–397

Gianessi, L. 2009. Solving Africa's weed problem: Increasing crop production & improving the lives of women. Available at http://www.croplifefoundation.org/Africa/News%20Conference/Press%20Conference%20Powerpoint.pdf (accessed December 3, 2010).

Hoffmann, G., C. Diarra, I. Ba, and D. Dembele. 1997. Parasitic plant species of food crops in Africa: Biology and impact, study in Mali. 1. Identification and biology of parasitic plants. 2. Impact of parasitic plants based on the results of a study in Mali (1991–1994). *Agriculture et Développement* 13:30–51.

Holou, R.A.Y. 1998. Etude de la dégradation par embroussaillement des pâturages artificiels, subnaturels et naturels en zone guinéenne au Bénin: Cas des stations d'Abomey-Calavi et d'Allada. Memoire Diplôme d'Etude Agricole Tropicale, Production Vegetale, Lycée Agricole Médji de Sékou, Allada, Bénin, 125p.

Holou, R.A.Y. 2002a. *Evaluation of rangeland monitoring at Betecoucou and Samiondji ranchs in Benin.* M.Sc dissertation, Faculty of Agronomical Sciences, University of Abomey Calavi, Abomey Calavi, 136pp.

Holou, R.A.Y. 2002b. Quel avenir pour le pastoralisme face aux mauvaises herbes des pâturages. *Acacia (L')* 23:13–15.

Holou, R.A.Y. 2010. *A Continent in Tears: The Origin of Africa's Collapse and How to Reverse It.* Denver, CO: Outskirts Press.

Holou, R.A.Y., and A. Dakpogan. 2003. Impact de *Commelina benghalensis* sur l'économie des agriculteurs au Bénin. *Acacia (Revue de l'Agriculture Ecologique en Afrique [Review of the Ecological Agriculture in Africa])* 26:8–9.

Holou, R.A.Y., and B. Sinsin. 2001. Grazing lands of the Guinea zone in Bénin under the threat of *Chromolaena odorata. Center for Cover Crops Information and Seed Exchange in Africa Newsletter* 7:2–3.

Holou, R.A.Y., and B. Sinsin. 2002. Bushy weed overrunning in artificial and native pastures under grazing in guinea zone in Benin. *Annales des Sciences Agronomiques du Bénin (Annals of the Agronomic Sciences of Benin)* 3:40–66.

Houehounha, R., H.T. Avohou, O.G. Gaoue, A.E. Assogbadjo, and B. Sinsin. 2010. Weed removal improves coppice growth of *Daniellia oliveri* and its use as fuelwood in traditional fallows in Benin. *Agroforestry Systems* 78:115–125.

Houinato, M., B. Sinsin, and J. Lejoly. 2001. Impact des feux de brousse sur la dynamique des communautés végétales dans la forêt classée de Bassila (Bénin). *Acta Botanica Gallica* 148:237–251.

Javaid, A. 2010. Herbicidal potential of allelopathic plants and fungi against *Parthenium hysterophorus*–A review. *Allelopathy Journal* 25:331–344.

Kamidi, M., and J. Mulati. 1997. Evaluation of new herbicides for weed control in maize. In *Proceedings of the 16th Biennial Weed Science Society for Eastern Africa*, eds. E. Adipala, G. Tusiime, and P. Okori, pp. 171–173. Kampala, Uganda: Weed Science Society for Eastern Africa.

Kanyomeka, L., and W. Schmid. 1995. Weed control in commercial soybean production in Zambia. In *Proceedings of the 15th Biennial Weed Science Society Conference for Eastern Africa*, pp. 13–21. Morogoro, Tanzania: Sokoine University of Agriculture.

Kasembe, E., O.A. Chivinge, I.K. Mariga, and S. Mabasa. 2001. The effect of different cowpea (*V. unguiculata* (L.) Walp] cultivars on witch weed *Striga asiatica* (L.) Kuntze] suppression under drylands conditions. In *Proceedings of the 5th Biennial Conference of the African Crop Science Society–Agricultural Policy, Sustainable Crop Production and Poverty Alleviation in Africa*, p. 82. Lagos, Nigeria: African Crop Science Society.

Lamers, J., A. Buerkert, H.P.S. Makkar, M. Von Oppen, and K. Becker. 1996. Biomass production, and feed and economic value of fodder weeds as by-products of millet cropping in a Sahelian farming system. *Experimental Agriculture* 32:317–326.

Lanyasunya, T.P., H.R. Wang, S.A. Abdulrazak, E.A. Mukisira, and J. Zhang. 2006. The potential of the weed, *Commelina diffusa* L., as a fodder crop for ruminants. *South African Journal of Animal Sciences* 36:27–31.

Marfo-Ahenkorah, E., F.K. Kumaga, and K. Ofori. 2001. Assessment of the potential of three leguminous cover crops for soil nitrogen improvement and weed control for young oil palm plantations. *Journal of the Ghana Science Association* 3:15–23.

Mashingaidze, A.B., W. Van Der Werf, L.A.P. Lotz, J. Chipomho, and M.J. Kropff. 2009. Narrow rows reduce biomass and seed production of weeds and increase maize yield. *Annals of Applied Biology* 155:207–218.

Maundu, P., E.G. Achigan-Dako, and Y. Morimoto. 2009. Biodiversity of African vegetables. In *African Indigenous Vegetables*, eds. C.M. Shackleton, M.W. Pasquini, and A.W. Drescher, pp. 65–104. London: Earthscan.

Mburu, D.N., D.S.H. Drennan, and R.W. Michieka. 1997. Weed control in onion red creole, using pre-emergence herbicides applied at low doses plus supplementary hand weeding. In *Proceedings of the 16th Biennial Weed Science Society for Eastern Africa*, eds. E. Adipala, G. Tusiime, and P. Okori, pp. 187–194. Kampala, Uganda: Weed Science Society for Eastern Africa.

Mitchell, D.S. 1985. African aquatic weeds and their management. In *The Ecology and Management of African Wetland Vegetation*, ed. P. Denny, pp. 177–202. Dordrecht, Netherlands: W. Junk.

Mkunga, M.J. 1991. The efficacy of fenoxaprop-P-ethyl for annual grass weed control in bread wheat in northern Tanzania. In *Proceedings, Seventh Regional Wheat Workshop for Eastern, Central and Southern Africa*, 16–19 September, Nakuru, Kenya, eds. D.G. Tanner, and W. Mwangi, pp. 469–473. Mexico: CIMMYT.

Mloza-Banda, H.R. 1997. Changes in weed species composition in maize based cropping system in central Malawi. In *Proceedings of the 16th Biennial Weed Science Society for Eastern Africa*, eds. E. Adipala, G. Tusiime, and P. Okori, pp. 71–79. Kampala, Uganda: Weed Science Society for Eastern Africa.

Monaco, T.J., S.C. Weller, and F.M. Ashton. 2002. *Weed Science: Principles and Practices*. New York, NY: John Wiley.

Munyasi, J.W., and J.D. Nichols. 2007. Communities and contrasting values attached to pasture weeds: The case of the Maasai and Kamba peoples in south-east Kenyan rangelands. *Agroforestry Systems* 70:185–195.

Muthamia, J.G.N., C.N. Muchendy, and A.N. Micheni. 1997. Long term effect of tillage methods on maize and bean intercrop performance and weed dynamics in eastern highlands of Kenya. In *Proceedings of the 16th Biennial Weed Science Society for Eastern Africa*, eds. E. Adipala, G. Tusiime, and P. Okori, pp. 119–123. Kampala, Uganda: Weed Science Society for Eastern Africa.

Mwaja, V., and J. Masiuna. 1997. Sustainable weed management systems in agriculture: A review. In *Proceedings of the 16th Biennial Weed Science Society for Eastern Africa*, eds. E. Adipala, G. Tusiime, and P. Okori, pp. 21–32. Kampala, Uganda: Weed Science Society for Eastern Africa.

Ngesa, H.O., D.K. Alembi, and H.W. Mwangi. 1997. Determination of soil weed bank. In *Proceedings of the 16th Biennial Weed Science Society for Eastern Africa*, eds. E. Adipala, G. Tusiime, and P. Okori, pp. 65–69. Kampala, Uganda: Weed Science Society for Eastern Africa.

Okalebo, J.R., G. Odiambo, P.L. Woomer, M. Omare, and M. Njok. 2005. A preliminary investigation of the effect of soil fertility on striga weed infestation in maize fields in western Kenya. *African Crop Science Conference Proceedings* 7(1):495–498.

Okot, A. 1995. *Effects of planting methods on weed management in upland rice (Oryza sativa)*. B.Sc. Research report, Faculty of Agriculture, Makerere University, Kampala, Uganda.

Omuony, J.M. 2006. *Evaluation of sesame cultivars as trap crop for the control of Striga weed in a rotational cropping system involving maize and assessment of the importance of sesame as a crop in western Kenya*. M.Sc. thesis, Department of Botany and Horticulture, Faculty of Science, Maseno University, Maseno, Kenya.

Orkwor, G.C., O.U. Okereke, F.O.C. Ezedinma, S.K. Hahn, H.C. Ezumah, and I.O. Akobundu. 1994. Response of yam (*Dioscorea rotundata* Poir.) to various periods of weed interference in an intercropping with maize (*Zea mays* L.), Okra (*Abelmoschus esculentus* L. Moench), and sweet potato (*Ipomoea batatas* L. Lam). In *Proceedings of Ninth Symposium of International Society for Tropical Root Crops*, October 20–26, 1991, eds. F. Ofori, and S.K. Hahn, pp. 349–354. Accra, Ghana: International Society form Tropical Root Crops.

Oumourou, M., B.A. Aboh, S. Babatounde, M. Houinato, and B. Sinsin. 2010. Valeur pastorale, productivité et connaissances endogènes de l'effet de l'invasion, par *Hyptis suaveolens* L. Poit., des pâturages naturels en zone soudano-guinéenne (Bénin). *International Journal of Biological and Chemical Sciences* 4:1262–1277.

Perret, F., P. Marnotte, T. Le Bourgeois, and A. Carrara. 1997. Practical identification of some Convolvulacae species, weeds found in central and West Africa. Agriculture et Développement 16:38–52.

PROTA. 2010. *Updated list of species and commodity grouping. Plant Resources of Tropical Africa (PROTA)*. Wageningen, Netherlands: PROTA Foundation/CTA, p. 391.

Rao, V.S. 2000. *Principles of Weed Science*. New Delhi, India: Oxford and IBH.

Rich, P.J., and G. Ejeta. 2008. Towards effective resistance to Striga in African maize. *Plant Signaling and Behavior* 3:618–621.

Rodenburg, J., and M. Demont. 2009. Potential of herbicide-resistant rice technologies for sub-Saharan Africa. *AgBioForum* 12:313–325.

Rodenburg, J., and D.E. Johnson. 2009. Weed management in rice-based cropping systems in Africa. *Advances in Agronomy* 103:149–218.

Rupende, E., O.A. Chivinge, and I.K. Mariga. 1995. Effect of period of cattle manure curing on survival of weed seeds and nutrient release. In *Proceedings of the 15th Biennial Weed Science Society Conference for Eastern Africa*. Morogoro, Tanzania: Sokoine University of Agriculture.

Sebuliba-Mutumba, R. 1995. *Weed effect on growth and yield of cowpea (Vigna unguiculata L. Walp.).* M.Sc. dissertation, Faculty of Agriculture, Makere University, Kampala, Uganda.

Smith, M.A.K., and I.K. Ojo. 2007. Influence of intra-row spacing and weed management system on gap colonization of weeds, pod yield and quality in okra [*Abelmoschus esculentus* (L.) Moench]. In *Proceedings of African Crop Science Conference*, 27–31 October, eds. K.Z. Ahmed. M.M. Abdel-Hakim, S.I. Shabali, A. El-Morsi, and H.A.M. Ismael, pp. 313–317. Minia, Egypt: African Crop Science Society.

Van Gils, H., J. Delfino, D. Rugege, and L. Janssen. 2004. Efficacy of *Chromolaena odorata* control in a South African conservation forest. *South African Journal of Science* 100:251–253.

Wanjala, B.W. 1997. Weed control methods in Pyrethrum production in Kenya: A review. In *Proceedings of the 16th Biennial Weed Science Conference for Eastern Africa*, 15–18 September, eds. E. Adipala, G. Tusiime, and P. Okori, pp. 57–64. Kampala, Uganda: Weed Science Society of Eastern Africa.

Williams, H.E., and L.G. Madire. 2008. Biology, host range and varietal preference of the leaf-feeding geometrid, *Leptostales ignifera*, a potential biocontrol agent for *Lantana camara* in South Africa, under laboratory conditions. *BioControl* 53:957–969.

Woomer, P.L., M. Bokanga, and G.D. Odhiambo. 2008. Striga management and the African farmer. *Outlook on Agriculture* 37:277–282.

Zewdu, K. 1985. *Evaluation of the traditional weed control practices and some herbicides in maize (Zea mays L.) culture at Bissidimo, Alemaya and its surrounding.* M.Sc. thesis. Addis Abeba University, Addis Abeba, Ethiopia, 98pp.

Zimmermann, H.G., and V.C. Moran. 1982. Ecology and management of cactus weeds in South Africa. *South African Journal of Science* 78:314–320.

chapter fourteen

Geospatial tools for identifying and managing invasive plants

Emma C. Underwood, Allan D. Hollander, and James F. Quinn

Contents

14.1   Introduction

Despite the enormous economic (Pimentel et al. 2005, Keller et al. 2007) and ecological (Lodge et al. 2006) damage caused by invasive species, the resources available for practical management of invasives on a landscape scale are far less than needed. As a result, land managers of all kinds have the challenge of identifying the species and locations at the most risk of environmental and ecological damage and concentrating their scarce resources accordingly. In particular, national policy emphasizes early detection and a rapid response (ED/RR) to identify and control or eradicate new infestations of particularly invasive or harmful species before they have a chance to spread to the point that full control or eradication is impossible or impractically expensive (FICMNEW 2003). Unfortunately, managers rarely have the field staff or resources to systematically search for new occurrences. As a result, it is important to both deploy the best available indirect evidence of occurrence and identify locations especially suitable to new invasions, to increase the probability that new infestations will be detected and controlled when they are small (Stohlgren and Schnase 2006).

Three major approaches have been effective in a number of settings: (1) statistical modeling of species distributions, (2) detection by remote sensing, and (3) enabling citizen scientists to better locate high-priority species and sites. All can be made more effective and complement one another within a well-designed geospatial information framework. Essential information includes good digital base maps, including at least elevation, major landforms, and waterways; human infrastructure (roads, buildings, and facilities); natural habitats; land use categories; soils and geology; and climate variables such as rainfall and temperature patterns. In the United States, downloadable GIS (geographic information system) datasets addressing all of these elements are available from national libraries (e.g., Geospatial One Stop—http://geo.data.gov) and from similar services offered by states, universities, libraries, and an increasing number of commercial organizations, sometimes but not always displaying greater local detail.

Invasive species occurrence records, per se, come in many forms, ranging from national maps displaying occurrences by county (Stohlgren et al. 2005, Graham et al. 2007) to a variety of more localized datasets with precisely mapped individuals or vegetation patches. Internationally, a few countries (e.g., Ireland's National Invasive Species Database) have comparable data centralized and available online, but in many regions, the data have not been collected or are not freely available for security or proprietary reasons. National datasets are also highly variable in their field methods and classification standards, making cross-border assessments challenging (Simpson et al. 2009). For particular applications, a variety of other datasets, for example, fire history and risk, livestock stocking rates, or recreational access, may also need to be assembled. In practice, readily available digital maps may or may not show sufficient detail for particular invasive species assessments, and it is often necessary to construct better map information from raw data, particularly aerial or satellite imagery.

Other essential information to assist in managing invasives, particularly if volunteers or other nonexperts help gather field data, include species lists for the region; accurate, well-illustrated identification guides; expert verification for problematic identifications; and well-organized repositories for photographs and specimens. In many places, museums and herbaria provide most of these services, but unaffiliated local experts, often dedicated amateurs, may be critical sources of localized information. Over the scale of a landscape, state, or region, effective early detection programs for invasive plants necessarily involve multiple eyes in the field, so an effective information system also needs to coordinate participants and provide standardized and easy-to-use ways for people to both learn about information needs and record what they have found. In practice, systems that do not make the information immediately available through interactive maps and searchable interfaces tend to lose participants. It is also critical to credit information providers for their contributions and, increasingly, to track reuse of the information (often termed "provenance") so that analysts and managers can track the ultimate sources of the information they rely on.

A number of web portals have arisen with the goal of providing many of these services to particular invasive-species-management communities. The Global Invasive Species Information Network (GISIN) (http://gisin.org) project has compiled a list of over 250 online invasive species information resources (Sellers et al. 2004 plus online updates). Most represent either identification and control information about invasive species of concern in particular countries or regions or lists of species found in entire jurisdictions (countries, states, counties, national parks, etc.). Many have generalized range maps. However, complete location-specific maps of actual invasive species occurrences are online for fewer than 30 countries, and relatively few of those have more than several thousand records, spread over hundreds of species (Simpson et al. 2009). A number of international initiatives, including the Global Invasive Species Programme (http://www .gisp.org) and working groups begun under the Convention on Biodiversity (http://www .cbd.int/), have promoted international data sharing on invasive species occurrences, in part because the best single predictor that an invader may escape and cause serious damage is that it has already done so elsewhere in the world (Mack et al. 2000). Current intergovernmental efforts are coordinated by GISIN, with pilot studies among the countries in America (http:// www.oas.org/en/sedi/dsd/iabin/) and a proposed data protocol based on standards adopted by the Global Biodiversity Information Facility (see Graham et al. 2007). Within the United States, several coordinated programs have mapped aquatic invasive species on a national level and have been linked by the Smithsonian Environmental Research and the United States Geological Survey (USGS) into a national clearinghouse for aquatic nonindigenous species (http://nisbase.org, see Steves et al. 2004). For terrestrial invasive plants, there are several regional collaborative invasive species mapping programs that are sufficiently established to have developed useful data for regional species distribution modeling, for example, the Invasive Plant Atlas of New England (Simpson et al. 2009) and the Invasive Plant Atlas of the Mid-South (Madsen and Ervin 2007). The National Institute for Invasive Species Science (http://niiss.org) is beginning to amalgamate data from U.S. regional efforts.

In some cases, occurrences of invasive plants can be inferred directly from imagery, as discussed in Section 14.5, particularly if a series of images has been taken over time, and the analyst can detect a sudden or unusual change in the appearance of a patch of vegetation. Alternatively, an analyst may estimate the probability that an aggressive or harmful invader will be found in a particular place, then conduct fieldwork to confirm, and initiate any needed control measures. Obviously, better predictions will lead to more detections and more cost-effective control. Quantitative estimates tend to draw on one or more of a set of related statistical procedures drawn from a rapidly growing literature on how to best map the distribution of species, which we discuss in Section 14.2. To run these procedures, analysts need digital maps of environmental variables that they assume will be good predictors of whether a species will survive and reproduce in a particular spot and locations of a number of documented occurrences of the species, so that the landscape can be statistically explored for unsampled environments that are similar to places the species is known to occur. Museum and herbarium specimen records are widely used for this purpose and are increasingly available in online datasets (see http://www.gbif.org/), but many invasive species are haphazardly collected or may not have been present when historical collections were made, so managers often have to assemble contemporary datasets by conducting surveys. Recommended survey protocols and data standards have been developed by multiple organizations, with varying degrees of standardization and rigor (e.g., see DiPietro et al. 2002, NAWMA 2002, Morse et al. 2004, McNaught et al. 2006), using field methods that vary from quick surveys by teams of volunteers to a series of permanent, balanced-design multiscale vegetation plots (Barnett et al. 2007). However, any dataset that contains the journalist's classic "Who? Where? What? When?" information—that is, species identities, location, and date, documented with information on the collector and method (and, preferably, photographs or specimens)—can be useful (Graham et al. 2007, Simpson et al. 2009). Many of the statistical methods also benefit from data on where the species are known not to be present as well as confirmed occurrences—data that are not typically available from museum collections (Guisan and Zimmermann 2000).

In the following sections, we discuss the development of predictive models (Sections 14.2, 14.3, and 14.4) and remote sensing (Sections 14.5, 14.6, and 14.7) approaches that can assist in delineating the extent of invasions, ascertain their severity, and determine areas at high risk of future invasion.

14.2   Predictive modeling of invasive plants

An ideal predictive system for invasive plants would enable identification of which species would likely spread invasively if introduced and would forecast the total geographic extent of the species spread. Such a system would be extremely beneficial from a management point of view as it enables targeting of limited resources: for instance, by facilitating searches for particular species at ports of entry or by being able to direct monitoring and eradication efforts to vulnerable habitats and geographic areas. A predictive system would operate at three different points in the invasion cycle: identifying the likelihood of arrival of a species in a new locality, identifying the likelihood of establishment in a new location, and identifying the likelihood of invasion in the new habitat or range (Mack et al. 2002). A spatial mapping component can enter into all three of those points. First, in assessing whether the plant has a wide geographic range already, which would increase the likelihood of arrival (Ehrlich 1989). Second, in determining whether the locations of species introductions have a similar climate to its current geographic range. And finally, predicting the extent of the invadable range in the locality and hence being able to infer the risk of introduction of the species.

As an ecological phenomenon stretching over time and space, the study of plant invasions has seen much modeling interest, especially in terms of spatial spread and population demographics (Hengeveld 1994). Classically, invasive spread has been treated through a number of theoretical approaches, including reaction-diffusion models, cellular automata models, and metapopulation models (Higgins and Richardson 1996). Yet, successful application of this genre of models may be limited to small-scale regions, owing to the difficulties of being able to parameterize long-distance dispersal processes (Pyšek and Hulme 2005). In what follows, we focus on the geographic rather the temporal component of predicting invasive plant distributions, that is, we discuss techniques used to identify the potential distribution of an invading plant given what is known about its biogeography.

14.3   Types of predictive modeling

Predictive models of biogeographic distributions have taken two forms: mechanistic and correlative models (Robertson et al. 2003). In a mechanistic model, detailed knowledge of a species' ecophysiology and life history is used to assess whether environmental conditions are suitable for a species. Correlative models are more indirect: associations between species distribution records and environmental conditions are used to indicate environmental suitability. Most likely because detailed ecophysiological information is difficult to obtain, correlative models are much more commonly seen than mechanistic models in modeling invasive distributions (see Sutherst and Bourne 2009).

The rise of statistically based correlative distribution models, sometimes known as ecological niche models, has been an important technical development in biogeography (Guisan and Zimmermann 2000). Ecological niche models have been applied to a variety of problems in biogeography, including the prediction of rare species (e.g., Williams et al. 2009), predicting shifts in the distribution of organisms due to climate change (Bakkenes et al. 2002, Hannah 2008), and analyzing phylogeographic patterns in the evolution of a species (Kozak et al. 2008). Starting in the late 1990s, these have been applied to predict potential spread of invasive species (Peterson 2003, Peterson et al. 2003). The principle of these models is as follows. Beginning with a set of observations of a species, one examines its pattern of association with a set of colocated environmental variables. Typically, the observations are represented in a GIS as point data, which are overlaid on a stack of environmental data layers to obtain the values of the environmental data at each point. This pattern of association is used to develop a model for the distribution of the species, which is then applied to predict the species distribution. In summary, these models take observations and spatially associated environmental variables in geographic space, translate these into niche space to build the predictive model, and then apply the predictive model back in geographic space (see Figure 14.1 and Table 14.1 for summary of applications).

An early and simple example of this approach is given by the BIOCLIM model (Busby 1986, 1991, Nix 1986). In BIOCLIM, observations of a species are overlaid on a stack of climate surfaces, including variables such as annual precipitation, mean temperature in the coolest month, and mean temperature in the warmest month, which are physiological predictors of plant distributions. The model is then constructed by determining the climatic range limits based on the set of observations. For instance, the observations may all fall between 200 and 350 mm annual precipitation and within 27°C and 35°C mean annual temperature (in practice, however, the limits are taken from the 5 to 95 percentile of all the observations). The resulting geographic range pattern would be determined by intersecting all areas on the map that are within the annual precipitation limits and the annual mean temperature limits. This model can be applied over more than two climatic dimensions simply by intersecting all of the climatic range limits. In geometric terms, the BIOCLIM model in niche space is taken to be the hypercube bounding all of the climatic

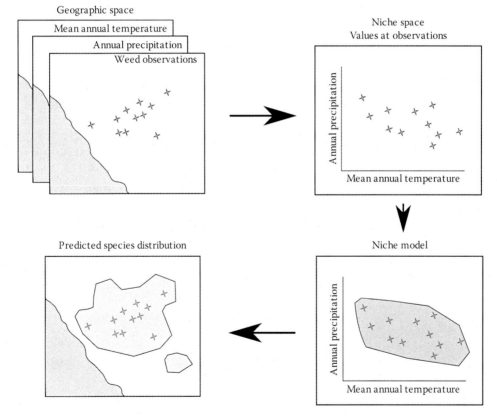

Figure 14.1 Workflow for ecological niche modeling. At upper left, species observations are overlaid in geographic space on a data stack of environmental variables. The relationships between these environmental variables become evident when plotted in niche space on the right, and statistical algorithms may be used to create a niche space model for the species at lower right. This model is then brought back into geographic space at lower left to map out the potential distribution for the species.

Table 14.1 Selection of Predictive Modeling Applications for Invasive Plant Species Mapping

Modeling family	Specific modeling algorithms	Target species	Location	Notes	Reference
Correlative presence–absence model	GARP	Hydrilla (*Hydrilla verticillata*), Russian olive (*Elaeagnus angustifolia*), sericea lespedeza (*Lespedeza cuneata*), and garlic mustard (*Alliaria petiolata*)	North America	Based predictions on native range of species	Peterson et al. (2003)
Correlative presence–absence model	GARP	Lehmann lovegrass (*Eragrostis lehmanniana*)	Southwestern United States, especially Arizona	Based predictions on native range of species	Mau-Crimmins et al. (2006)
Correlative presence–absence model, ensemble modeling	GLM, GAM, GBM, Artificial Neural Networks, Random Forests, Classification Tree Analysis, Mixture Discriminant Analysis, and Surface Range Envelope	Hawkweed (*Hieracium aurantiacum, Hieracium murorum*, and *Hieracium pilosella*)	Europe, North America, and Australia	Compared calibration of predictions using native range versus native and invaded range	Beaumont et al. (2009)
Correlative presence–absence model	GLM, GAM, and GBM	Ragweed (*Ambrosia artemisiifolia*)	Austria	Incorporated two measures of propagule pressure	Dullinger et al. (2009)
Correlative presence–absence model, dynamic spread model	GLM, GAM, and Interacting Particle System	Ragweed (*Ambrosia artemisiifolia*)	Austria	A combined model fitted the data better than the individual components	Smolik et al. (2010)
Correlative presence–absence model, ensemble modeling	Logistic Regression, Boosted Regression Trees, Maxent, MARS, and Random Forests	Dalmation toadflax (*Linaria dalmatica*), Musk thistle (*Carduus nutans*), Cheatgrass (*Bromus tectorum*), and White sweet clover (*Melilotus officinalis*)	Yellowstone, Grand Teton, Sequoia & Kings Canyon, and Denali National Parks, United States	Ensemble model outperformed individual models	Stohlgren et al. (2010)

GARP, Genetic Algorithm for Rule-set Production; GLM, Generalized Linear Models; GAM, Generalized Additive Models; GBM, Generalized Boosted Models.

range limits. This is a simple representation of Hutchinson's (1957) original "fundamental niche" concept, designating the range of habitats and conditions physically suitable for a species. It is important to note that ecological niche models, in general, are really "fundamental niche" models and that distributions are often limited by biotic interactions (resulting in a more restricted "realized niche"), which are almost never modeled. Hence, physiologically correct ecological niche models will usually overpredict a species' actual distribution.

From a statistical point of view, ecological niche modeling is an application of statistical or machine learning techniques (Berk 2008, Hastie et al. 2009). These range in complexity from very simple classifiers such as the BIOCLIM hypercube-based approach, through regression models, tree-based models, and finally ensemble models that aggregate results from simpler models. Ecological niche models can also be divided into presence-only models and presence-absence models (Hirzel et al. 2002). All that is needed to compute a presence-only model is a set of known occurrences for the species. In contrast, there is a family of techniques that produce a predictive model by statistically establishing the differences between known presences and known absences of the species and is certainly a preferable approach if the occurrence locations are limited in number or are likely not to be representative of suitable locations, as in many museum collections. In cases where no absence observations are available, presence–absence methods often use "pseudo-absences," which are frequently points scattered randomly across the geographic domain of interest.

The BIOCLIM technique is an example of a presence-only distribution model. There are also presence-only methods that are more complex than BIOCLIM. For instance, GARP (Genetic Algorithm for Rule-Set Production) is an ensemble method that randomly generates different rules for predicting distributions and uses a genetic algorithm to combine these rules into a more optimum model (Stockwell 1999). Another technique is ecological niche factor analysis (Hirzel et al. 2002), a multivariate technique that computes the specialization of a species in niche space with respect to the environmental conditions of the entire study area. A final presence-only technique that has recently become popular is Maxent, which is based on finding a probability distribution that maximizes entropy (hence the name) subject to constraints that in the case of species distribution modeling is the environmental variables associated with each species observations (Phillips et al. 2006). Maxent has been found to perform particularly well in modeling occurrences of invasive plant species (Evangelista et al. 2008).

There are also many presence-absence techniques used in ecological niche modeling. In statistical terms, these are termed "supervised learning models" (Hastie et al. 2009), and any such supervised learning approach could potentially be used in ecological niche modeling. One such modeling technique is logistic regression. This is a commonly used regression technique for modeling binary category data, which is one component of statistical techniques called generalized linear models (GLMs) that are sometimes applied to niche modeling. Logistic regression differs from ordinary least squares regression in that it uses a logit transform to constrain the output of the regression function to lie between 0 and 1, thus making it suitable for probability measures (Venables and Ripley 2003). A second example of a presence/absence method used in predictive distribution modeling is multivariate adaptive regression splines (MARS). This is a variant on linear regression that allows for changes in the slope of the fitted surface at critical breakpoints of the independent variables (Hastie et al. 2009). In distribution modeling, this technique has an advantage over logistic regression of allowing the model to capture different functional relationships over different portions of environmental space (Leathwick et al. 2005).

A third presence/absence technique, random forests, is based on repeated applications of using classification trees as predictors. A classification tree is a statistical learning technique that creates a decision tree by recursively partitioning the data observations based along split points of a single independent variable (Breiman et al. 1984). These classification trees are easy to interpret, work well for combinations of categorical and numerical variables, but often have only moderate predictive accuracy. Random forests are a recent variant on the classification tree technique that tries to improve accuracy by building an ensemble set of trees. Each of the trees in this set is a random variant, the randomness arriving in a couple of ways. First, the algorithm only selects among a randomly chosen small subset of the independent variables. Second, the observations for each of these trees are randomly sampled with replacement from the original set of observations (Breiman 2001). The method of random forests is an increasingly popular statistical learning technique because of its high predictive skill over a wide range of types of inputs (Hastie et al. 2009). The method is increasingly being used in ecology for problems such as species distribution modeling (Cutler et al. 2007, Williams et al. 2009).

Another recent development in predictive modeling for invasive species is the use of ensemble or consensus techniques. Of the many statistical approaches for ecological niche modeling described in this section, no one technique has emerged as superior to the rest (Segurado and Araujo 2004), although some methods, for instance GARP, seem to underperform (Meynard and Quinn 2007, Kumar et al. 2009). If no overall best technique can be identified, it follows that combining the results from different modeling techniques using algorithms such as averaging may produce an ensemble model with higher overall accuracy (Marmion et al. 2009). Stohlgren et al. (2010) suggest that ensemble modeling may be particularly useful in situations where robustness is especially valued such as in automated mapping workflows and also may be well suited for modeling recently arrived species that have not spread to enough habitats to make the species–environment relationships clear. Stohlgren et al. used ensemble techniques to predict the distribution of four species of invasive plants at four different sites (Grand Tetons National Park, Sequoia and Kings Canyon National Park, Yellowstone National Park, and Alaska) in the United States, evaluating five different statistical modeling techniques including logistic regression, boosted regression trees, Maxent, MARS, and random forests. The different modeling techniques varied in their performance across the four sites and species, but the ensemble model outperformed the individual models at two of the four sites and performed almost as well as the top-performing individual model at a third site (Figure 14.2).

14.4   *Applications and considerations of ecological niche modeling to invasive plants*

Peterson et al. (2003) present an early example of applying ecological niche modeling to invasive plant species. In this study, they used the GARP method to predict the potential distribution of four invasive plants in North America, hydrilla (*Hydrilla verticillata*), Russian olive (*Elaeagnus angustifolia*), sericea lespedeza (*Lespedeza cuneata*), and garlic mustard (*Alliaria petiolata*). They developed their models based on occurrence records from the native ranges of the species in Europe and Asia, using a suite of climate and topographic layers as predictor variables. To assess accuracy within North America, they compared the predicted distribution with known occurrence records from counties and hydrological units. The GARP method predicted fairly narrow distribution ranges for hydrilla and

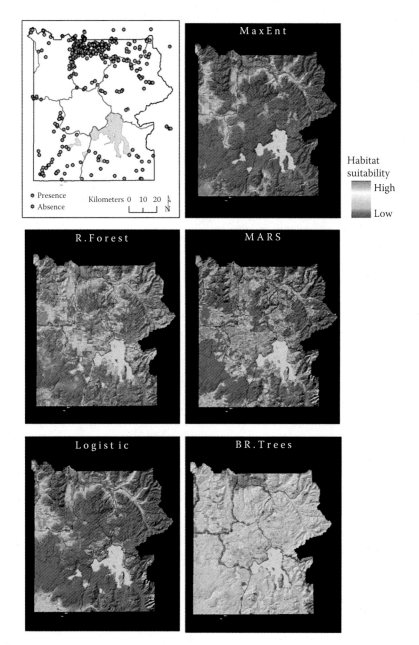

Figure 14.2 **(See color insert.)** Point distribution of the invasive plant *Linaria dalmatica* (Dalmatian toadflax) in Yellowstone National Park, Wyoming, United States, and habitat suitability maps generated using five common species distribution models: Maxent, random forest (R. Forest), multivariate adaptive regression splines (MARS), logistic regression (Logistic), and boosted regression trees (BR. Trees). All models predicted suitable habitat in the north-central part of the park, which overlapped the majority of reported presence locations, otherwise the models accentuated different locations, for example, the boosted regression model highlighted road corridors compared to the MARS model, which emphasized habitat suitability in the park's interior. (Reprinted from Stohlgren, T.J. et al., Ensemble habitat mapping of invasive plant species. *Risk Analysis*. 2010. 30:224–235. Copyright Wiley. Reproduced with permission.)

sericea lespedeza and relatively wide ranges for Russian olive and garlic mustard, compared to records from a random selection of counties. One difficulty Peterson et al. report is that it was very time consuming to collect observation points from the native ranges of the species, an order of magnitude longer than to accumulate the digital occurrence data from the invaded ranges in North America.

The Peterson et al. (2003) study raises an interesting question of theory: in modeling the potential distribution of an invasive species, should one base the model on occurrence records from the native range of the species or from initial occurrences in the invaded region? Both choices have their difficulties. Basing a model on occurrences in a still-expanding invaded range runs the risk of underestimating the potential niche space the invader may occupy. Conversely, relying on a species' native range to develop an ecological niche model overlooks the possibility that the niche of the species may shift in the process of invasion (Beaumont et al. 2009), perhaps due to genetic selection and recombination by the invader (Lavergne and Molofsky 2007). Many invasive species are uncommon or unaggressive in their native ranges, perhaps limited by parasites, predators, or diseases, yet proliferate in new settings, maybe because they arrive without natural enemies. For this reason, patterns of occurrence and spread in other nonnative regions may be the best predictor of the dynamics of new invasions (Mack et al. 2002).

A study that compared modeling an invasive plant using occurrences from either the invaded or the native ranges examined the perennial bunchgrass *Eragrostis lehmanniana* in Arizona (Mau-Crimmins et al. 2006). This grass is unusual as an invasive species in that its introduction is particularly well documented: it was brought to Arizona from southern Africa in the 1930s to be planted for erosion control, the seeds being selected from a single cultivar. In this study, the distribution model built from occurrence points chosen from the native range of the species had a much wider geographic extent than the distribution model built from occurrences in the invaded range. Specifically, the first model predicted the species to be able to spread to central California and western Texas, whereas the predicted range in the second model was largely restricted to southeastern and northwestern Arizona. Based on field experiments and expert opinion, the results from the model fitted from occurrences from the invaded range of these species seem more plausible. The authors also modeled the distribution of the grass in its native range in southern Africa. The model built from occurrence points from southern Africa, not surprisingly, covered the entire native range of the species. In contrast, the model built from occurrences in the invaded range in Arizona had a very restricted geographic extent, which, interestingly, occupied a region with lower temperatures than that of the original cultivar. This could represent one or more of nonrepresentative genetics of the invasive propagules (a classic founder effect); chance introduction in a nonrepresentative location (a geographic disequilibrium); a distribution not limited by temperature; or impacts of a different community of competitors, predators, and disease (a shift in realized niche relative to the fundamental niche). The authors suggest that invasive species without much genetic variation and with a well-known history of introduction may be well suited to build predictive distribution models from the invaded range.

As illustrated by the *E. lehmanniana* example, a basic problem in building predictive distribution models for invasive plants is that, almost by definition, invasive species are most often not at equilibrium with respect to their (new) geographic range. The general framework outlined in Figure 14.1 for producing a species distribution model assumes an equilibrium condition: that is, for the model to work, the known occurrences of the species should effectively capture the niche dimensions of the species. In contrast, there have been some efforts to incorporate nonequilibrium conditions in distribution modeling for invasive plants. Dullinger et al. (2009) examine the case of ragweed (*Ambrosia artemisiifolia* L.)

invasion in Austria. Ragweed is native to the central United States, but has invaded many parts of the world including central Europe where it has spread greatly since the middle of the twentieth century. Dullinger et al. collated records for ragweed in Austria, using the database of the Floristic Mapping Project of Austria as a primary source. In addition to using temperature, precipitation, topography, and geology as environmental variables, in their modeling, Dullinger et al. incorporated two measures of propagule pressure, namely, the length of major roads within a raster cell and the presence of ragweed in an adjacent raster cell. They also considered as a variable whether a population was naturalized or not in a raster cell, judging naturalization by whether at least 100 reproductive individuals were present in the cell. They found that the models using records from only the naturalized populations were much more accurate than the ones that used the total set of records, and the incorporation of propagule pressure only slightly improved the models.

In another study using the example of *A. artemisiifolia* in Austria, Smolik et al. (2010) combine the ecological niche modeling approach used in the Dullinger et al. (2009) study with a dynamic spread modeling approach. There has been research on mathematically modeling the spread of invasive species for many decades (Hastings et al. 2005), but it is unusual in that approach to incorporate geographic heterogeneity, which is basic in ecological niche modeling. Smolik et al. incorporate an interacting particle system (IPS) model to build a combined spread model for *A. artemisiifolia*. IPSs are spatially explicit models that are a generalization of cellular automata in that they can incorporate spatial variation and stochasticity in their transition rules. Smolik et al. use a time series of *Ambrosia* spread from 1990 to 2005 to fit three models, one based solely on the time pattern of spread, one based solely on the pattern of habitat associations, and one model that combined the time pattern of spread with habitat associations. They found that the spread-alone model did the poorest job of capturing the pattern of invasion, frequently predicting the occupation of unsuitable raster cells. The habitat-alone model considerably improved the prediction of the invasion, and the combined model offered a slight improvement over the habitat-alone model. The authors expected a more substantial improvement with the combined model and believed that if tested over a larger extent the spatial spread component would be relatively more important to the success of the model. At any rate, incorporation of spread information in models necessitates having an accurate time series of the initial stages of a plant invasion, data that are quite often lacking.

The choice of a predictive modeling algorithm may end up being a trade-off between complexity and practicality. Studies that have examined the performance of species distribution models indicate that algorithms that finely differentiate species occurrences from background conditions (e.g., ensemble methods such as random forests or methods that use breakpoints to highlight different ecological conditions such as MARS) often outperform those with coarser characterizations of the environment (e.g., BIOCLIM) (Elith et al. 2006). However, the simpler methods may be more practical to implement in terms of GIS-based predictive modeling, avoiding the need to deeply embed statistical models within a GIS framework. For example, it is simple to take the coefficients from a logistic regression model and write a GIS script to output a predictive map. It is much more difficult to write a GIS script to produce a predictive map from a random forest model composed of several hundred deeply rooted classification trees, even though the accuracy of the random forest model is likely to be higher than the logistic regression model. Moreover, the choice of algorithm is less critical than the quality of the information that goes into the predictive modeling process, in terms of both environmental data layers and the occurrence records for the invading species. It is often challenging to map many of the environmental variables most proximal to the dispersal and reproductive success of the invading plant, such as disturbance conditions on the ground or the number of passages across a location by dispersal vectors. The quality of the modeling also depends on the spatial representativeness of the occurrence records that are used to build the models. Efforts to systematically coordinate the search for invasive plants, as discussed in Section 14.1, and the use of remote sensing techniques, as discussed in Section 14.5, address this issue of spatial bias and are important for improving the accuracy of predictive modeling.

14.5   Remote sensing of invasive plants

Imaging through remote sensing can be described, at its simplest, as any method of observing the Earth's surface without being in direct contact with it. More specifically, images record information about the electromagnetic spectrum. Remote sensing is one of a suite of geospatial tools, used together with GIS and GPS, which offers potential advantages over field-based detection for mapping invasive plant species. For instance, an entire area can be mapped simultaneously and more quickly than field-based approaches alone, while repeat acquisitions of images over the same area can assist in monitoring trends in invasion or assess the effectiveness of control activities.

A primary distinction can be made between types of remote sensing imagery based on the energy source used. Passive sensors capture solar illumination reflected from the Earth's surface and typically collect information in three parts of the spectrum—the visible, near infrared, and middle infrared, at wavelengths between 0.40 and 14 μm (Turner et al. 2003, Figure 14.3). In contrast, active sensors emit a pulse of energy and then capture the energy returned to the sensor. Radar (radio detection and ranging) sensors emit microwave pulses and LiDAR (light detection and ranging) sensors emit laser light. Physical characteristics of the Earth's surface such as information on tree height, biomass, and leaf area of vegetation are described by the intensity and timing of the returned signals. Active sensors such as LiDAR operate in the visible to near-infrared wavelengths while Radar collects data in longer microwave wavelengths from 1 mm to 1-m (Turner et al. 2003, Figure 14.3).

Remotely sensed images can also be characterized by their spatial, spectral, temporal, and radiometric resolutions. Spatial resolution refers to the size of the smallest picture element (pixel) in the scene, for example, Landsat 7 Enhanced Thematic Mapper Plus (ETM+) has 30-m pixels. The spectral resolution of the image describes the width of the bands of the electromagnetic spectrum detected by the sensor (Steininger and Horning 2007). A higher spectral resolution means more bands of narrower width—for example,

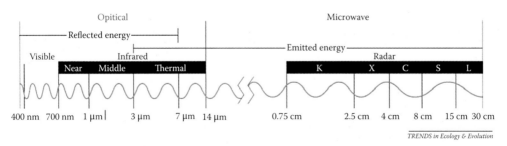

Figure 14.3 The electromagnetic spectrum. The figure illustrates that optical and microwave sensors occupy distinct regions of the spectrum and are detected using different technologies. (Reprinted from *Trends in Ecology and Evolution*, 18, Turner, W. et al., Remote sensing for biodiversity science and conservation, 306–314, Copyright 2003, with permission from Elsevier.)

hyperspectral sensors capture information in up to 224 bands compared to Landsat ETM+ with seven broad bands. As the spectral resolution increases, so does the ability to detect unique spectral signatures of target invasive species.

Remote sensing offers two general approaches for detecting invasive plants. First, it can assist directly by mapping the invasive species or the invaded ecological community. Direct detection relies on exploiting unique features of the target invader such as phenological or structural characteristics and the spatial patterning of the target compared to the surrounding mosaic of species. Second, remote sensing can indirectly help identify areas of potential invasion using surrogates such as areas of human-related disturbance (e.g., bare ground near construction sites) or features that are likely to encourage the dispersion of invasive plants (e.g., hiking and off-road vehicle trails). Furthermore, image-derived environmental parameters, such as topographical variables, are important inputs for developing predictive models to determine areas susceptible to invasion (see Cohen and Goward 2004, Andrew and Ustin 2009). In this section, we review applications of different types of imagery to mapping invasive plants (see Table 14.2 for summary) and discuss some of the advantages and disadvantages of each (see also Huang and Asner 2009).

14.6   Applications and considerations of remote sensing to invasive plants

Early airborne sensors for remote sensing include aerial photography and videography with high spatial (<10-m) but low spectral resolution (~5 bands), which provide a relatively simple approach for mapping invasive plants over small spatial areas. Successful identification of the target invader rests on the species having unique spectral characteristics compared to surrounding vegetation, often relating to a specific phenological stage or having a distinct spatial pattern. For example, Everitt et al. (1995) found conventional color aerial photographs acquired in early summer could successfully distinguish leafy spurge (*Euphorbia esula*) in North Dakota and Montana, owing to the higher visible reflectance of the bright yellow bracts compared to surrounding brush and herbaceous species. However, sparse stands of leafy spurge with less than 25% canopy cover were generally undetectable in photos. In another study, color infrared photos successfully detected the invasive redberry juniper (*Juniperus pinchotii*) in rangelands in northwest Texas (Everitt et al. 2001), its dark evergreen foliage in winter results in lower visible reflectance compared to surrounding dormant species, while its dense structure results in higher near-infrared reflectance. The appeal of aerial photography and videography is their relatively low cost and, often, the availability of archival images. Limitations of the technology include ensuring image acquisition at the optimal phenological time and the large amount of manual processing that is involved (e.g., digitizing and georegistration) (Anderson et al. 1993).

Since the 1970s, satellite sensors have been orbiting the Earth. Two commercial multispectral satellites with high spatial resolution are Quickbird (2.4-m resolution) and IKONOS (4-m resolution), each with five spectral bands. Both have been applied to mapping invasive species. For example, IKONOS was used to map the distribution of invasive *Melaleuca* (*Melaleuca quinquenervia*) in southern Florida using a supervised neural network classifier (Fuller 2005). However, success was limited to moderate to dense stands; consequently, small outlying populations, which may be instrumental for future expansion, may have gone undetected. Quickbird has been used around the Hudson River in New York State to generate a land cover map (20 classes), which included three invaded communities—purple loosestrife (*Lythrum salicaria*), common reed (*Phragmites australis*), and water chestnut (*Trapa natans*) (Laba et al. 2005), with overall mapping accuracies of 65–74%. Again, owing to the coarse

Table 14.2 Selection of Remote Sensing Applications to Invasive Plant Species Mapping Highlighting Different Spatial, Spectral, and Temporal Resolutions

Sensor	Target species	Habitat type	Location	Basis and method of identification	Reference
High spatial resolution technologies					
Color and color infrared aerial photos, color videography (0.12–0.5-m; 3–4 bands)	Leafy spurge (*Euphorbia esula*)	Mixed herbaceous species, mixed brush	North Dakota & Montana, United States	Phenology (associated with yellow bracts); known locations (GPS) used to determine mean values	Everitt et al. (1995)
Color infrared aerial photo (0.4-m; 3 bands)	Redberry juniper (*Juniperus pinchotii*)	Rangelands	Texas, United States	Phenology (lower visible and higher infrared reflectance); known locations (GPS) used to determine mean values	Everitt et al. (2001)
Color infrared aerial photo (0.3-m; 3 bands)	Blackberry (*Rubus fruticosus*), Montpellier broom (*Genista monspessulana*), and desert ash (*Fraxinus angustifolia*)	Eucalypt woodlands	Mount Lofty Ranges, South Australia	Field data used in supervised classification	Crossman and Kochergen (2002)
Airborne videography (1.4–2.3-m; 3 bands)	Chinese tamarisk (*Tamarix chinensis*)	Riparian areas	Southwestern United States	Phenology (color of leaves prior to falling); known locations (GPS) used to determine mean values	Everitt et al. (1996)
IKONOS (4-m, 4 bands)	*Melaleuca* (*Melaleuca quinquenervia*)	Woody vegetation, prairie	Florida, United States	Texture and phenology (NDVI) used in a neural network classifier	Fuller (2005)
Quickbird (0.61 cm and 2.4-m; 4 bands)	Purple loosestrife (*Lythrum salicaria*), common reed (*Phragmites australis*), and water chestnut (*Trapa natans*) land cover classes	Tidal wetlands	Hudson River, New York, United States	Field data and aerial photo interpretation used as training data	Laba et al. (2005)
Low spatial and high temporal resolution technologies					
AVHRR (1 km, 6 bands)	Cheatgrass (*Bromus tectorum*)	Semiarid grasslands	Great Basin, United States	Phenology; presence of cheatgrass linked with interannual variability of greenness (NDVI)	Bradley and Mustard (2005)

Sensor	Species	Habitat	Location	Method	Reference
MODIS (250-m, 2 bands at 250-m)	Lehmann lovegrass (*Eragrostis lehmanniana*)	Semidesert grasslands	Arizona, United States	Phenology; time series of NDVI linked with gradient of infestation	Huang et al. (2009)
Moderate spatial and moderate spectral resolution technologies					
Landsat 7 ETM+ (30-m, 7 bands)	Cheatgrass (*B. tectorum*)	Shrubs or coniferous trees with bunch grass understory	Nevada, United States	Phenology; NDVI calculated from images acquired in two seasons and used as input in to model	Peterson (2005)
Landsat 7 ETM+	Cheatgrass (*B. tectorum*)	Sagebrush steppe	Southern Idaho, United States	Phenology; four scenes from different dates acquired and stacked, linear spectral unmixing conducted on MNF bands	Singh and Glenn (2009)
Landsat 7 ETM+	Tamarisk (*Tamarix chinensis*)	Riparian areas	Arkansas River, Colorado, United States	Phenology; time series analysis of six scenes, with tasseled cap transformation for soil/vegetation wetness proving one of best variables in Maxent model	Evangelista et al. (2009)
Landsat 5 TM and Landsat 7 ETM+	Amur honeysuckle (*Lonicera maackii*)	Deciduous forest	Ohio and Indiana, United States	Phenology; NDVI best predictor to map this understory species	Wilfong et al. (2009)
Landsat 7 ETM+	Bitter bush (*Chromolaena odorata*)	Lowland *Shorea robusta* forest	Nepal	Light intensity and forest canopy density; neural networks used to determine these and then inverted to indicate seed productivity of this understory species	Joshi et al. (2006)
High spectral resolution technologies					
AVIRIS (4-m, 224 bands)	Jubata grass (*Cortaderia jubata*)	Chaparral	California, United States	Cellulose and lignin absorption features; maximum likelihood classification of selected MNF bands	Underwood et al. (2007)

(Continued)

Table 14.2 Selection of Remote Sensing Applications to Invasive Plant Species Mapping Highlighting Different Spatial, Spectral, and Temporal Resolutions (*continued*)

Sensor	Target species	Habitat type	Location	Basis and method of identification	Reference
High spectral resolution technologies					
AVIRIS	Pickelweed (*Salicornia virginica*)	Marsh	Petaluma Marsh, California, United States	Water-absorption features; regression of field canopy water content measurements with image measurements	Sanderson et al. (1998)
Hymap (3-m, 126 bands)	Submerged aquatic vegetation	Aquatic system	Sacramento-San Joaquin Delta of California, United States	Threshold value of NDVI; spectral angle mapper and spectral mixture analysis used in decision tree	Hestir et al. (2008)
AVIRIS and Hymap	Perennial pepperweed (*Lepidium latifolium*)	Marsh	San Francisco Bay and Sacramento-San Joaquin Delta of California, United States	Spectrally unique white flowers; 19 physiological indexes generated and used in CART	Andrew and Ustin (2006)
Hyperion (30-m, 220 bands)	Common reed (*Phragmites australis*)	Coastal wetlands	Great Lakes, Wisconsin, United States	Field reference spectra of pure stands used with Spectral Correlation Mapper	Pengra et al. (2007)
Active sensor technologies					
LiDAR	Spartina (*Spartina alterniflora*) and hybrid (*S. alterniflora x foliosa*)	Marsh	San Francisco Bay, California, United States	First and last returns detected differences in canopy	Rosso et al. (2006)

spectral resolution, the target species still requires distinct phenological characteristics or spatial patterning in order to be detected and the cost of the imagery is relatively expensive.

In contrast to high spatial resolution imagery is low spatial (but high temporal) resolution imagery associated with sensors such as AVHRR (Advanced Very High Resolution Radiometer) and MODIS (Moderate Resolution Imaging Spectroradiometer). Although these satellites have spatial resolutions in excess of 1 km, they have the advantages of regular repeat acquisitions (daily), have a large spatial footprint, and are available at no cost. Bradley and Mustard (2005) generated the Normalized Difference Vegetation Index (NDVI) from AVHRR data in the Great Basin, United States, and found that cheatgrass-dominated (*Bromus tectorum*) areas had a greater response to interannual rainfall, which was distinct from the native shrub and grasses. Similarly, NDVI and brightness calculated from MODIS imagery (at 250-m spatial resolution) was used to detect phenological differences across grasslands with differing levels of invasion by Lehmann lovegrass (*E. lehmanniana*) (Huang et al. 2009).

Satellite images with moderate spectral and spatial resolution are available from the Landsat ETM+ sensor with 7 bands and 30-m resolution, repeat acquisition over the same site occurs every 16 days, and imagery is available at no cost. Applications for invasive species include a study by Peterson (2005), which used six Landsat 7 ETM+ scenes from two dates in Nevada to map invasive cheatgrass (*B. tectorum*) using differences in phenology. An image from the spring established a baseline of green vegetation when cheatgrass and surrounding native species are photosynthetically active, which was compared to an image in early summer when the cheatgrass had senesced compared to the surrounding green vegetation. These changes were summarized by the NDVI generated from reflectance values in the red and near-infrared regions of the spectrum. After using regression methods to describe the seasonal growth and senescence patterns, phenology, elevation, and late-season green band were found to be statistically significant for accurately determining cheatgrass distribution. In some cases, researchers have improved the "spectral" resolution of the sensor by stacking images from different dates. For example, Singh and Glenn (2009) stacked four Landsat 7 ETM+ scenes over a growing season and utilized techniques commonly associated with hyperspectral image processing to detect cheatgrass from the surrounding vegetation in southern Idaho. Results indicated that 85% (producer's accuracy) of known cheatgrass locations were detected.

Although the detection of invasive plants is often reliant on the target species dominating the vegetation canopy, there are some examples of successful mapping when the invasive is in the understory. Wilfong et al. (2009) mapped the understory invasive Amur honeysuckle (*Lonicera maackii*) occurring in deciduous forests in the midwestern United States. Images from Landsat TM and ETM+ in different years were acquired and NDVI was used to determine the greener invasive compared to the senescent vegetation of the canopy. Joshi et al. (2006) also used Landsat ETM+ to locate the invasive bitter bush (*Chromolaena odorata*) in the forest understory of Nepal. The authors derived information on forest canopy density and light intensity to indirectly determine areas of bitter brush seed production, which thrive in areas with high light intensities (and correspondingly reduced canopy density).

High spectral (hyperspectral) resolution sensors, either airborne (e.g., the Airborne Visible/Infrared Imaging Spectrometer, AVIRIS, with 224 bands and 4-m or 20-m spatial resolution) or spaceborne (e.g., Hyperion with 220 bands and 30-m resolution), detect radiation across a continuous spectrum. The wealth of radiometric information offers great potential for species-level mapping through subtle variations in reflection and absorption patterns (Vane and Goetz 1993, Aspinall et al. 2002). Studies have successfully mapped invasive species by exploiting unique biochemical and structural characteristics such as photosynthetic pigment, cellulose and lignin absorption, water, and nitrogen. Underwood et al. (2007) mapped jubata grass (*Cortaderia jubata*) based on distinct absorption features associated with the cellulose and lignin in senesced leaves using imagery acquired by AVIRIS in coastal California. Alternatively, the water content of succulent leaves can result in strong water-absorption features that are also detectable using hyperspectral imagery (around 0.90 μm). Sanderson et al. (1998) successfully identified pickleweed (*Salicornia virginica*) and Underwood et al. (2003) mapped ice plant (*Carpobrotus edulis*) at different densities using this feature (Figure 14.4). In contrast, Asner et al. (2008b) found differences across the full range of the spectrum in the reflectance properties of introduced and invasive species in Hawaii. The authors found highly invasive trees could be distinguished from introduced (nonproliferating) species based on differences in leaf pigment (chlorophyll and carotenoids), nutrients (nitrogen and phosphorus), structural properties, and canopy leaf area index.

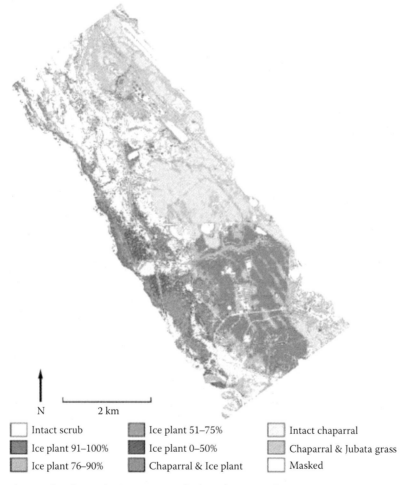

☐ Intact scrub	■ Ice plant 51–75%	☐ Intact chaparral
■ Ice plant 91–100%	■ Ice plant 0–50%	▨ Chaparral & Jubata grass
▨ Ice plant 76–90%	■ Chaparral & Ice plant	☐ Masked

Figure 14.4 **(See color insert.)** A supervised classification of an AVIRIS flight line in coastal California using bands generated from a minimum noise fraction procedure. Invasive ice plant (*Carpobrotus edulis*) was detected based on its water-absorption features in the spectral profile (around 0.9 μm) and different depths of this feature permitted four classes of ice plant density to be mapped with a high level of accuracy.

As with detecting invasive species using lower spectral resolution sensors, phenological differences between target invasive species and surrounding vegetation have again been leveraged. For example, perennial pepperweed (*Lepidium latifolium*) could be separated spectrally from co-occurring marsh vegetation using AVIRIS images in the delta of central California (Andrew and Ustin 2006) (Figure 14.5a). The authors used physiological indexes, or ratios of different wave bands, to encapsulate information-related specific spectral features, which are particularly useful with hyperspectral imagery when adjacent wavelengths are highly correlated (Andrew and Ustin 2006). The spectral uniqueness is primarily conferred by the perennial pepperweed's dense spray of white flowers, resulting in bright, relatively uniform reflectance throughout the visible wavelengths (0.45–0.70 μm, Figure 14.5b).

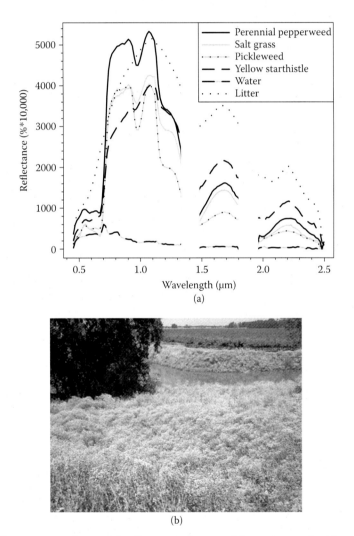

Figure 14.5 (a) Comparison of spectral reflectance of perennial pepperweed with co-occurring species (noisy bands removed). Note the bright reflectance of perennial pepperweed in the visible wavelengths (from 0.45 to 0.7 μm). (Adapted from Andrew, M. E., and S. L. Ustin, *Weed Science* 54, 1051–1062, 2006.) (b) **(See color insert.)** Photograph of perennial pepperweed (*Lepidium latifolium*) highlighting the white flowers. (Photo courtesy of M. Andrew.)

The spaceborne Hyperion sensor (220 bands) has a similarly high spectral resolution to AVIRIS but has been used less often for mapping invasive species probably because of its lower spatial resolution (30-m) (Huang and Asner 2009). However, the spaceborne hyperspectral sensor offers repeatable collections (every 16 days) of data and is relatively inexpensive compared to airborne data (Huang and Asner 2009). Pengra et al. (2007) used Hyperion data to map the extent of common reed (*P. australis*) in coastal wetlands of the Great Lakes in Wisconsin, United States, with an overall accuracy of 81%, although the spatial resolution made detection of small and linear *Phragmites* stands difficult. A comparison between the ability of spaceborne Hyperion to detect invasive plants and of airborne AVIRIS with higher spatial resolution was made by Root et al. (2002) in the Theodore Roosevelt National Park in North Dakota. The authors found they could map infestations of leafy spurge (*E. esula*) of 160 m² using images from AVIRIS (at 17-m) compared to images from Hyperion, which could only detect patches of 500 m² or greater.

Some of the limitations of high spectral and spatial resolution data are the expense, with only a few airborne image providers, and it requires experienced users to process the imagery (Turner et al. 2003). Indeed, with hundreds of potential data points for each pixel and often strong correlations among patterns in nearby bands, it can be challenging to filter and transform the data to a useful number of relatively independent spectral predictors. In addition, the relatively narrow width of airborne imagery (approximately 3 km for 4-m resolution AVIRIS compared to the 185-km swath of Landsat TM) makes its use over large spatial areas challenging, for example, mosaicking imagery can cause radiometric inconsistencies between flight lines due to sun-sensor geometry, which can cause classification issues. Even though the higher spectral resolution offers greater potential over multispectral sensors for species-level mapping, the detection of non-canopy invasives is, in general, still challenging. Lower mapping accuracies for leafy spurge (*E. esula*), for example, were recorded in more wooded, closed canopy sites compared to open prairies (Williams and Hunt 2002). Similarly, the identification of weeping lovegrass (*Eragrostis curvula*), an understory species, was greatest when it occurred in canopy gaps and other openings at road edges (Andrew and Tom, unpublished data).

Finally, active sensors have had some limited applications to invasive species mapping. Rosso et al. (2006) used LiDAR data to detect the invasive *Spartina* (both S. alterniflora and its hybrid with the native *S. foliosa*) in salt marshes of San Francisco Bay, California. The authors followed sediment dynamics and colonization of the mudflats using interannual LiDAR data and were able to separate and identify native and invasive *Spartina* based on structural differences in the canopies that were recorded by the first and the last returns in the LiDAR images. Other researchers have fused LiDAR or Radar data with high spatial resolution data from passive sensors (e.g., hyperspectral or IKONOS), which have the potential for improving the detection of invasive species. A recent study in a semiarid rangeland classified sagebrush using hyperspectral imagery and coregistered LiDAR data found a 14% increase in mapping accuracy, with the advantage of providing more detailed stand-specific descriptions than possible through either dataset used independently (Mundt et al. 2006). Asner et al. (2008a) deployed the Carnegie Airborne Observatory (CAO), which integrates high-fidelity imaging spectrometers (HiFIS) with LiDAR sensors to identify individual crowns of five highly invasive trees in Hawaiian forests. Limitations of the image fusion approach include the additional processing time and expertise required. Also, when images have been acquired from different aircraft, they need to be coaligned for analysis, with possible misalignment errors of up to four pixels (Asner et al. 2007) and the possible need to resample both images to a lower spatial resolution to accommodate geometric error.

14.7   Future directions of remote sensing to invasive species

There are two areas of remote sensing where development could benefit invasive species mapping. First, when acquiring imagery (particularly hyperspectral) in a new study area, spectral reflectance measurements of the target invasive species, surrounding vegetation, and other types of land cover (e.g., roads, buildings, or bare ground) are required for processing. Ideally, spectral reflectance measurements should be acquired at the same time of image acquisition to match the phenological stage of the vegetation. When this is not possible (e.g., processing archival imagery), a spectral library consisting of multiple examples of the reflectance of the target species across a range of conditions would increase data processing efficiency. Some extensive spectral libraries have been developed for soils (Baumgardner et al. 1985) or geology (Jet Propulsion Lab 2002, Clark et al. 2003), but until recently, there have been few examples of spectral libraries of plant species. The concept was first presented for the Santa Monica Mountains in southern California (Roberts et al. 1998), but a recent study of salt marsh species from various coastal wetlands in the Gulf of Mexico and California showed species consistently expressed similar spectral patterns (Zomer et al. 2009); consequently, the library can now be used for wetland studies across a broad geographic range. The development of libraries specifically for a suite of invasive species across different habitats and seasons is important for improving our capacity to utilize the full mapping potential from hyperspectral images.

Second, one of the main concerns voiced by the user community is the infrequency of satellite data overpasses, which, when coupled with frequent cloud cover in tropical and boreal regions, limits the availability of these data. One prominent concept to increase the temporal frequency of data availability is to launch several identical satellites in different orbital configurations. The simplest concept is to have several satellites follow in the same orbit ("train" concepts) that collect data at various time intervals. A related concept is to launch a cluster of satellites ("constellation" concepts) for the same purpose. These are potentially expensive options and to date, with the exception of the two MODIS sensors on the Terra and Aqua platforms, have not been implemented.

14.8   Summary

Our goal in writing this review is to suggest that geospatial analysis methods applied to locating infestations of invasive species have now reached the point of sophistication and reliability that it is plausible to deploy these technologies routinely for managing invasive plant species over large areas. For most land managers and conservation practitioners, the most effective strategies will necessarily involve a mix of methods. Traditional ground surveys, mapping confirmed point occurrences (and absences) of invaders, which are preferably precisely geo-located, are still the most fundamental data for species management, and the <90% accuracies of invasive plant identifications from the best of imagery studies suggest that fieldworkers will not be replaced by technology anytime soon. However, fieldwork can be made much more efficient by directing survey or weed control teams to high priority locations, which could be identified by distribution models as places very likely to harbor new invasions (i.e., very high suitability as predicted by the local mix of environmental properties, coupled with proximity to known source populations, or vectors and pathways). Alternatively, high-risk sites could be identified from a spectral signature, the timing of seasonal change (green-up brown-down, and flower/fruit colors), or by automated change detection of rapid shifts in spectral properties. Combinations may be even more effective. For example, remotely sensed metrics from higher-resolution imagery can be used as proxies for direct physiological measures or plants (wetness, chlorophyll levels, soil properties, and productivity), are available at much higher resolution than the digital maps (e.g., temperature, soil, and rainfall) used for most published niche models, and, as a result, appear to be more effective predictors of invasive species occurrences.

Field surveys can also be made more effective by using technology to enable volunteers and citizen scientists to do fieldwork previously done by professional biologists. Citizen science methods under development include directing volunteers to sites where confirming the presence or absence of a species would greatly decrease uncertainties in the modeled distributions of those species elsewhere (critical tests) or just locations where a confirmation could trigger an ED/RR control effort. They include field keys to make identifications more reliable, methods to log photographic confirmations, and could include triggering mechanisms to identify particularly interesting observations and immediately involve experts in confirming and responding to them. The emergence of GPS-enabled handheld computers and smartphones seems to have prompted a burst of innovation in citizen-science applications, but making the most use of them will involve embedding the resulting data in full GIS and statistically assessing them for geographic context and regional predictive power using the kinds of methods described in this review.

Many challenges remain. While niche models can, in principle, predict locations where an aggressive invader might be newly established and rare, the imagery-based methods depend on unique phenological or biochemical properties or unique spatial patterning. Given this, remotely sensed detection of invasive species is best suited to established, widespread populations that dominate the vegetation canopy, rather than the early detection of new populations. Costs are still a serious issue, although some will certainly drop dramatically over time (Turner et al. 2003). While 30-m pixel and coarser satellite data (e.g., Landsat and AVHRR) are now readily available for free or at low cost over long time periods, higher resolution satellite data are mostly proprietary and priced outside the typical budgets of land managers, as are airborne hyperspectral and LiDAR instruments. However, higher spatial resolution data may not always be needed to accurately map invasive plants or invaded communities. One standardized study of six community types in central California (including three invaded ones) recorded highest mapping accuracies using full resolution AVIRIS hyperspectral data, followed by an image with hyperspectral resolution but downgraded spatial resolution (from 4-m to 30-m), suggesting that spectral resolution can be more important than spatial resolution (Underwood et al. 2007). In addition, all remote sensing analyses require capable computers, specialized software, and large libraries of ancillary data for processing. In practice, future invasive species teams will ideally need one or more GIS and remote sensing specialists in addition to field biologists expert with the species and ecosystems (Turner et al. 2003). On the contrary, invasive species cause tens of billions of dollars per year in economic damage in the United States alone, are one of the foremost threats to native ecosystems, and will only get worse with increasing globalization and climate shifts. In comparison, even large investments in technology and geospatial science seem well justified.

Acknowledgments

Special thanks to Susan Ustin, Margaret Andrew, Tom Stohlgren, Peter Ma, Ned Gardiner, and Woody Turner. Partially funded by the National Science Foundation and the United States Geological Survey.

References

Anderson, G. L., J. H. Everitt, A. J. Richardson, and D. E. Escobar. 1993. Using satellite data to map false broomweed (*Ericameria austrotexana*) infestations on south Texas rangelands. *Weed Technology* 7:865–871.

Andrew, M. E., and S. L. Ustin. 2006. Spectral and physiological uniqueness of perennial pepperweed (*Lepidium latifolium*). *Weed Science* 54:1051–1062.

Andrew, M. E., and S. L. Ustin. 2009. Habitat suitability modelling of an invasive plant with advanced remote sensing data. *Diversity and Distributions* 15:627–640.

Asner, G., D. Knapp, M. Jones, T. Kennedy-Bowdoin, R. Martin, J. Boardman, and C. Field. 2007. Carnegie Airborne Observatory: In-flight fusion of hyperspectral imaging and waveform light detection and ranging (wLiDAR) for three-dimensional studies of ecosystems. *Journal of Applied Remote Sensing* 1:013536.

Asner, G. P., R. F. Hughes, P. M. Vitousek, D. E. Knapp, T. Kennedy-Bowdoin, J. Boardman, R. E. Martin, M. Eastwood, and R. O. Green. 2008a. Invasive plants transform the three-dimensional structure of rain forests. *Proceedings of the National Academy of Sciences of the United States of America* 105:4519–4523.

Asner, G. P., M. Jones, R. Martin, D. Knapp, and R. Hughes. 2008b. Remote sensing of native and invasive species in Hawaiian forests. *Remote Sensing of Environment* 112:1912–1926.

Aspinall, R. J., W. A. Marcus, and J. W. Boardman. 2002. Considerations in collecting, processing, and analyzing high spatial resolution hyperspectral data for environmental investigations. *Journal of Geographical Systems* 4:15–29.

Bakkenes, M., J. R. M. Alkemade, F. Ihle, R. Leemans, and J. B. Latour. 2002. Assessing effects of forecasted climate change on the diversity and distribution of European higher plants for 2050. *Global Change Biology* 8:390–407.

Barnett, D. T., T. J. Stohlgren, C. S. Jarnevich, G. W. Chong, J. A. Erickson, T. R. Davern, and S. E. Simonson. 2007. The art and science of weed mapping. *Environmental Monitoring and Assessment* 132:235–252.

Baumgardner, M. F., L. F. Silva, L. L. Biehl, and E. R. Stoner. 1985. Reflectance properties of soils. *Advances in Agronomy* 38:1–44.

Beaumont, L. J., R. V. Gallagher, W. Thuiller, P. O. Downey, M. R. Leishman, and L. Hughes. 2009. Different climatic envelopes among invasive populations may lead to underestimations of current and future biological invasions. *Diversity and Distributions* 15:409–420.

Berk, R. A. 2008. *Statistical Learning from a Regression Perspective*. New York: Springer.

Bradley, B. A., and J. F. Mustard. 2005. Identifying land cover variability distinct from land cover change: Cheatgrass in the Great Basin. *Remote Sensing of Environment* 94:204–213.

Breiman, L. 2001. Random forests. *Machine Learning* 45:5–32.

Breiman, L., J. Friedman, C. J. Stone, and R. Olshen. 1984. *Classification and Regression Trees*. 1st edition. Boca Raton, Florida: Chapman & Hall/CRC.

Busby, J. R. 1986. A biogeoclimatic analysis of *Nothofagus cunninghamii* (Hook.) Oerst. in southeastern Australia. *Austral Ecology* 11:1–7.

Busby, J. R. 1991. BIOCLIM – A bioclimate analysis and prediction system. In *Nature Conservation: Cost Effective Biological Surveys and Data Analysis*, eds. C. R. Margules and M. P. Austin, pp. 64–68. Melbourne: CSIRO.

Clark, R. N., G. A. Swayze, R. Wise, K. E. Livo, T. M. Hoefen, A. R. F. Kokaly, and S. J. Sutley. 2003. *USGS Digital Spectral Library (splib05a)*. Open File Report 03-395. U.S. Geological Survey.

Cohen, W. B., and S. N. Goward. 2004. Landsat's role in ecological applications of remote sensing. *BioScience* 54:535–545.

Crossman, N. D., and J. Kochergen. 2002. Mapping environmental weeds in the Mount Lofty Ranges, South Australia, using high resolution infrared aerial photography. In *Proceedings of the Thirteenth Australian Weeds Conference*, eds. S. H. Jacob, J. Dodd, and J. H. Moore, pp. 1–4. Perth, Australia: Plant Protection Society of Western Australia.

Cutler, D. R., T. C. Edwards, K. H. Beard, A. Cutler, K. T. Hess, J. Gibson, and J. J. Lawler. 2007. Random forests for classification in ecology. *Ecology* 88:2783–2792.

DiPietro, D., M. Kelly, S. Schoenig, D. Johnson, and R. Yacoub. 2002. *The California Weed Mapping Handbook*. California: California Department of Food & Agriculture. (http://cain.ice.ucdavis.edu/weedhandbook).

Dullinger, S., I. Kleinbauer, J. Peterseil, M. Smolik, and F. Essl. 2009. Niche based distribution modelling of an invasive alien plant: Effects of population status, propagule pressure and invasion history. *Biological Invasions* 11:2401–2414.

Ehrlich, P. R. 1989. Attributes of invaders and invading processes: Vertebrates. In *Biological Invasions: A Global Perspective*, eds. J. Drake, F. di Castri, R. Groves, F. Kruger, H. A. Mooney, M. Rejmánek, and M. Williamson, pp. 315–328. New York: John Wiley & Sons.

Elith, J., C. H. Graham, R. P. Anderson, M. Dudik, S. Ferrier, A. Guisan, R. J. Hijmans, F. Huettmann, J. R. Leathwick, A. Lehmann, J. Li, L. G. Lohmann, B. A. Loiselle, G. Manion, C. Moritz, M. Nakamura, Y. Nakazawa, J. M. Overton, A. T. Peterson, S. J. Phillips, K. Richardson, R. Scachetti-Pereira, R. E. Schapire, J. Soberon, S. Williams, M. S. Wisz, and N. E. Zimmermann. 2006. Novel methods improve prediction of species' distributions from occurrence data. *Ecography* 29:129–151.

Evangelista, P., S. Kumar, T. Stohlgren, C. Jarnevich, A. Crall, J. Norman, and D. Barnett. 2008. Modelling invasion for a habitat generalist and a specialist plant species. *Diversity and Distributions* 14:808–817.

Evangelista, P., T. J. Stohlgren, T. Morisette, and S. Kumar. 2009. Mapping invasive Tamarisk (*Tamarix*): A comparison of single-scene and time-series analyses of remotely sensed data. *Remote Sensing* 1:519–533.

Everitt, J., D. Escobar, M. Alaniz, M. Davis, and J. Richerson. 1996. Using spatial information technologies to map Chinese tamarisk (*Tamarix chinensis*) infestations. *Weed Science* 44:194–201.

Everitt, J. H., G. L. Anderson, D. E. Escobar, M. R. Davis, N. R. Spencer, and R. J. Andrascik. 1995. Use of remote sensing for detecting and mapping leafy spurge (*Euphorbia esula*). *Weed Technology* 9:599–609.

Everitt, J. H., C. Yang, B. J. Racher, C. M. Britton, and M. R. Davis. 2001. Remote sensing of redberry juniper in the Texas rolling plains. *Journal of Range Management* 54:254–259.

FICMNEW. 2003. *A National Early Detection and Rapid Response System for Invasive Plants in the United States*. Washington, D.C.: Federal Interagency Committee for the Management of Noxious and Exotic Weeds.

Fuller, D. 2005. Remote detection of invasive Melaleuca trees (*Melaleuca quinquenervia*) in South Florida with multispectral IKONOS imagery. *International Journal of Remote Sensing* 26:1057–1063.

Graham, J., G. Newman, C. Jarnevich, R. Shorya, and T. J. Stohlgren. 2007. A global organism detection and monitoring system for non-native species. *Ecological Informatics* 2:177–183.

Guisan, A., and N. E. Zimmermann. 2000. Predictive habitat distribution models in ecology. *Ecological Modelling* 135:147–186.

Hannah, L. 2008. Protected areas and climate change. *Annals of the New York Academy of Sciences* 1134:201–212.

Hastie, T., R. Tibshirani, and J. Friedman. 2009. *The Elements of Statistical Learning: Data Mining, Inference, and Prediction*. 2nd edition. Berlin: Springer.

Hastings, A., K. Cuddington, K. F. Davies, C. J. Dugaw, S. Elmendorf, A. Freestone, S. Harrison, M. Holland, J. Lambrinos, U. Malvadkar, B. A. Melbourne, K. Moore, C. Taylor, and D. Thomson. 2005. The spatial spread of invasions: New developments in theory and evidence. *Ecology Letters* 8:91–101.

Hengeveld, R. 1994. Small-step invasion research. *Trends in Ecology & Evolution* 9:339–342.

Hestir, E., S. Khanna, M. E. Andrew, M. Santos, J. Viers, J. Greenberg, S. Rajapakse, and S. Ustin. 2008. Identification of invasive vegetation using hyperspectral remote sensing in the California Delta ecosystem. *Remote Sensing of Environment* 112:4034–4047.

Higgins, S. I., and D. M. Richardson. 1996. A review of models of alien plant spread. *Ecological Modelling* 87:249–265.

Hirzel, A. H., J. Hausser, D. Chessel, and N. Perrin. 2002. Ecological-niche factor analysis: How to compute habitat-suitability maps without absence data? *Ecology* 83:2027–2036.

Huang, C., E. L. Geiger, W. J. D. Van Leeuwen, and S. E. Marsh. 2009. Discrimination of invaded and native species sites in a semi-desert grassland using MODIS multi-temporal data. *International Journal of Remote Sensing* 30(4):897–917.

Huang, C. Y., and G. P. Asner. 2009. Applications of remote sensing to alien invasive plant studies. *Sensors* 9:4869–4889.

Hutchinson, G. E. 1957. Concluding remarks. *Cold Spring Harbor Symposia on Quantitative Biology* 22:415–427.

Jet Propulsion Lab. 2002. *Spectral Library*. http://speclib.jpl.nasa.gov/.

Joshi, C., J. de Leeuw, J. van Andel, A. Skidmore, H. Lekhak, I. van Duren, and N. Norbu. 2006. Indirect remote sensing of a cryptic forest understory invasive species. *Forest Ecology and Management* 225:245–256.

Keller, R. P., D. M. Lodge, and D. C. Finnoff. 2007. Risk assessment for invasive species produces net bioeconomic benefits. *Proceedings of the National Academy of Sciences* 104(1):203–207.

Kozak, K. H., C. H. Graham, and J. J. Wiens. 2008. Integrating GIS-based environmental data into evolutionary biology. *Trends in Ecology & Evolution* 23:141–148.

Kumar, S., S. A. Spaulding, T. J. Stohlgren, K. A. Hermann, T. S. Schmidt, and L. Bahls. 2009. Potential habitat distribution for the freshwater diatom *Didymosphenia geminate* in the continental US. *Frontiers in Ecology and the Environment* 7:415–420.

Laba, M., F. Tsai, D. Ogurcak, S. Smith, and M. E. Richmond. 2005. Field determination of optimal dates for the discrimination of invasive wetland plant species using derivative spectral analysis. *Photogrammetric Engineering & Remote Sensing* 71:603–611.

Lavergne, S., and J. Molofsky. 2007. Increased genetic variation and evolutionary potential drive the success of an invasive grass. *Proceedings of the National Academy of Sciences* 104:3883–3888.

Leathwick, J., D. Rowe, J. Richardson, J. Elith, and T. Hastie. 2005. Using multivariate adaptive regression splines to predict the distributions of New Zealand's freshwater diadromous fish. *Freshwater Biology* 50:2034–2052.

Lodge, D. M., S. Williams, H. J. Macisaac, K. R. Hayes, B. Leung, S. Reichard, R. N. Mack, P. B. Moyle, M. Smith, D. A. Andow, J. T. Carlton, and A. Mcmichael. 2006. ESA Report: Biological invasions: Recommendations for U.S. policy and management. *Ecological Applications* 16:2035–2054.

Mack, R. N., S. C. H. Barrett, P. L. Defur, W. L. MacDonals, L. V. Madden, D. S. Marshall, D. G. McCullough, P. B. McEvoy, J. P. Nyrop. S. E. Hayden Reichard, K. J. Rice, and S. A. Tolin. 2002. *Predicting Invasions of Nonindigenous Plants and Plant Pests*. 194pp. Washington, D.C.: National Academy Press.

Mack, R. N., D. Simberloff, W. M. Lonsdale, H. Evans, M. Clout, and F. A. Bazzaz. 2000. Biotic invasions: Causes, epidemiology, global consequences, and control. *Ecological Applications* 10:689–710.

Madsen, J. D., and G. N. Ervin. 2007. Integrating effects of land use change on the invasive plant species distribution into an invasive plant atlas for the Mid-South (IPAMS). *Proceedings of the Southern Weed Science Society* 60:197.

Marmion, M., M. Parviainen, M. Luoto, R. K. Heikkinen, and W. Thuiller. 2009. Evaluation of consensus methods in predictive species distribution modelling. *Diversity and Distributions* 15:59–69.

Mau-Crimmins, T. M., H. R. Schussman, and E. L. Geiger. 2006. Can the invaded range of a species be predicted sufficiently using only native-range data?: Lehmann lovegrass (*Eragrostis lehmanniana*) in the southwestern United States. *Ecological Modelling* 193:736–746.

McNaught, I., R. Thackway, L. Brown, and M. Parsons. 2006. *A Field Manual for Surveying and Mapping Nationally Significant Weeds*. Canberra, Australia: Bureau of Rural Sciences (http://www .weedcenter.org/management/docs/Aust_mapping.pdf).

Meynard, C. N., and J. F. Quinn. 2007. Predicting species distributions: A critical comparison of the most common statistical models using artificial species. *Journal of Biogeography* 34:1455–1469.

Morse, L. E., J. M. Randall, N. Benton, R. Hiebert, and S. Lu. 2004. *An Invasive Species Assessment Protocol: Evaluating Non-Native Plants for Their Impact on Biodiversity*. Version 1. Arlington, Virginia: NatureServe (http://www.natureserve.org/library/invasiveSpeciesAssessmentProtocol.pdf).

Mundt, J. T., D. R. Streutker, and N. F. Glenn. 2006. Mapping sagebrush distribution using fusion of hyperspectral and lidar classifications. *Photogrammetric Engineering & Remote Sensing* 72:47–54.

NAWMA. 2002. *North American invasive plant mapping standards*. North American Weed Management Association (http://science.nature.nps.gov/im/monitor/Protocols/inv_NAWMAMapStds.pdf).

Nix, H. A. 1986. A biogeographic analysis of Australian elapid snakes. In *Snakes of Australia*, ed. R. Longmore, pp. 4–15. Canberra, Australia: Australian Government Publishing Service.

Pengra, B. W., C. A. Johnston, and T. R. Loveland. 2007. Mapping an invasive plant, *Phragmites australis*, in coastal wetlands using the EO-1 Hyperion hyperspectral sensor. *Remote Sensing of Environment* 108:74–81.

Peterson, A. T. 2003. Predicting the geography of species' invasions via ecological niche modeling. *Quarterly Review of Biology* 78:419–433.

Peterson, A. T., M. Papes, and D. A. Kluza. 2003. Predicting the potential invasive distributions of four alien plant species in North America. *Weed Science* 51:863–868.

Peterson, E. B. 2005. Estimating cover of an invasive grass (*Bromus tectorum*) using tobit regression and phenology derived from two dates of Landsat ETM plus data. *International Journal of Remote Sensing* 26:2491–2507.

Phillips, S. J., R. P. Anderson, and R. E. Schapire. 2006. Maximum entropy modeling of species geographic distributions. *Ecological Modelling* 190:231–259.

Pimentel, D., R. Zuniga, and D. Morrison. 2005. Update on the environmental and economic costs associated with alien-invasive species in the United States. *Ecological Economics* 52:273–288.

Pyšek, P., and P. E. Hulme. 2005. Spatio-temporal dynamics of plant invasions: Linking pattern to process. *Ecoscience* 12:302–315.

Roberts, D., M. Gardner, R. Church, S. Ustin, G. Scheer, and R. Green. 1998. Mapping chaparral in the Santa Monica Mountains using multiple endmember spectral mixture models. *Remote Sensing of Environment* 65:267–279.

Robertson, M. P., C. I. Peter, M. H. Villet, and B. S. Ripley. 2003. Comparing models for predicting species' potential distributions: A case study using correlative and mechanistic predictive modelling techniques. *Ecological Modelling* 164:153–167.

Root, R., S. Ustin, P. Zarco-Tejada, C. Pinilla, R. Kokaly, G. Anderson, K. Brown, K. Dudeck, S. Hager, and E. Holroyd. 2002. Comparison of AVIRIS and EO-1 Hyperion for classification and mapping of invasive leafy spurge in Theodore Roosevelt National Park. In *Proceedings of the 11th Earth Science Airborne Workshop*, ed. R. O. Green. Jet Propulsion Laboratory Publication 03-4. Pasadena, California: Jet Propulsion Laboratory, and National Aeronautics and Space Administration.

Rosso, P., S. Ustin, and A. Hastings. 2006. Use of lidar to study changes associated with *Spartina* invasion in San Francisco Bay marshes. *Remote Sensing of Environment* 100:295–306.

Sanderson, E. W., M. Zhang, S. L. Ustin, and E. Rejmankova. 1998. Geostatistical scaling of canopy water content in a California salt marsh. *Landscape Ecology* 13:79–92.

Segurado, P., and M. B. Araujo. 2004. An evaluation of methods for modelling species distributions. *Journal of Biogeography* 31:1555–1568.

Sellers, E., A. Simpson, and S. Curd-Hetrick. 2004. List of Invasive Alien Species (IAS) Online Information Systems. *A preliminary draft document, prepared for the Experts Meeting Towards the Implementation of a Global Invasive Species Information Network (GISIN)*, April 6–8, 2004. Baltimore, Maryland (http://www.niiss.org/cwis438/websites/GISINDirectory/DatabaseDirectory_Table.php).

Simpson, A., C. Jarnevich, J. Madsen, R. Westbrooks, C. Fournier, L. Mehrhoff, M. Browne, J. Graham, and E. Sellers. 2009. Invasive species information networks: Collaboration at multiple scales or prevention, early detection, and rapid response to invasive species. *Biodiversity* 10(2–3):5–13.

Singh, N., and N. F. Glenn. 2009. Multitemporal spectral analysis for cheatgrass (*Bromus tectorum*) classification. *International Journal of Remote Sensing* 30:3441–3462.

Smolik, M., S. Dullinger, F. Essl, I. Kleinbauer, M. Leitner, J. Peterseil, L. Stadler, and G. Vogl. 2010. Integrating species distribution models and interacting particle systems to predict the spread of an invasive alien plant. *Journal of Biogeography* 37:411–422.

Steininger, M., and N. Horning. 2007. *The Basics of Remote Sensing*. Sourcebook on Remote Sensing and Biodiversity Indicators Convention on Biological Diversity Technical Series no. 32. Montreal, Canada: Secretariat of the Convention on Biological Diversity.

Steves, B., S. Dalton, P. Fuller, and G. Ruiz. 2004. NISBase: A distributed database system for nonindigenous species information. In *Proceedings of Experts Meeting Towards the Implementation of a Global Invasive Species Information Network (GISIN)*, April 6–8, 2004. Baltimore, Maryland.

Stockwell, D. 1999. The GARP modelling system: Problems and solutions to automated spatial prediction. *International Journal of Geographical Information Science* 13:143–158.

Stohlgren, T. J., D. Barnett, C. Flather, J. Kartesz, and B. Peterjohn. 2005. Plant species invasions along the latitudinal gradient in the United States. *Ecology* 86:2298–2309.

Stohlgren, T. J., P. Ma, S. Kumar, M. Rocca, J. T. Morisette, C. S. Jarnevich, and N. Benson. 2010. Ensemble habitat mapping of invasive plant species. *Risk Analysis* 30:224–235.

Stohlgren, T. J., and J. L. Schnase. 2006. Risk analysis for biological hazards: What we need to know about invasive species. *Risk Analysis* 26:163–173.

Sutherst, R., and A. Bourne. 2009. Modelling non-equilibrium distributions of invasive species: A tale of two modelling paradigms. *Biological Invasions* 11:1231–1237.

Turner, W., S. Spector, N. Gardiner, M. Fladeland, E. Sterling, and M. Steininger. 2003. Remote sensing for biodiversity science and conservation. *Trends in Ecology & Evolution* 18:306–314.

Underwood, E., S. Ustin, and D. DiPietro. 2003. Mapping nonnative plants using hyperspectral imagery. *Remote Sensing of Environment* 86:150–161.

Underwood, E. C., S. L. Ustin, and C. M. Ramirez. 2007. A comparison of spatial and spectral image resolution for mapping invasive plants in coastal California. *Journal of Environmental Management* 39:63–83.

Vane, G., and A. F. H. Goetz. 1993. Terrestrial imaging spectrometry: Current status future trends. *Remote Sensing of Environment* 44:117–126.

Venables, W., and B. Ripley. 2003. *Modern Applied Statistics with S.* 4th edition. Berlin: Springer.

Wilfong, B. N., D. L. Gorchov, and M. C. Henry. 2009. Detecting an invasive shrub in deciduous forest understories using remote sensing. *Weed Science* 57:512–520.

Williams, A. E. P., and E. R. Hunt. 2002. Estimation of leafy spurge cover from hyperspectral imagery using mixture tuned matched filtering. *Remote Sensing of Environment* 82:446–456.

Williams, J. N., C. Seo, J. Thorne, J. K. Nelson, S. Erwin, J. M. O'Brien, and M. W. Schwartz. 2009. Using species distribution models to predict new occurrences for rare plants. *Diversity and Distributions* 15:565–576.

Zomer, R. J., A. Trabucco, and S. L. Ustin. 2009. Building spectral libraries for wetlands land cover classification and hyperspectral remote sensing. *Journal of Environmental Management* 90:2170–2177.

chapter fifteen

Improving restoration to control plant invasions under climate change

Qinfeng Guo and Steve Norman

Contents

15.1 Introduction

Native forests and grasslands worldwide have been converted to developed lands or invaded by exotic species due to human activities. These pressures are predicted to increase with population growth and climatic stress in coming decades, escalating concerns for the viability of native species and communities that are affected. Ecological restoration is frequently offered as a partial solution to these changes because less stressed ecosystems may be more tolerant of novel changes in the environment (Temperton et al. 2004, Clewell and Aronson 2008). In this sense, restoration could provide a strategy for enhancing ecological resilience, given escalating problems associated with invasives and a changing climate (see Hobbs and Norton 1996).

Traditional restoration efforts have been concerned with matching restored systems to historical or nearby natural habitats, but climate change and biotic invasions can alter the viability of historically based objectives (Figure 15.1). In many places, historically based restoration has become impossible, particularly where development dominates, restoration conflicts with other objectives, or persistent invasive diseases and pathogens have removed dominant or keystone species (Clewell and Aronson 2008). Elsewhere, where restoration has been attempted, restored conditions can only be maintained over the long-term through vigilant monitoring and costly maintenance (Aronson and van Andel 2006). Most restoration efforts are rarely absolute as aggressive invasive species simply cannot be eradicated given current technology. Our management of these latter sites would benefit

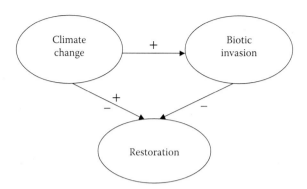

Figure 15.1 A joint framework showing the interactive effects of climate change and biotic invasions on restoration discussed in this chapter. Although climate change may have either positive or negative effects on restoration depending on the species used and climate change as a growing disturbance agent can promote species invasions, both climate change and biotic invasions would make restoration efforts more difficult.

if ecosystems were more self-restoring or self-sustaining—if they were more capable of resisting new invasions or climate-related stress. In recent years, both theoretical and field-based ecological research have provided insights into the means by which stability may be enhanced amid climate change and biotic invasion.

In systems that have been invaded, exotic–native interactions can be complex, and complete removal of exotics may be impossible and, in some rare cases, undesirable, particularly when exotics check the spread of invasives or favorably alter habitat, competition, or food web relationships for natives (D'Antonio and Mack 2001, Ewel and Putz 2004, Vander Zanden et al. 2006). These fundamental biological processes of inhibition and facilitation, broadly conceived to incorporate both natives and exotics, can be manipulated to help meet a broad range of management objectives, some historical and some practical.

Strategies to address climate change and biotic invasions are normally devised and discussed separately rather than in an integrated way (Berger 1993, Perry and Galatowitsch 2003, Price and Weltzin 2003, Bakker and Wilson 2004, Harris et al. 2006, Guo 2007, Holsman et al. 2010). Both factors affect ecological outcomes and can determine restoration success. Therefore, managers and researchers would benefit when these drivers of potential ecological change are considered jointly as integrated strategies in both research and practice. The purpose of this chapter is to review how these joint strategies can improve restoration and promote ecological resilience and stability in a changing environment.

15.2   Setting restoration goalposts

When restoration appears as a management goal, it implies that some measurable change to an ecosystem has occurred and that all or key aspects of the system can be returned toward an earlier condition or dynamic. Today, restoration constitutes an explicit goal of land management to sustain desired attributes or to redirect successional change to a more natural range of variability. For example, the removal of exotics, particularly invasives that aggressively compete with natives, has led to the restoration of species composition to historical assemblages. Similarly, the return of historical fire regimes after decades of exclusion has led to both compositional and structural restoration in forests that developed with fire (Keeley 2006). In both situations, the long-term viability of ecosystems is thought to be enhanced by arresting and reverting the evolutionary or ecological novelty imposed by exotics or the anomalous competitive or demographic dynamics from altered disturbance regimes.

In practice, few restoration efforts can be expected to wholly arrest and recover from novel biotic change (see Hobbs 2007). In exotic eradication efforts, it is often only those species that are most invasive and able to be removed that serve as indicators of restoration success. From the reverse perspective, it is usually the density of only the most valuable native species (i.e., threatened or endangered, prominent, commercially important) that measures success. There is no guarantee that such indictors are reliable measures of the recovery or resilience of the ecosystem over the long-term, and so restoration success is often ambiguous.

The success of ecological restoration is also difficult to measure because the attributes of ecosystems are mediated by broader forces (i.e., climate variability and disturbance regimes). For example, restoration may be successful only when climate and disturbance regimes work in its favor, and decades of efforts may be lost during a drought or other disturbance events (Figure 15.1). With such a complex broad-scale and long-term driver at work, restoration success must often be measured by the ephemeral condition of the environment, such as native species abundance.

Given the trend toward more and more invasives and an ecosystem in flux, restoration in practice normally perceives goals as inflexible. Success is measured incrementally as progress toward a predefined goal. A range of strategies are acceptable, and fixed goals clearly demarcate a range of possible acceptable outcomes. Having this combination of clear goals and a range of acceptable future conditions that are informed by historical insight adds flexibility for managers much more than do impractical notions of restoration as mimicry.

The two most tangible aspects of ecosystem restoration—composition and structure—may be less important than a more abstract use of the word. The restoration of ecosystem resilience prioritizes those processes, structures, and compositions that enhance system inertia and elasticity during periods of stress or recovery. For example, returning stem density to historical conditions could reduce moisture stress and increase the vigor of dominants. In other situations, resilience may be engineered above historical levels by modifying historical structure or composition (Ren et al. 2012). Such efforts may be warranted to successfully resist the undesirable effects of invasives and changing climate conditions.

15.3   Cumulative effects of novel changes

As mentioned earlier, traditional approaches to restoration rely on species composition and structure from some reference period of the past as a model for management. Yet, under certain climate change and invasive scenarios, these historical species and ecosystems may no longer be viable on site. Under less extreme scenarios, the cumulative effects of climate and biotic stressors are likely to make restoration objectives harder to achieve (Figure 15.2).

Acting individually or together, climate change and biotic novelty due to species invasion could eliminate or reduce habitat for some species while creating more favorable habitat for others. With a warming climate, the most similar habitat to that of the recent past and present is generally at higher latitudes and altitudes. Migration may be difficult for native species. Many invasives are highly aggressive colonizers and so they could slow down or even thwart successful migration of natives to these sites without intensive management. Within landscapes that have been highly fragmented by development, the

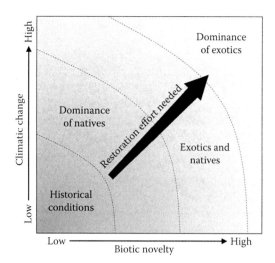

Figure 15.2 The degree of effort needed to restore an ecosystem to historical conditions generally increases with climate change and biotic novelty. Restoration may not be feasible under extreme climate scenarios, but restoration becomes easier where or when climate conditions resemble that of the past.

most mobile species may have a selective advantage. Exotics, particularly invasives, have already been pre-selected for their broad environmental tolerance and dispersal abilities. This may put natives at an even greater disadvantage (Kohli et al. 2008, Ricklefs et al. 2008, Qian and Guo 2010). Given these hurdles that may be faced by native species in the presence of climatic range shifts and aggressive exotic competitors, enhancing the resilience of species and their habitats at or near their current locations may be a worthwhile investment. Stress from moderate climate extremes or disturbances is tolerable for species that have adapted to those conditions. However, the cumulative effects of this stress in the presence of invasives can result in a loss of resilience. Restoration efforts that reduce stress from invasives could make the system more tolerant to climate change, and vice versa.

15.4   *Engineering resistance*

Biological invasion can alter the structure and dynamics of entire communities and ecosystems while threatening natives more directly because of their typically high productivity and competitive strength (Berger 1993). In a successional sense, invasives provide classic examples of inhibition. At a community level, existing exotic species can also facilitate or inhibit the arrival of later invaders or they can be absorbed into the species mix with no discernable effects. The same can be said for natives. Understanding these interspecies relationships is critical for identifying if a need for biotic restoration exists or if an increase in species richness might actually enhance resistance to serious invaders. Certain exotic species may be of minor ecological consequence and a few may have management value (Figure 15.3).

One common restoration strategy is to increase resistance to biological invasions through niche occupation and a high rate of biomass accumulation of native species. Habitat resistance to invasives is likely to be affected by the combination of both richness and biomass of existing (native and exotic) species, among other factors (Guo and Symstad 2008). Diverse plantings could help identify the most suitable species that can compete with existing invasive species in local or surrounding habitats. This can help guide future

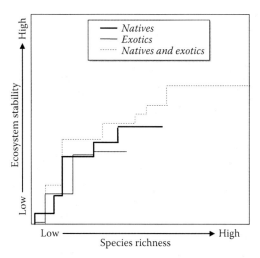

Figure 15.3 Native and exotic species may not vary substantially in their ability to alter the functional stability of ecosystems unless the latter group constitutes highly invasives. In a system where species are not saturated (i.e., niches still available for invading exotics and both natives and exotics can coexist), ecosystem functions may be enhanced due to increased species richness. Identification of natives and exotics as "drivers" of stability or "passengers" that serve complimentary or redundant roles provides powerful tools for restoration. The curve levels off because additional species become complimentary or redundant, whether exotic or native, depending on their individual attributes. The scales for both axes and for natives and exotics may or may not be proportional to each other. (Modified from Peterson, G., C. Allen, and C.S. Holling, *Ecosystems*, 1, 6–18, 1998.)

large-scale restoration efforts, where the habitats may or may not have been invaded. If designed and practiced correctly, the manipulation of species richness and cover could help buffer sites from the effects of highly undesirable invasives.

Restorers need to make sure that native or exotic species used do not facilitate the arrival of new invasives by altering the habitat or through mutualisms. An often neglected yet critical issue in restoration ecology is to make sure the species being utilized for restoration is native but not invasive (i.e., the so-called "home-grown" invasives) (Cox 1999). Further, landscape-level restoration and management are needed to maximize the neighborhood propagule density of natives while minimizing those of invasives (see Larson 2009).

In highly disturbed habitats, we may no longer rely on seed banks and natural seed dispersal processes for revegetation by native species; seed planting or assisted migration of needed species become necessary (Wang et al. 2009). Because of the threats from increasing impacts of climate change and biotic invasions, restoration techniques are often aimed to maximize vegetation growth rate and to inhibit exotic invasion. In a given area, the number and type of species to be planted in order to maximize restoration rates are critical questions that need to be resolved. The importance of germination and survival of the species planted will be used to measure the success of restoration.

However, even though such questions are being addressed, many practical challenges lie ahead. For example, seeds of many native species are very costly, and therefore, many native plantings are seeded with relatively few species. Although certain practices save money in the short-term, it may lead to inferior results or restoration failure. This could lead to more costly management in the long-term. For this and many other reasons, the number of successful efforts appears to be few, and many restorations are not sustainable.

15.5   Lessons from biodiversity experiments

In most experimental studies, the purpose is to experimentally examine the response of restoration rates and vegetation structure and dynamics to different numbers of planted species. Experimental research to date shows that (1) high-diversity planting did increase the habitat productivity during ecosystem early development (cf. Roy 2001), (2) high-diversity planting could reduce the habitat invasibility by introduced species (e.g., Tracy et al. 2004), and (3) however, there could be "optimal" number of species to be planted in a given area as very high-diversity planting can be wasteful and may cause adverse effects on restoration rates. Results of the experiment will advance our understanding of plant species interactions and their potential role in habitat restoration. The study will also furnish useful information to managers involved in seeding native species.

Diversity, biomass, productivity, and stability are key variables in community and ecosystem ecology, but only diversity and biomass could be directly manipulated in restoration and management and other variables change accordingly (Guo 2007, Larson 2009). High diversity has many other benefits for overall ecosystem health including higher nutrient use and carbon sequestration (CO_2 uptake) efficiency, higher litter decomposition rate, higher community and ecosystem stability, and better habitat and product quality. Diversity indirectly plays the role of resisting invasion through facilitating biomass accumulation. Small-scale planting of more native species could enhance the resistance of community to invasions by exotic species (Figure 15.4). For example, Bakker and Wilson (2004) found that restoration can act as a filter for constraining invasive species and both diversity and species identity can be important.

Experimental studies offer useful insights to improve restoration outcomes (Roy 2001, Guo 2003, Schmid and Hector 2004). However, unlike biodiversity experiments that often seed multiple species on bare soils, restoration in the real world often involves planting native species (1) on preexisting vegetation, (2) across large scales, and (3) with long-term perspectives (see Grace et al. 2007). Thus, applying knowledge from seeding experiments on bare-soil, small-scale, and often short-term requires caution. Experiments show the benefits of high-diversity planting. But what experiments to date have not been able to offer is when, where, and what species to plant first and what species to follow (e.g., different successional stages, multitrophic levels).

In general, productivity increased with the number of planted species but only up to a certain level, and invasibility may therefore decline. The decline could be due to competition, allelopathy, and functional redundancy, among others (Guo et al. 2006). Thus, practical details regarding what species, how many species, how much (number or weight) seeds (or seedlings) for all and each species, and how to arrange the relative abundance among species are all potential issues that need to be resolved before action. Given the costs of seeds and sources of species (which may extend to much larger regions, thus some of the species may not actually be native to the particular habitat or landscape to be restored), planting appropriate number of species and seeds is still the major challenge.

New initiations of restoration efforts, identifying the most successful native species, not only those that can resist invasion by exotic species, but also the ones that can better adapt to future climate changes, become an urgent need in practice (Guo 2007). As mentioned earlier, species selection should also take both inhibition and facilitation effects into account. Inhibition may occur either biochemically, as with allelopathy, or because the invader exhibits faster or denser growth or is in some way more competitive for the available water, light, or nutrients at a site. Inhibition can be successful by lowering the rates of establishment, growth, reproduction, or mortality. For many high-profile invasives, the

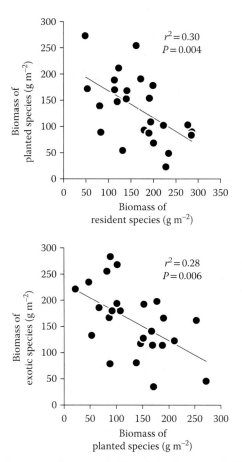

Figure 15.4 Experimental demonstration of biodiversity effects on habitat invasibility. The plots with higher biomass of resident species suppressed the growth of seeded species (as invading species; Top), and the plots with higher biomass of seeded species had lower biomass of exotic species (Bottom). However, when number of species was used, the number of seeded species was marginally negatively related to the number of exotic species ($r^2 = 0.15$, $p = .057$; not shown). (Adapted from Guo, Q., T. Shaffer, and T. Buhl, *Ecology Letters*, 9, 1284–1293, 2006.)

key conservation concern is that invasives become too dominant and inhibit biodiversity. They seemingly fill all or most of the available niches and reduce the number of available niches (cf. Grime 2002). In contrast, less aggressive species may extirpate only a few competitors, simply reduce their population densities, or fill some unoccupied space. In some situations, these latter effects may be managerially useful, as niches filled by relatively benign exotics may reduce the risk of dominance by aggressive invasives. Theoretically, such exotics or transplanted natives could enhance the resilience of ecosystems dominated by native species.

On the other hand, some exotics, especially invasives, can facilitate successional changes that are either compatible or incompatible with site objectives. When invasives reduce site diversity, they may not close niches entirely as much as selectively remove vulnerable species. This can retain opportunities for new species that are more tolerant of the invader. Alternatively, an invader may alter site conditions or disturbance processes in a way that favors native species or invasives that were formerly at a disadvantage. For example, exotic species have proven useful for gradual restoration of native-dominant mangroves (Ren et al. 2008). Also, increased fire frequency from invasive grasses can facilitate the recovery of fire-tolerant natives or the spread of certain invasives (Drewa et al. 2001). As a third example, nonnative trees are often used for reclamation after strip mining with the long-term expectation of a more diverse forest, even when it is not similar to what was historically there (Wikipedia, 2012).

During restorational planting, it is likely that not every species in any multispecies seeding can germinate and establish. Therefore, planting greater numbers of native species may lead to high diversity and restoration rate as measured by biomass accumulation. This is because diverse species planting can ensure that the niches in the habitat could be fully occupied in case of failures among the most productive species, especially in early stages of restoration when the habitat is quite open or the targeted habitat is quite heterogeneous (Tracy et al. 2004). High-diversity planting could increase the chances that rare or endangered species residing in the natural habitats may be included. By planting diverse species, we may identify the most suitable species that are most likely to be successful and persistent and those that are least competitive and might disappear in subsequent years. At the same time, we would not miss other potentially adaptive and productive native species in the restored habitats. Historical vegetation data may be used to identify highly productive species but might not be enough, because disturbance regimes and other factors such as climate might have changed over time. Finally, a high-diversity planting may form persistent seed banks, which would help protect the area from the effects of further disturbances and by ensuring that species important for early succession stages are present in the seed banks.

What has gained less attention in traditional restoration is that logical plantation and management regimes should follow the evolutionary forces that favored the initial success, for example, best adaptation to climate change and disturbances such as fire. If phylogenetically close relatives could better resist invasions by closely related species because of using the similar resources (Cadotte et al. 2009), planting congeners would be an effective way to resist invasions of closely related exotic species. On the other hand, recent studies suggest that some hybrids between native and introduced species often show greater invasiveness (Gaskin et al. 2009). Thus, avoiding planting the native species or preventing possible introductions of close relative or sister exotic species that may form hybrid invasive species have become increasingly important.

15.6   Bottom-up, top-down, and multitrophic (or food-web) restoration

Most previous and ongoing restorations are focusing on a single trophic level, mostly in plant communities (or primary producer). However, many plant species are relying on animals that serve as agents of dispersal, pollination, or reproduction that may require a broad view of restoration. Other critical elements to be restored include fungi and soil microbes that some plants need. Therefore, the more effective ways for restoration could be to restore multiple trophic levels as they depend on each other for long-term survival. If herbivores' existence relies on their native predators, the latter will need to be introduced as well ("the trophic cascade hypothesis" or the food web theory) (see Hastings 1988). In this regard, the effective way could only be the "bottom-up" approach.

The most common approach of restoration in the past and at present is the bottom-up producer. In contrast, an alternative and new approach proposed would be, First, the much needed elements for producers (plants) such as below-ground communities (e.g., arbuscular mycorrhizal fungi—as a symbiotic supporting system) (Perry and Amaranthus 1990, van der Heijden 2004) may need to be restored with producers and other elements such as pollinators should follow immediately. Second, the top-down approach (restoration of consumers) should also follow when producers have established. For example, often top predators are needed to regulate lower-level predators and primary producers to maintain the established systems.

15.7   Suggested restoration pathways

Sites that have different levels of destruction or degradation may benefit from different restoration pathways. For example, in highly destructed grassland or forests, restoration often involves direct seeding of native species after pretreatment of soils to eliminate the seed banks of exotic species. Yet, in highly degraded habitats invaded by exotics, restorationists usually seed or plant native species on preexisting vegetation. Often, exotics are also useful for creating temporal favorable conditions, for example, for soil erosion control in early stages of succession (e.g., California chaparral and grassland after fires) or for creating initial habitats suitable for natives in forests (Williams 1997, Guo 2003, Ren et al. 2008). Although actual restoration processes may be complex and may need flexible justification from time to time, here we propose a simplified pathways or scenarios for consideration. First, in barren areas or highly degraded (destroyed) habitats where natives might not survive and establish, planting facilitating exotics would be necessary to ensure later colonization of native species (Figure 15.5). In such cases, natives are not suitable to be used, and suitable but noninvasive exotics are used at the initial stages because they can fix soil erosion or create benign conditions for natives (Path I in Figure 15.5) (Ren et al. 2008). Second, in most cases where exotic species already exist, restorationists are likely to plant native species directly onto preexisting vegetation and, therefore, forming a mix of both natives and exotics. In such conditions, however, exotics may be dominant, especially in early stages (Path II) (Guo et al. 2006). Third, in other cases where bare soils exist or can be fully treated to remove the seed banks of exotics, native species are planted directly to form native communities without exotics (e.g., many controlled seeding experiments in grasslands), although certain exotics might invade subsequently in the future (Path III). However, in all scenarios, given the ongoing (and even accelerating) species invasions and the fact that completely eradicating exotics is virtually impossible, an eventual habitat condition with both native and exotic coexistence may be acceptable. This is especially the case because restoration sites are most likely at places with great human disturbances and have been already invaded by exotic species. Intense management of exotics at such sites is needed to ensure the dominance of natives (Figures 15.5 and 15.6).

15.8   Biomass manipulation and tradeoffs of enhancing biodiversity

While most physical factors are almost completely beyond human control, careful manipulation of other factors such as fire, grazing, nutrient addition, and continued seeding is important for restoration and ecosystem performance (Bradshaw 1987). One needs to realize that restoration is a continuing process, rather than one-time event. Because much needed species, especially those that fail to establish after initial planting, will need to be planted until viable populations are established.

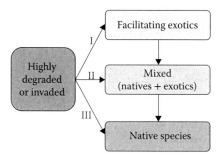

Figure 15.5 A sample of possible pathways in terms of species selection in restoration process. Highly degraded or invaded area includes all habitats that need to be restored (e.g., newly created bare grounds such as landslides and chemically treated areas for removing exotic seed banks). Path I: in barren areas or highly degraded (destroyed) habitats where natives might not survive and establish, planting facilitating exotics would be necessary to ensure later colonization of native species. In such cases, natives are not suitable to use, noninvasive exotics may be used at the initial stages because they can fix soil erosion or create suitable conditions for natives. Path II: in most cases where exotic species already exist, restorationists are likely to plant native species directly onto preexisting vegetation, therefore, forming a mix of both natives and exotics. Path III: in some cases where bare soils exist or can be fully treated to remove the seed banks of exotics, native species are planted directly to form native communities without exotics. However, in all scenarios, given the ongoing (and even accelerating) species invasions and the fact that completely eradicating exotics is impossible, an eventual acceptable habitat with both native and exotic coexistence may result. Proper control of exotics is needed to ensure the dominance of natives.

Figure 15.6 An example of restoration by seeding native species onto preexisting prairie grassland that had been heavily invaded by several exotics such as Canada thistle (*Cirsium arvense*), alfalfa (*Medicago sativa*), and smooth brome (*Bromus inermis*) near Jamestown, North Dakota, U.S.A. (photo by J. Schatz). During the first few years after seeding, natives and exotics appeared to coexist. (Adapted from Guo, Q., T. Shaffer, and T. Buhl, *Ecology Letters*, 9, 1284–1293, 2006.)

In management practices, biomass is the most easily and frequently manipulated variable (Huston 2004, Guo 2007). It is likely that after the community biomass reaches a certain level, some of the planted species will likely disappear because of increased competition or other environmental changes. Periodic removal of above-ground biomass through varying frequency, intensity, and timing of burning and grazing can often increase species diversity and habitat productivity. However, the optimal frequency and intensity of some of these activities in various habitats are often debated.

In habitats invaded by nonnative species, when total elimination of invasive species is not feasible, techniques that can effectively remove their biomass should be developed. However, although invasive species management usually adopts burning, grazing, physical or chemical treatment, and biocontrol agents, these practices are also disturbances and their effects on native species need to be evaluated. In addition, below-ground biomass, a frequently neglected factor, could also be considered in future restoration and management plans.

Management tradeoffs may develop from strategies designed to control invasives and the need for native species to migrate in response to climate change. Although enhanced species diversity at a site may slow the spread of problematic invasives in the near term, this change may close niches that are necessary for native species to effectively respond to environmental changes at the landscape scale. In other words, while increasing diversity may enhance resistance to invasives, it may *reduce* the ability of species to migrate or alter population densities at the landscape scale. Such factors need to be considered in restoration planning.

15.9 Perspectives

The effects of biomass on biodiversity and the effects of diversity on productivity are two closely linked foci in both basic and applied research. Most related studies have dealt with these two issues separately and their practical implications have not been given enough attention (Guo 2007). Experiments show the benefits of initial high-diversity planting. But what experiments to date have not been able to offer is when and where to plant what species first and what species (e.g., multitrophic) and where to follow. Such questions are at least equally important, if not more. Future restoration should focus on these issues in planning and actual efforts.

Human-driven changes are reflected in vastly disturbed habitats worldwide. This suggests that the high-end of the classic disturbance–diversity curve may be soon reached in these places, thereby threatening diversity (i.e., the intermediate disturbance hypothesis). How to reduce the effects from human activities amid ongoing efforts of economic growth and development is the major challenge in maintaining global or regional biodiversity and ecosystem health.

Many restored ecosystems seem unsustainable, especially those without continued management and subsequently invaded by highly invasive species (SER 2004). This is in part due to our lack of understanding of basic ecology needed in restoration and in part due to the lack of timely and efficient communication and proper application (Palmer et al. 1997). To effectively control biotic invasions, management and restoration must go hand in hand. In addition to the practical requirements for restoration, continuing and sometimes intensive management is necessary to effectively constrain biotic invasions and adept climate change. For example, to preserve biodiversity, periodic thinning, burning, and grazing are needed to reduce biomass and competitive exclusion by invasive species.

Aside from biomass manipulation, restoration is also about how diversity manipulation (e.g., seeding, planting) affects biomass and productivity. For invasive species management, restoration is to build up enough abundance (biomass or cover) to constrain invasion. In most cases, optimal management often involves regulating biomass so that high diversity and productivity or other preferred habitat characteristics can be achieved and maintained, while restoration usually involves planting or seeding a certain number or combination of native species so that the native structure and function of the habitat can be restored and degraded ecosystems can recover faster.

15.10   Conclusions

In the past, the concept of restoration was useful to highlight the degree to which ecosystems had changed and how past conditions or dynamics often provided a better means for sustaining a wide assortment of ecosystem values. These have not changed. What has and will change is the degree to which conditions have been altered due to forest loss and fragmentation, invasive species, and climate change (Kohli et al. 2008). While historical conditions may have been useful in the past, trends toward increased novelty may require creative management strategies that are less and less guided by the past.

To retain its usefulness, the restoration concept must evolve past its historical usage. Restoration will always invoke the past, but less so as a goal and more as a justification for maintaining and enhancing site resilience of species and ecosystems. The word is also useful in a more general sense, as in restoring habitat structures that may include novel components that are engineered using exotic species. Our best options for achieving species conservation given the cumulative threats from invasives and climate change may be incompatible with restoration, even if it were possible.

Although many physical factors are clearly beyond our control, diversity and biomass can be manipulated through management. There is evidence that the optimal management that maintains higher biodiversity may to some degree buffer the adverse physical effects, especially those from catastrophic events. However, no matter how well designed, the newly restored ecosystems are most likely to be novel or different from the conditions if the habitats had not been altered by human activities. Thus, not only do we need historical information and past experience to help guide us in restoration and management efforts, but we also need to be creative to counter the new challenges that may emerge in the future (Falk et al. 2006).

In restoration, planting or encouraging the right species and an optimal level of native species richness would help achieve and maintain habitat productivity and stability. Such niche occupation and a high biomass or cover of native species would increase the resistance to biological invasion by nonnative species. In some cases, even using certain exotics with certain genetic or life history traits could help stabilize the systems and initially create more suitable habitats for natives. Future insights from the basic research on the relationships among diversity, biomass, and productivity and those between diversity and stability or resilience can offer better guidelines for efficient restoration and for reducing the effects of invasive species.

In short, while there is still no quick fix in restoration amid biotic invasions, continuing effective and proper management especially through biomass manipulation is crucial. We should keep in mind that restoration is a continuing, long-term effort and often involves intense follow-up management. A carefully designed and restored ecosystem may not be the ideal system a habitat might support, but it could become the best system that is practically achievable. In future efforts, projected changes in climate and from exotic invasions must be considered in restoration guidelines to ensure the desired outcomes and to avoid unintended consequences.

Acknowledgments

We thank Corinne Diggins, Yude Pan, and anonymous reviewers for helpful comments, and many individuals who assisted in the experimental restoration project. This study was in part supported by the USDA Forest Service.

References

Aronson, J., and J. van Andel. 2006. Challenges for restoration theory. In *Restoration Ecology: The New Frontier*, eds. J. van Andel, and J. Aronson, pp. 223–233, Oxford, UK: Blackwell.

Bakker, J., and S. Wilson. 2001. Competitive abilities of introduced and native grasses. *Plant Ecology* 157:117–125.

Bakker, J.D., and S.D. Wilson. 2004. Using ecological restoration to constrain biological invasion. *Journal of Applied Ecology* 41:1058–1064.

Berger, J.J. 1993. Ecological restoration and nonindigenous plant species: A review. *Restoration Ecology* 1:74–82.

Bradshaw, A.D. 1987. The reclamation of derelict land and the ecology of ecosystems. In *Restoration Ecology*, eds. W.R. Jordan III, M.E. Gilpin, and J.D. Aber, pp. 53–74. Cambridge, UK: Cambridge University Press.

Cadotte, M.W., M.A. Hamilton, and B.R. Murray. 2009. Phylogenetic relatedness and plant invader success across two spatial scales. *Diversity and Distributions* 15:481–488.

Clewell, A.F., and J. Aronson. 2008. *Ecological Restoration: Principles, Values, and Structure of an Emerging Profession*. Washington, DC: Island Press.

Cox, G.W. 1999. *Alien Species in North America and Hawaii*. Washington, DC: Island Press.

D'Antonio, C.M., and M.C. Mack. 2001. Exotic grasses potentially slow invasion of an N-fixing tree into a Hawaiian woodland. *Biological Invasions* 3:69–73.

Drewa, P.B., D.P.C. Peters, and K.M. Havstad. 2001. Fire, grazing, and honey mesquite invasion in black gramma-dominated grasslands of the Chihuahuan Desert. In *Proceedings of the Invasive Species Workshop: The Role of Fire in the Control and Spread of Invasive Species*, eds. K.E.M. Galley, and T.P. Wilson, pp. 31–39. Tallahassee: Tall Timbers Research Station.

Ewel, J.J., and F.E. Putz. 2004. A place for alien species in ecosystem restoration. *Frontiers in Ecology and the Environment* 2:354–360.

Falk, D.A., M.A. Palmer, and J.B. Zedler (eds.) 2006. *Foundations of Restoration Ecology*. Washington, DC: Island Press.

Gaskin, J.F., G.S. Wheeler, M.F. Purcell, and G.S. Taylor. 2009. Molecular evidence of hybridization in Florida's sheoak (*Casuarina* spp.) invasion. *Molecular Ecology* 18:3216–3226.

Grace, J.B., T.M. Anderson, M.D. Smith, E. Seabloom, S.J. Andelman, G. Meche, et al. 2007. Does species diversity limit productivity in natural grassland communities? *Ecology Letters* 10:680–689.

Grime, J.P. 2002. Declining plant diversity: Empty niches or functional shifts? *Journal of Vegetation Science* 13:457–460.

Guo, Q. 2003. Disturbance, life history and optimal management for biodiversity. *AMBIO* 32:428–430.

Guo, Q. 2007. The diversity-biomass-productivity relationships in grassland management and restoration. *Basic and Applied Ecology* 8:199–208.

Guo, Q., and A. Symstad. 2008. A two-part measure of degree of invasion for cross-community comparisons. *Conservation Biology* 22:666–672.

Guo, Q., T. Shaffer, and T. Buhl. 2006. Community maturity, species saturation, and the variant diversity-productivity relationships in grasslands. *Ecology Letters* 9:1284–1293.

Harris, J.A., R.J. Hobbs, E. Higgs, and J. Aronson. 2006. Ecological restoration and global climate change. *Restoration Ecology* 14:170–176.

Hastings, A. 1988. Food web theory and stability. *Ecology* 69:1665–1668.

Hobbs, R.J. 2007. Setting effective and realistic restoration goals: Key directions for research. *Restoration Ecology* 15:354–357.

Hobbs, R.J., and Norton, D.A. 1996. Toward a conceptual framework for restoration ecology. *Restoration Ecology* 4:93–110.

Holsman, K.K., P.S. McDonald, P.A. Barreyro, and D.A. Armstrong. 2010. Restoration through eradication: Removal of an invasive bioengineering macrophyte (*Spartina alterniflora*) restores some habitat function for a native mobile predator (*Cancer magister*). *Ecological Applications* 20:2249–2262.

Huston, M.A. 2004. Management strategies for plant invasions: Manipulating productivity, disturbance, and competition. *Diversity and Distributions* 10:167–178.

Keeley, J.E. 2006. Fire management impacts on invasive plants in the western United States. *Conservation Biology* 20:375–384.

Kohli, R.K., S. Jose, D.R. Batish, and H.P. Singh. (eds.). 2008. *Invasive Plants and Forest Ecosystems*. Boca Raton, FL: CRC/Taylor and Francis.

Larson, D.L. 2009. Evaluation of restoration methods to minimize Canada thistle (*Cirsium arvense*) infestation. U.S. Geological Survey. Open-File Report 2009–1130. Reston, U.S. Geological Survey.

Palmer, M.A., R.F. Ambrose, and N.L. Poff. 1997. Ecological theory and community restoration ecology. *Restoration Ecology* 5:291–300.

Perry, D.A., and M.P. Amarathus. 1990. The plant-soil bootstrap: Microorganisms and reclamation of degraded ecosystems. In *Environmental Restoration: Science and Strategies for Restoring the Earth*, ed. J.J. Berger, pp. 94–102. Washington, DC: Island Press.

Perry, L.G., and S.M. Galatowitsch. 2003. A test of two annual cover crops for controlling *Phalaris arundinacea* invasion in restored sedge meadow wetlands. *Restoration Ecology* 11:297–307.

Peterson, G., C. Allen, and C.S. Holling. 1998. Ecological resilience, biodiversity and scale. *Ecosystems* 1:6–18.

Price, C.A., and J.F. Weltzin. 2003. Managing non-native plant populations through intensive community restoration in Cades Cove, Great Smoky Mountains National Park, USA. *Restoration Ecology* 11:351–358.

Qian, H., and Q. Guo. 2010. Linking biotic homogenization to habitat, invasiveness, and growth form of non-native plants in North America. *Diversity and Distributions* 16:119–125.

Ren, H., H.F. Lu, J. Wang, N. Liu, and Q.F. Guo. 2012. Forest restoration in China: Advances, obstacles, and perspectives. *Tree and Forestry Science and Biotechnology* 6 (Special Issue 1):7–16.

Ren, R., S. Jian, H. Lu, Q. Zhang, W. Shen, W. Han, Z. Yin, and Q. Guo. 2008. Natural colonization of native mangroves into exotic mangrove plantations in Leizhou Bay, South China. *Ecological Research* 23:401–407.

Ricklefs, R.E., Q.F. Guo, and H. Qian. 2008. Growth form and distribution of introduced plants in their native and non-native ranges in Eastern Asia and North America. *Diversity and Distributions* 14:381–386.

Roy, J. 2001. How does biodiversity control primary productivity? In *Terrestrial Global Productivity*, eds. J. Roy, B. Sangier, and H.A. Mooney, pp. 169–186. San Diego., USA: Academic Press.

Schmid, B., and A. Hector. 2004. The value of biodiversity experiments. *Basic and Applied Ecology* 5:535–542.

SER. 2004. *The SER International Primer on Ecological Restoration, Version 2*. Society for Ecological Restoration International Science and Policy Working Group. Available online at: http://www.ser.org/reading_resources.asp.

Temperton, V.M, R.J. Hobbs, T. Nuttle, and S. Halle (eds). 2004. *Assembly Rules and Restoration Ecology: Bridging the Gap between Theory and Practice*. Washington, DC: Island Press.

Tracy, B.F., I.J. Renne, J. Gerrish, and M.A. Sanderson. 2004. Effects of plant diversity on invasion of weed species in experimental pasture communities. *Basic and Applied Ecology* 5:543–550.

van der Heijden, M.G.A. 2004. Arbuscular mycorrhizal fungi as support systems for seedling establishment in grassland. *Ecology Letters* 7:1–11.

Vander Zanden, M.J., J.D. Olden, and C. Gratton. 2006. Food-web approaches in restoration ecology. In *Foundations of Restoration Ecology*, eds. D.A. Falk, M.A. Palmer, and J.B. Zedler, pp. 165–189. Washington, DC: Island Press.

Wang, J., H. Ren, L. Yang, D. Li, and Q. Guo. 2009. Soil seed banks in four 22-year-old plantations in South China: Implications for restoration. *Forest Ecology and Management* 258:2000–2006.

Wikipedia. Surface mining. http://enwikipedia.org/wiki/Surface_mining. Last accessed November 12, 2012.

Williams, C.E. 1997. Potential valuable ecological functions of nonindigenous plants. In *Assessment and Management of Plant Invasions*, eds. J.O. Luken, and J.W. Thieret, pp. 26–36, New York: Springer.

chapter sixteen

Converting invasive alien plant stands to natural forest nature's way

Overview, theory, and practice

Coert J. Geldenhuys

Contents

16.1   Introduction

The general view of invasive alien plants (IAPs) in the environment is negative and that they are a threat to the native vegetation and species, that is, the biodiversity and productivity of any natural system (Macdonald et al. 1986, Henderson 2001, Kohli et al. 2009). A part of this view is true when one considers the replacement of grassland and shrubland in many areas by invasive alien species, and also by native pioneer species and vegetation. Sometimes people question the fuzz about alien invaders and the associated emotion around them, and they attribute any kind of environmental problem to the presence of alien species. This results in the general approach to get rid of alien invaders at all costs, but the high costs spent on controlling invasions are often not effective. But often the presence of invasive alien species or pioneer species in general is an indication of environmental degradation of some or other kind.

Different situations require different solutions, and general observations in different areas show that particularly in the natural evergreen forest environment, invasive alien species may need to be dealt with in a different way (Parrotta et al. 1997, FAO 2003). Some of the emotions around invasive alien species demonstrate many wrong perceptions because of a poor understanding of the natural disturbance processes and the associated recovery processes and how invasive alien species fit such a perspective.

Disturbance–recovery processes determine vegetation types and biodiversity patterns in the natural forest environment. Fires in particular determine the forest–woodland–shrubland–grassland mosaics in the landscape (Geldenhuys 1994). There are also many wrong perceptions that forest clearing by people is the main causal factor of the very fragmented forest distribution, although land use is a contributing factor to this pattern (Feely 1987). The question is, therefore, whether all IAP species in the natural evergreen forest environment are indeed a problem, or do some at least facilitate natural forest rehabilitation?

The management of IAPs covers a wide range of topics and involves different disciplines (Macdonald et al. 1986, Kohli et al. 2009). There is a need for a framework within which to coordinate the research and management efforts within any region to facilitate focused and cost-effective effort and reduced overlap and to introduce more differentiation in alien plant control measures. Sustainable development needs to consider the frequent conflict between conservation and development, specifically in terms of invasives and the balancing of their use value with their impacts on biodiversity and productivity of the land. The root cause of many of the alien invaders is not the species as such but poor land management—in some areas, the invasion problem is a symptom of the problem of unsustainable land use practices. The framework should provide for a land use policy to guide sustainable and integrated land management at the landscape level for conservation of biodiversity and ecological processes, productive agriculture, productive forestry, and socio-economic development in each eco-region of the area.

In South Africa, the focus of the National Working for Water program is to clear alien plant invasions in water catchments to make water more available downstream. The program is organized in such a way that it contributes to job creation through training of project teams. The tasks for clearing invaded land are determined by the density and type of invader plant species. The application uses basically the same approach in Mediterranean *sclerophyll fynbos shrublands, dry Karoo shrublands*, seasonal grassland and woodland, and natural forests, that is, total clearing of alien plant stands and burning the debris where the woody components cannot be used for secondary industries. The expectation is to recover the natural vegetation biodiversity of the different natural systems in the catchments. A question often asked is whether IAP species in the forest environment should be managed like the fire-adapted grassland, shrubland, and so on. Observations in cleared alien stands in the forest environment showed that their clearing diverted the pressure for collection of poles, fuel wood, and so on from the woody alien stands back to the natural forest species and also cause dense regrowth stands of the alien tree stands.

Some people call every introduced species an invasive alien because of the potential for such species to invade a natural area. For the purposes of this chapter, an IAP species in the natural evergreen forest environment is considered an introduced plant species that increases its spread, numbers, and/or biomass to the detriment of the forest substance and life-support systems through various processes such as continuous regeneration, prevention, and/or suppression of natural forest regeneration, upsetting and/or displacing the reproductive processes (which could be positive or negative), suppressive growth rate, and suppressive crown development, and/or upsetting or changing the natural disturbance regime (Geldenhuys 1996). For example, the fast growth combined with poor root development and large emergent crowns of the introduced Australian blackwood (*Acacia melanoxylon*) trees make them more vulnerable to strong gusty winds in the Southern Cape Afrotemperate forests to cause more and larger windfalls than the indigenous large-crowned Ironwood (*Olea capensis* subsp *macrocarpa*). The larger gaps can be more susceptible to regeneration of and domination by blackwood trees, preventing forest pioneer species to regenerate in the gaps (Geldenhuys 1996). However, the same study showed that blackwood trees substitute the harvesting of indigenous timber species because of their good timber and that the blackwood stands nurse the regeneration of natural forest species. It is therefore necessary to find a balance between the useful properties of the species and the costs associated with its control (Geldenhuys 1986, 1996, 2002).

The primary objective of this chapter is to use the South African situation as an example of developing a strategic framework for the control of invasive alien species in the natural forest environment of a country or region. This is done by

1. Considering a short overview of the natural evergreen forests in South Africa and the history of the introduction of alien species that became invasive
2. Extracting the key issues highlighted during a national survey of invasive alien species in and around the natural forests
3. Developing a conceptual understanding of the invasion process and the options it offers to develop sustainable IAP control and associated forest rehabilitation practices
4. Developing some guidelines for the practical control of IAP species in the rehabilitation of natural forest

16.2   Natural evergreen forests in South Africa

The natural evergreen forest in South Africa is generally a multilayered vegetation unit dominated by trees (largely evergreen or semi-deciduous) whose combined strata have overlapping crowns (crown cover >75%), where grasslike plants in the herbaceous stratum (if present) are generally rare, and in which fire does not play a major role in normal forest function or dynamics except at the fringes (Shackleton et al. 1999). The stand height ranges from over 30 m in high forests to just over 3 m in scrub forests.

The forests form the smallest, most widespread, and highly fragmented vegetation formations in South Africa (Mucina and Geldenhuys 2006). They occur scattered along the Great Escarpment, mountain ranges, and coastal lowlands of eastern and southern South Africa, from the Soutpansberg Mountains (22°40′S) and Maputaland coast (27°S) in the northeast to the Cape Peninsula in the southwest (34°S). Most forest patches are very small (<10 ha) to small (<100 ha) and embedded within a matrix vegetation of fire-prone and -adapted grassland, sclerophyll shrubland (fynbos), or deciduous woodland (Cooper 1985, Geldenhuys 1991, Berliner 2009). Few forests are larger than 1000 ha. Total forest cover is 3000 to 4000 km², which forms less than 0.1% of the country. Forests persist in areas with winter or all-year rainfall >525 mm and summer rainfall >725 mm and on a wide range of geological substrates (Rutherford and Westfall 1986, Mucina and Geldenhuys 2006). The small size and high margin-to-interior ratio of the forest patches make them very vulnerable within the fire-adapted vegetation and accentuates the importance of forest margins in forest survival in relation to regular fires and cultivation. The forest interior is subject to regular small-scale disturbances, causing gaps less than 0.1 ha in size (Mucina and Geldenhuys 2006).

The natural forests contain a high biodiversity and are considered the second richest biome in South Africa containing 5.35% of all South African species with a relatively rich 0.514 species/km², with many widespread but also endemic or fragmented species (Mucina and Geldenhuys 2006). The use value of the forests is high in terms of timber and nontimber forest products (Lawes et al. 2004). Most forests occur in areas of high population density. Forests surrounded by affluent societies expand with a small impact from infrastructure development (Geldenhuys and MacDevette 1989). Forests surrounded by developing poor rural communities are often degraded by traditional subsistence practices to satisfy their livelihood needs for building material, fuel wood, food and medicine, and other household goods (Geldenhuys and MacDevette 1989, Cawe and Geldenhuys 2007). However, they could not provide in the timber, fiber, and energy needs of South Africa. Today, commercial timber plantations and cultivated land surround natural forests in many areas (Geldenhuys 2002).

16.3 Historical perspective on alien plant introductions in the South African natural forest environment

The number of introduced species planted in South Africa followed an exponential increase over the years (Geldenhuys et al. 1986). European colonists settling in South Africa since 1652 brought with them trees and shrubs and other life forms to plant in their new environment to provide in their needs for timber, poles, fiber and firewood. Planting such species for their aesthetic value was a secondary objective. Jan van Riebeeck, who started the European settlement in South Africa, planted the first alien tree, *Quercus robur*, in 1656; initially it was planted for its timber, but later for shade and shelter. Commercial afforestation was started in the Cape Peninsula in 1884, mostly with *Pinus* and *Eucalyptus* species. Over the years, a wider range of species was introduced. Some species were planted in overexploited portions of the natural forests, either in gaps or on the forest margin where they began to spread. As the Europeans moved further inland and along the east coast, they tried more and more species. Sometimes individual trees were planted in the forest gaps but later they were planted in groups, and eventually many species were planted outside the natural forests to start a plantation forestry industry, particularly with species such as various *Pinus* species (mainly *P. radiata*, *P. canariensis*, *P. patula*. *P. elliottii*, *P. taeda*, and *P. halepensis*), various *Eucalyptus* species (mainly *E. globulus*, *E. diversicolor*, *E. saligna*. *E. grandis*, and *E. camaldulensis*), various *Acacia* species (mainly *A. mearnsii*, *A. decurrens*, and *A. melanoxylon* for timber and fibre and *A. cyclops* and *A. saligna* for dune reclamation), and various *Populus* species. Some other species were planted on a smaller scale for their timber or as windbreaks such as *Cinnamomum camphora*, *Toona australis*, *Grevillea robusta*, various *Cupressus* species, *Cedrus* species, *Sequoia sempervirens*, and many others.

During 1870–1910, over 300 tree and shrub species, mostly aliens, were sold from forestry nurseries (Geldenhuys et al. 1986). Since then many more species may have been introduced, particularly with the many private nurseries around, with at least one in every town and many more in larger towns and cities. Many species have been introduced for their value in landscaping or as fruit trees, including species such as *Solanum mauritianum*, *Lantana camara*, and various species of *Cestrum*, *Leptospermum*, *Ligustrum*, *Morus*, *Pittosporum*, and *Psidium*.

The plantation forestry industry expanded to cover an area of 1.33 million ha of commercial plantations by 2006, with over 80% in the Mpumalanga, KwaZulu-Natal, and Eastern Cape Provinces, producing >20 million cubic meters of roundwood worth an estimated 5.1 billion South African Rand (about US$0.65 billion) (Table 16.1; adapted from DWAF 2007). Most of the commercial plantations were established in grassland, open woodland, and shrubland (fynbos) areas (very small areas in cleared natural forest areas) in the higher rainfall areas, in mainly summer rainfall areas, but also in areas of all-year and winter rainfall.

Table 16.1 Area by Ownership, Species, and Purpose of Commercial Forestry Plantations in South Africa during 2005/2006

Category	Area (ha)	% of total area
Total area (ha)	1,281,519	100.0
Ownership		
Private ownership	978,299	76.3
Public ownership	303,219	23.7
Main species groups		
Softwoods, mainly *Pinus* species	688,313	53.7
Hardwoods of *Eucalyptus* species	478,191	37.3
Hardwoods of *Acacia* species (mainly *A. mearnsii*)	104,821	8.2
Hardwoods of other species	10,194	0.8
Main product groups		
Sawtimber Softwood	471,586	36.8
Sawtimber Hardwood	20,198	1.6
Pulpwood Softwood	209,570	16.3
Pulpwood Hardwood	470,343	36.7
Mining timber Hardwood	71,841	5.6
Other roundwood products: Softwood	7,157	0.6
Other roundwood products: Hardwood	30,824	2.4

Source: DWAF (2007).

16.4   National survey of invasive alien species in South African forest environment

During 1985, a questionnaire survey was completed at eight regional forestry offices throughout South Africa by knowledgeable foresters to collect information on IAPs occurring in the forest environment (inside the forest, in forest gaps, and along the forest margin) (Geldenhuys et al. 1986). Respondents were asked to distinguish between passive invaders (plants that grew and reproduced successfully in the adopted habitat, but not yet widely dispersed) and active invaders (plants widely dispersed). This information was supplemented with information from various publications dealing with IAP species. The data were summarized by forest eco-region, growth form (herbs, soft shrubs, woody shrubs, lianas, and trees), extent of spread, aggression of spread (passive or active), and forest condition (only in large forest gaps and along forest margins or also in small gaps and closed forest) (Appendix; Geldenhuys et al. 1986).

A total of 82 perceived alien invader plant species were recorded, but the majority (77%) were confined to forest margins and larger gaps inside the forest. Trees, shrubs, and woody lianas comprised 88% of the species, but less than 44% were regarded as invaders. Only 10 species of herbs and soft shrubs were listed probably because they are generally less conspicuous or less known. The majority (84%) of the species recorded from closed forest were considered as passive invaders. The forests were considered particularly resistant to alien plant invasion. Some species that were considered active invaders of closed forests were not able to establish under those conditions, such as *Acacia longifolia, A. mearnsii, A. melanoxylon, Cedrela toona, Grevillea robusta, Jacaranda mimosifolia, Pinus patula,, Populus canescens, Chromolaena odorata,* and *Rubus cuneifolius*. Their presence in closed forests was more likely the result of their establishment in former large gaps. Only plants that are able to reproduce and establish in the closed forest are more likely to become problem plants. Threatening invader plant species are shade-tolerant and bird-dispersed or small-seeded species such as *Cinnamomum camphora*, *Pittosporum undulatum*, various *Ligustrum* spp and *Cestrum* spp and, to some extent, *Psidium guajava* and *Solanum mauritianum*, the thorny liana *Pereskia aculeata*, and lianas that attach to tree stems to reach to the light of the canopy such as *Macfadyena unguis-cati*. No follow-up survey has been conducted since the initial survey of Geldenhuys et al. (1986).

Assessment of the characteristics of IAP species in the natural forest environments showed that most of the perceived invasive alien species are pioneer trees, shrubs, or climbers in their natural environments that establish in disturbed sites (Geldenhuys et al. 1986). They are generally intolerant of shade, cannot establish under their own canopies, and depend on disturbance to become established. The fruits or seeds of many invader tree species are either small, wind-dispersed (eucalypts, pines, and poplars), or have long-living hard-coated seeds (acacias, some of which are dispersed by birds, e.g., *Acacia melanoxylon*). They are generally easily raised and fast-growing. By contrast, the majority of the natural forest species have short-lived seeds, are bird- or mammal-dispersed, and are tolerant of shade.

Several of the species that are considered invasive species have three or more uses (Geldenhuys 1986). Many of them are introduced commercial timber species used in plantation forestry such as pines, eucalypts, and acacias (Table 16.1). A recent study in a rural area showed the use value of the invader species in the rural households and how they contribute to the conservation of the natural forest by providing alternative resources for poles and firewood (Table 16.2) (Cawe and Geldenhuys 2007). A traditional household would have three or more traditional round houses. The laths are used to fill in the space between the poles before the wall is plastered with mud and the roof is covered with grass. Stems of eucalypts, pines, acacias, and inkberry (*Cestrum laevigatum*) (for laths) became an important substitute in many of the houses and have also been collected for firewood (for cooking, warmth, and social gatherings). The current alien clearing program has to reconsider total clearing of those alien species that have potential value for rural livelihoods to prevent their clearing reverting the pressure back to the natural forests for construction material and firewood.

The information obtained from the survey suggested that the different alien plant species in the natural forest environment require different approaches for their control: (1) The majority of alien species (trees, shrubs, and nonwoody species) are just present and pose no threat to the forest ecology and biodiversity; (2) several species that appear to be invasive are in reality allies to the conservation and rehabilitation of the natural forest because of their use value and/or their pioneering characteristics (they are sometimes more active

Table 16.2 Average Wood Usage in the Wall and Roof of One Traditional Round House, in the Fences of One Household and for Firewood Used by the Household in the Rural Wild Coast of South Africa

Usage	Dimensions	Poles	Laths
Walls	21 m × 2.1 m	29 (2.6 m × 6–8 cm)	721 (1.5 m × 3–4 cm)
Roof	–	27 (3.3 m × 6 cm)	168 (1.5 m × 3–4 cm)
Garden fence	281 m	214 (1.7 m × 8 cm)	–
Livestock fence	40 m	28 (1.7 m × 10 cm)	–
Firewood	2 head loads/day; 0.4 m³; 2 m × 0.3 m		

Source: Cawe and Geldenhuys (2007).

or aggressive pioneer trees or plants than the native pioneer species); and (3) relatively few species pose a real threat to the integrity and biodiversity of the forest function because of their shade tolerance and their fruit characteristics.

16.5 Conceptual understanding of the invasion process of shade-intolerant alien species

The key concept presented here for a better understanding of the perceived invasion of natural forests by shade-intolerant invasive alien species lies in the natural disturbance-recovery processes at the landscape level. Many examples of the degradation of natural forests exist. These include human disturbance activities of resource use, crop cultivation, mining, and changed land uses (IAPs, grazing, increased or decreased fire frequency, and changed regimes). They also include natural disturbances by fire, lightning, cyclones or hurricanes, windfalls, flooding, drought, and so on. When we take a closer look, we find many similarities and also major differences between natural and human impacts through such disturbances and the recovery processes, which include alien plant invasions of degraded landscapes.

Natural forests have intrinsic adaptations to recover from diverse disturbance regimes (fires, cyclones, landslides, volcanic eruptions, windfalls, etc.) through suites of species dominating stands in different development stages toward recovery to mature forest. The total forest biodiversity of a particular forest system is therefore a function of the habitat and the suite of recovery stages after diverse disturbances (Van Gemerden et al. 2003, Geldenhuys 2009). Different species in such a forest system function optimally in different development stages to mature forest, such as pioneer species in disturbed sites to shade-tolerant species in mature forest, and they require such conditions to maintain healthy populations. Our challenge is to recover the biodiversity and productivity of degraded and cleared natural forests and in areas invaded by invasive shade-intolerant invasive species, cost-effectively.

In many other areas with natural evergreen forest, the fragmented forest location pattern is mostly not because of clearing by people although land use is a contributing factor to different degrees in different areas (Feely 1987, Geldenhuys 1994). In South Africa, the area with more than 800 mm annual rainfall cover 7% of the country, that is, potential area for natural forests, whereas the total actual area of natural forests cover about 0.1% of the country (Geldenhuys 1994). Feely (1987) has shown that the extensive areas of natural grassland in the areas potentially suitable for natural forests existed long before the destructive activities of the Iron age smelters degraded many natural forests and that the forests still persist in such areas. The fire regime, in particular, determines the forest–woodland–shrubland–grassland mosaics in the broader landscape where the annual rainfall intensity and seasonality is high enough to maintain natural evergreen forest (Geldenhuys 1994). This provides a basis for understanding how to manage the majority of IAPs in the forest environment.

South Africa, for example, has experienced a long period of fire ignition by lightning and people. The fires driven by prevailing winds in the dry season determine the forest location pattern as hot air moves up, a fire cannot climb down a cliff but jumps over the edge, and air movement over a ridge or edge creates an eddy on the leeward side of a ridge (Figures 16.1 and 16.2). Fire sparks carried by the wind jumping across a ridge can start a spot fire, which could run with an eddy up the ridge from the leeward side. Commercial timber plantations are generally planted in the fire pathway in grassland, fynbos shrubland, or woodland (fire-adapted systems), and very seldom in cleared natural forest. The timber plantations require costly fire protection systems, but during extreme

Figure 16.1 **(See color insert.)** Fire flows (solid lines) along ridges and jumps (dotted lines) across gullies to contribute to the fragmented forest location pattern within fynbos shrubland near Swellendam (left) and within grassland in the Catherdral Peak area, KwaZulu-Natal Drakensberg (middle and right), South Africa. (Adapted from Geldenhuys, C.J., *Journal of Biogeography*, 21, 49–62, 1994.)

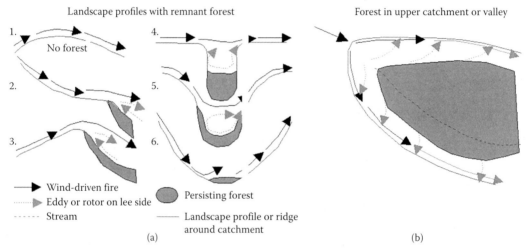

Figure 16.2 Generalized model of forest location pattern in fire shadow areas with fire-adapted fynbos shrubland or grassland in the fire pathways within broken landscapes. (Adapted from Geldenhuys, C.J., *Journal of Biogeography*, 21, 49–62, 1994.)

conditions some plantations get regularly destroyed by fire (Figure 16.3). The natural forests persist in mountainous landscapes surrounded by fire-adapted grassland, shrubland, and woodland in topographic shadow areas in relation to wind-driven fires in the dry season (Geldenhuys 1994).

Fire is withdrawn from the landscape directly with the intensive protection measures for commercial timber plantations and indirectly through crop agriculture, infrastructure and urban development, and changed fire regimes (cool manageable fires) for grazing and conservation areas. Pioneer species, often IAP species, establish in dense stands in such protected landscapes to start the recovery process through natural succession (Figure 16.4).

The fire-determined natural forest boundaries are more or less fixed in the landscape because they are maintained naturally by the fire–topography interactions, in which the changes in slope provide natural dynamic buffers between fire-sensitive forest and fire-adapted other vegetation (Figures 16.5 and 16.6). The structure of the ecotone vegetation between the mature forest and the adjacent fire-adapted vegetation changes gradually from dense low vegetation on the side of the fire-adapted vegetation to the less dense tall forest, that is, a soft edge. When planted tree stands or naturalized pioneer tree stands of

Figure 16.3 Commercial timber plantations planted in grassland or fynbos within the fire zone were destroyed by severe fires, while the natural forests were not destroyed at all, at Langverwacht plantation near Kokstad (left) and Witelsbos plantation near Humansdorp (right) in South Africa.

(a) (b) (c)

Figure 16.4 **(See color insert.)** Natural pioneer species and invasive alien species become established in the natural forest environment with the protection of the landscape against fire: (a) Natural regeneration of *Juniperus procera* on the Nyika Plateau, Malawi, inside the firebreak around a mixed evergreen natural forest with a few large trees of this species; (b) Invasion of grassland by the native legume pioneer tree *Acacia karroo* after the private nature reserve near East London, South Africa, was protected against fire; (c) Invasion of fynbos shrubland vegetation above the natural forest edge by *Pinus radiata* trees when the total area was protected against fire to safeguard the plantation forestry estate behind the photographer.

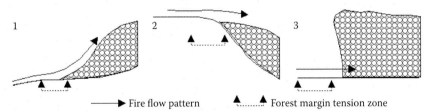

⟶ Fire flow pattern   ▲⋯⋯▲ Forest margin tension zone

Figure 16.5 Profiles of forest margins with gradual change (soft edges) along fire-prone vegetation such as shrubland and grassland (diagrams 1 and 2) or with an abrupt change (hard edge; diagram 3).

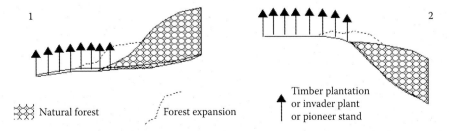

⬡ Natural forest   ⋯ Forest expansion   ↑ Timber plantation or invader plant or pioneer stand

Figure 16.6 Profiles of forest margins with timber plantation, invader plant, or pioneer tree stands establishing on the forest margin with fire exclusion in the landscape, and forest expansion into these newly established stands.

indigenous or alien species on the forest edge are removed, they form an abrupt or hard forest edge, which makes the forest edge much more vulnerable to penetration by fire. If a pioneer tree stand on the forest edge is manipulated to develop into mixed forest, then the ecotone adopts the natural soft edge.

In natural pioneer stands, after ending the disturbance (*Virgilia divaricata* after fire in Southern Cape Afrotemperate forest in South Africa [Van Daalen 1981] or *Musanga cecropioides* after slash & burn traditional agriculture in tropical moist forest in the African Congo Basin [Mala 2009]), the monospecific shade-intolerant pioneer tree (nurse) stands become "invaded" by shade-tolerant forest species as the nurse stand naturally undergoes thinning. As the nurse stand matures and becomes less dense, more shade-tolerant natural forest species of middle to late successional stages become established and an increasingly more diverse regrowth natural forest develops (Geldenhuys 2009). A similar process can be observed in planted stands of pines, eucalypts, and acacias over time with alien plant stand development (Parrotta 1995, Geldenhuys 1996, 1997, Loumeto and Huttel 1997, Parrotta and Turnbull 1997). The nurse stands buffer the more variable and extreme conditions of regularly disturbed grassland, shrubland, and agricultural fields and add litter to the site that ameliorate the nutrient and moisture content of the upper soil layer. This is a typical succession process that could take many years toward the recovery of the mixed species mature forest stand. Geldenhuys (1994) estimated the age of a regrowth forest stand after a fire in the Southern Cape Afrotemperate forest to be at least 230 years old. Most of the tree species in the regrowth forest canopy had a similar composition than the adjacent mature forest that was not damaged by the fire. However, the understory shrub and herbaceous vegetation in the regrowth forest had not recovered to the understory vegetation of the adjacent unburnt stand, although many pole-sized trees of the canopy species were present. This forest succession process through the nurse stands of natural pioneer species or of commercial timber species or of naturalized invasive alien tree stands can be manipulated through selective thinning to speed up the recovery process toward mixed-species forest (Figure 16.7) at relatively low costs.

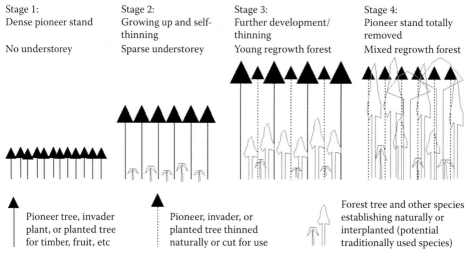

Figure 16.7 Conceptual model of the relationship between stand development stage and the establishment of understory vegetation of natural forest tree regeneration. (Adapted from Geldenhuys, C.J. 1999. In *Gestion Agrobiologique des sols et des systèmes de culture*, eds. F. Rasolo, and M. Raunet, pp. 543–551. *Proceedings of an International Workshop*, Antsirabe, Madagascar, March 23–29, 1998.)

This management approach is possible because most natural forest "invaders" are shade-intolerant but most natural forest species are shade-tolerant, and the nurse stand facilitates the recovery of the natural forest system and species. This is in contrast to the process in the fire-adapted systems where the invasive alien tree and shrub species compete directly with the grassland or shrubland systems and species. The alien species in the fire-adapted systems, therefore, need to be cleared or eliminated before the natural vegetation systems and species can recover. In the forest environment, such clearing of the alien cover will maintain the stand development in stages 1 and 2 (Figure 16.7), that is, perpetuate the invasion problem. Clearing alien invaders and subsequent planting of indigenous tree species to rehabilitate natural forest is expensive. Personal practical experience of such an action in different areas has shown that natural regeneration outgrew the planted seedlings. The cover of the nurse stand systems (particularly the alien tree cover) can be removed from the system through gradual manipulation (thinning) of the nurse stand (Stages 3 and 4 in Figure 16.7) until no more alien trees are present. Planting of indigenous tree or other species within the regrowing forest system will only be necessary for species that would not easily disperse there.

16.6   Practical management of shade-intolerant alien invaders in forest

The areas with stands of invasive shade-intolerant species are usually a degraded forest or grassland or shrubland that became invaded with alien tree stands because of the lack of fire. Some of the invader plant stands have to be restored back to grasslands or shrublands but in the forest environment has to be restored back to forest. The rehabilitation area needs to be managed through the following four steps (Geldenhuys 2008, Geldenhuys and Bezuidenhout 2008):

Step 1: Broad zonation of the rehabilitation area in terms of the end-points in alien tree removal, such as productive farmland, continued timber production areas, broader riparian zones, and stream bank areas. This should differentiate between areas that need abrupt total clearing of alien vegetation, such as grassland or shrubland areas, and areas that need gradual removal of alien stands such as "forest" and riparian zones. The purpose is to maintain "natural" forest–other vegetation mosaics and "development" or land use zones. Not all areas suitable for forests could be converted to forests sustainably. Rehabilitation should be confined to "fire-shadow" areas outside of the fire pathways, unless the area is within a built-up environment or agricultural land or forestry estate, which is directly or indirectly protected against fire. In terms of the fire patterns (Figure 16.3), the riparian or riverine zones in a potential forest environment could be considered "forest" zones. In natural areas, the conversion should be confined to near the forest margin or on sites known to be cleared of forest. Old aerial photographs of the area could be used to determine safe sites for forest rehabilitation. Other conversion constraints and rehabilitation costs should be determined for the different forest areas. In some areas, it may also be necessary to determine the user needs of stakeholders for potential species (forest, shrublands, other land uses) through directed user-need surveys to satisfy the needs in combination with the rehabilitation actions. For example, poles of the invader trees could be used for house construction in rural areas or for making small garden chairs for sale in urban areas as small enterprises. Some regenerating forest species could be a source for plants that could be used for traditional medicine that would need special management in the rehabilitation process.

Step 2: Zone the "forest" and riparian sites identified for rehabilitation to regrowth natural forest by stand development stages (see Figure 16.7). The criteria used for the differentiation include stands of different overhead composition (species) and "age" (size) and development stages of the indigenous woody vegetation in the understory. The following criteria have been developed for assessment of the stand development stages in the natural forest zone (Geldenhuys and Bezuidenhout 2008):

Development stage 1: Dense, young nurse stands with many small stunted stems of the nurse stand species. Gradually as this stand grows taller, many stems die because of competition from the faster-growing stems. Very few to no indigenous tree species, or scattered individuals that existed before, may be present. Manual stand manipulation needs to be delayed until the stand develops through self-thinning. Where forest tree seedlings and saplings do appear, some invader plants can be removed (hand-pulled, cut, or ring-barked). The stands need to be inspected from time to time.

Development stage 2: The nurse stand is still relatively dense, but taller. Forest tree seedlings and saplings and species of other growth forms start to establish in the understory, sometimes in small clusters. These species often include shade-tolerant tree species from a nearby natural forest, but they are small and develop slowly because of the relatively low light conditions, even though they are tolerant of some shade. The focus in stand manipulation should be on removal of invader plants to facilitate growth of the forest tree seedlings into saplings and poles.

Development stage 3: Natural stand thinning is in progress. More stems of the nurse stand die back to create more space in the understory with more light. This allows the development of more forest tree clusters and development of the establishing natural forest tree species into young tree stems amongst the fewer nurse stand trees. Tree clusters expand laterally, with new plants establishing away from the cluster core. The focus in stand manipulation should be to remove all nurse stand plants within the clusters and to create just enough open space adjacent to the cluster to facilitate expansion of the cluster.

Development stage 4: This is the advanced stage of forest recovery, with most or all of the nurse plant stems removed (or dead). The forest structure is developing toward a continuous forest canopy and tree stems in a range of stem diameters as more species become established and the young stems grow into trees and into the canopy. At this stage, the stand conditions are such that shade-intolerant invader plant species cannot become established, even if they are present in small numbers in the understory. The focus in stand manipulation should be to remove all invader plants in a gap or on the forest edge. It is a low-key activity that requires an occasional inspection to remove alien trees that may establish in gaps. Such stands could be used as a norm for recovery of stages 1 to 3.

Step 3: Implement stand manipulation in stand development stages 2 and 3. The intensity of rehabilitation activities will vary according to the development stage (canopy and understory) of each nurse stand. Selected unwanted trees of the nurse stand should be gradually removed with care to enable regeneration of natural forest species to become established naturally. The selected stems could be cut for direct use or ring-barked to die standing in the selective thinning (not clearing). The following five golden rules need to be considered:

1. The focus should be on the establishing forest species clusters.
2. A relatively closed nurse stand canopy should be maintained to prevent regeneration and regrowth of the nurse stand species. It should be opened above forest species clusters to help the forest seedlings, saplings, and poles to grow stronger and to flower and fruit earlier to enable fruit or seed dispersers to help with the process.
3. Small suppressed or stunted sub-canopy stems of the nurse stand species should be removed to provide more growing space for the regeneration and growth of forest species in the understory.
4. The ground cover (herbaceous plants, litter, dead branches, etc.) should be kept intact to prevent invader or pioneer species to become established.
5. Tree debris should not be burnt on the rehabilitation site: the generated heat will (1) kill seeds of forest species that may be present in the organic or litter layers, (2) destroy the soil organic layer and associated microorganisms involved in the decomposition and nutrient cycling processes (sterilize the soil), and (3) stimulate hard-coated seeds of particularly invasive *Acacia* species to regenerate in mass. This will push the recovery process back several years.

    This general process is for conversion of the alien stand to natural forest through selective cutting (thinning) of the invader plant (nurse) stand stems, with the nurse stand having relatively small stem diameters. However, sometimes the nurse stand is composed of large, branchy, and useful trees of one or more alien tree species that could be used for timber, poles, fuelwood, charcoal, and so on (Geldenhuys and Bezuidenhout 2008). The first priority would then be to remove all utilizable stems before the rehabilitation actions are implemented. The non-harvestable tree debris could be left on site but in sections not longer than 2 m (as bird perch sites, as microsites for seedling establishment and small animals and micro-organisms, and as a slow-release "fertilizer" through natural wood decomposition). However, if such an alien tree stand grows inside the riparian zone or on the banks of a river, then the first priority would be the safety and security of the river in terms of people's lives, infrastructure (bridges, pipelines, water pumps, etc.), and ecotourism activities on and along the river. Then all alien trees should be removed in one action to prevent stream blockage by drifting branch wood and deformed non-harvestable main stems with associated increased flood levels and infrastructure damage. If good native forest regeneration would be present in such an alien tree stand, then care should be taken to minimize damage to the native tree regeneration by the falling cut nurse stand trees. But even if the native tree regeneration is damaged by the cut trees, then most of the regenerating woody species will recover through relatively fast vegetative regrowth. In this latter case of removing large tree debris along the river, the non-harvestable tree debris could be burnt on site with care to keep the burnt spot footprint small and to prevent scorch of the regenerating plants of native woody species.

Step 4: Plant forest seedlings in spots without seedlings. Planting should be limited to spots in the stand with no forest seedlings or to specific species that do not disperse easily. Seedlings could be collected from clusters of seedlings (maximum 50% of the seedlings in dense clusters) from the nurse stand or the nearby natural forest. Such planting should be done during misty or rainy weather to ensure success. After the ash bed of burnt large-tree debris on sites along a stream bank has cooled down, sods of topsoil and small plants from the areas around the burnt spot could be placed on the burnt spot to ensure rapid recovery.

The results from the various rehabilitation actions according to this approach have shown that a pioneer or planted or naturalized nurse stand can be converted to natural forest, where this is not the case in the normal fire zone. Even in several *Chromolaena odorata*–covered disturbed sites in subtropical coastal forest areas in South Africa and in tropical moist forest areas in the Democratic Republic of Congo (DRC), the forest tree species eventually grew through the *C. odorata* cover. In the DRC, the *C. odorata* rapidly covered exposed sites after mining, but native pioneer tree species such as *Trema orientalis* and *Musanga cercopioides* established underneath *C. odorata* cover and then grew through it to outshadow *C. odorata*. This practice has benefits that make it better than the conventional clearing of invasive alien species. It can be done much cheaper because nature does a large part of the job for free (birds and mammals carry many of the seeds into the stand). The manipulative "interference" through selective thinning is simply to speed up the stand development process! A practice of repeated clearing of invader plants, for example, in the riparian zone, can be changed into a system of regrowing forest with much reduced maintenance costs. The removed stems could be used for laths, poles, firewood, timber, depending on the development stage of the nurse stand. Medicinal plant crops and other forest species could be harvested from such stands (Geldenhuys and Delvaux 2007). The practice, therefore, provides options for small business development parallel to the rehabilitation process. It is a different system, a different mindset, and a different approach, with many hidden values to the owner of the land, adjacent communities in many areas, and society as a whole.

16.7   Practical management of shade-tolerant alien invaders in natural forest

Shade-tolerant invasive species pose a real threat to the integrity of the mature forest, such as the camphor tree (*Cinnamomum camphora*), and sub-canopy and understory trees such as *Pittosporum undulatum* and *Ligustrum* species (Richardson and Brink 1985, Geldenhuys et al. 1986). In the Cape Peninsula (Cape Town), South Africa, the South African national tree and a nationally protected tree, *Afrocarpus (Podocarpus) falcatus*, grows about 250 km to the west and outside of its natural range and has become an invasive "alien" tree species. Its bat-dispersed seeds are spreading from a few scattered planted trees within Table Mountain National Park and also from the suburbs below the national park. Dense seedling banks are established in the understory of the small natural forest patches and have to be removed (with much debate over this controversial issue). In general, these species have a very confined, patchy occurrence in the natural forest. The best strategy is to remove the seed-bearing trees inside or close to the forest and to maintain a close observation or monitoring of such areas to prevent the seedlings from becoming dense and suppressive or growing into future seed-bearing trees. Fortunately, most shade-tolerant species, native or alien, grow relatively slower under the forest canopy.

The spiny and succulent climbing *Pereskia aculeata* appears to be the most serious IAP in coastal forest areas of South Africa. Currently, its distribution is still very patchy in disjunct localities. The plant climbs up the stem of a tree and forms dense tangles draping over and smothering the tree crown. A biological control agent, the Pereskia flea beetle (*Phenrica guérini*), has been released in some sites, but at present the action is still slow (Williams 2008). The best current approach is to cut the stems hanging to the ground as high up as possible above ground level. This causes the crown tangles to wither and die, but that can take some time. All Pereskia stems cut at ground level maintain growth and need to be sprayed with suitable herbicide. Those rooted in the ground sprout aggressively and those stem pieces lying on the ground develop roots and sprouts. The action is to accumulate all the cut stems and branches into small piles in forest gaps and to burn them once relatively dry. In forests where a patch is heavily infested and where the crowns are difficult to access, it may even be better to cut and burn the entire infested patch, including the native trees, and then use either native or alien pioneer species to form a nurse stand to manage the site back to natural regrowth forest as described earlier.

16.8   General discussion: A strategic framework for alien plant control in the forest environment

The development history of the informal adaptive management approach developed since the initial IAP survey of 1985 for the natural forest environment in the South African landscape provides a useful basis for developing a general strategic framework for alien plant assessment, control priorities and management practices around natural forests, in three phases (Figure 16.8).

In Phase 1, the alien plant species in different eco-regions in the forest environment need to be identified and characterized for an objective assessment of their invasive status. Generally, in most areas, there is much emotion, realism, or skepticism that varies amongst scientists, resource managers, and the general public, which needs to be considered for an objective assessment. Questionnaires need to be carefully designed to objectively obtain relevant information from the different types of stakeholders in each eco-region. The categories for the information to be supplied by the respondents need to cover the categories in parts 2 and 3 of Figure 16.8 to enable the initial categorization of the recorded alien plant species: functional attributes of the species such as typical growth or life form, fruit or seed type in relation to dispersal mechanisms, regeneration strategies, and ecological drivers for optimal functioning; behavior and spread of the species in relation to the forest stand conditions; impacts of the species in the forest environment; and potential benefits that will have to be considered in control of the invasiveness of the species. This information should provide a basis for rating the invasiveness of each recorded species of each eco-region and forest type and to list the species that need priority attention for control. The prioritization needs to be developed in terms of seriousness of threats, impacts, and perceived problems of control. The process of Phase 1 should preferably take 3 to 6 months, but not longer than 1 year. Eighty to ninety percent of the invasive species should be identified and the relevant information captured during this period through a well-coordinated effort between national coordinators and experts from the different eco-regions.

In Phase 2, consensus needs to be developed on the prioritization of the invasive alien species in terms of control. Initially, people from the different eco-regions (by their main vegetation types) need to prioritize the invasive species independently. Afterwards, national or even regional consensus needs to be reached at a relevant workshop in the country or region. Some smaller countries may share similar alien species and forest conditions and the process could be managed for the region as a whole. After consensus has been reached on the priority species by different ecological groups for the different eco-regions and forest types or conditions, it is useful to develop an information base and common ecological understanding of the priority of invasive alien species and how they behave under different conditions in the forest environment within the ecological groups. The forest types relate to forest dynamics based on species assemblages with common adaptations to specific disturbance regimes within different moist to dry environments,

---

PHASE 1: IDENTIFICATION AND INITIAL PRIORITIZATION OF INVASIVE ALIEN PLANT SPECIES

1. National survey of alien plant species in different natural forest eco-regions

2. List alien species by Eco-region and Categories of functional attributes

Growth/Life form	Fruits/Seeds & Dispersal	Regeneration strategy	Ecological drivers
Woody (tree, shrub, liana) Herbaceous (fern, vine, epiphyte, geophyte, graminoid, forb)	Fleshy/Dry Large/small Birds, Mammals, Wind, Water, Ballistic	Seedling/Vegetative Period to reproductive maturity Seed longevity	Shade tolerance Fire tolerance Cold tolerance Drought tolerance Seasonality

3. List and prioritize alien species by Forest habitat/condition, Species spread/abundance, Impacts and control issues, and potential Products/uses

Forest habitat/condition	Species spread/ abundance	Impacts and Control Issues	Use value and Other benefits
Moist/dry sites Closed forest Forest gaps (small/large) Forest margin Degraded forest (broken canopy, scattered trees, all trees removed, all vegetation removed)	Distribution in area (widespread, scattered, occasional) (clusters/groups, individuals) Abundance (abundant, intermediate, few)	Seriousness of impacts (type and duration of impact) Control problems (regular widespread seed spread, root suckering, aggressive vegetative regrowth, etc)	Timber & wood products Firewood/Charcoal Fibre for crafts/binding Traditional medicine Horticulture & Landscaping Food crops Wind breaks

PHASE 2: CONSENSUS PRIORITIZATION AND INFORMATION BASE OF INVASIVE ALIEN PLANT SPECIES

4. Workshops: Local and regional consensus on species prioritization

5. Develop information base and ecological understanding of priority species by Eco-region

Status & Behavior in different natural systems	Status & Behavior in artificial production systems	Research requirements and priorities
Forest development stage (pioneer, early regrowth, advanced regrowth, mature) Causes of degradation or changed gap size (fire, grazing, over-utilization, crop cultivation, etc) Conflicts with which native species (plants/animals)	Artificial production systems (Timber/tree plantation, fruit orchards, cultivated crops, grazing areas, etc) Stage of production system (site preparation, mature production stage, fallow period, etc)	History and status of invasion (how introduced, how long ago) Invasion stage (early, optimal, declining, naturally controlled) Specific disciplines (ecological, social, economic) (biological, mechanical, chemical controls)

PHASE 3: DEVELOP GUIDELINES FOR COST-EFFECTIVE INVASIVE ALIEN PLANT CONTROL

6. Implement invasion plant control, management, and monitoring procedures for each group of related priority species through adaptive management approaches

Alien species with no invasive tendencies	Invasive shade-intolerant alien species in disturbed forest	Invasive shade-tolerant alien species in closed-canopy forest
No action	Alien plant stand manipulation to facilitate forest regrowth	Total removal of all seed-bearing mature plants by various means: biological, mechanical, chemical.

Figure 16.8 Schematic outline of steps in a process to develop a strategic framework for the development and implementation of invasion plant control, management, and monitoring procedures for priority problem species.

which determine response patterns to different invasive alien species. It is also useful to understand how these alien species behave in the changed environments of artificial production systems such as agricultural crops, fruit orchards, timber plantations, grazing areas, slash and burn sites, and the development stage of the production system. Processes of invasion may be very different from natural systems and may relate to cultivation or management activities or may be a problem only when the particular crop is harvested and the area is left fallow, and management and control measures may be very different. The situations or conditions may affect different alien invader species differently. Such an information base will provide questions that need to be addressed in terms of research requirements and priorities for each identified invasive alien species or group of related species. For example, it may be useful to know how long the species had been present in a particular area, how long it took to become invasive, and in what invasion stage it currently is in the particular area: early invasion stage, optimal invasion stage, declining invasion stage (how far or close it is to become naturally controlled in the area).

In Phase 3, the information base and ecological understanding should be used to develop specific guidelines for cost-effective control and management of the invasive alien species, as described in the earlier sections of this paper. In broad terms, there are two main categories for management of invasive alien species: the shade-intolerant species in forest gaps, forest margins, and degraded forests, which function as nurse stands for the recovery of the shade-tolerant forest species, and the shade-tolerant invasive alien species in closed forest, which pose a real threat for the shade-tolerant forest species through competition for the same resources.

There are always many unknown aspects of the priority invasive species and also species that are currently not perceived as invasive. All these invasive alien species need to be monitored in terms of their impacts on the natural systems and their responses to different control actions. They should be managed in an adaptive management approach to learn from trying different approaches through monitoring results (Geldenhuys and Bezuidenhout 2008).

16.9   Concluding remarks

IAPs in the forest environment can be partners in forest rehabilitation. The natural succession process with the IAPs as the pioneer nurse stand can be used to reduce costs as follows:

- The aliens provide rapid ground cover, which prevents erosion and adds litter to a degraded site, which would help to restore the nutrient cycling process and increase the moisture content of the growing medium.
- The nurse stand provides shade and reduces strong wind to facilitate the establishment of shade-tolerant natural forest species and thereby recover the biodiversity as in any native pioneer stand.
- The nurse stand of alien species and the establishment of natural forest species can provide rural resource use needs (fuel wood, poles, construction material, and medicinal plant resources), which would reduce the resource use pressures on the natural forest.

However, resource managers need to be careful and watchful for the shade-tolerant alien invaders, which could gradually invade a natural seemingly uninvaded forest.

As indicated earlier, we need a changed mindset towards this cost-effective way of using alien invader plant stands to rehabilitate natural forests, to balance the impacts of an invader species with its potential benefits, and to control invaders where they pose a real threat in relatively few sites.

Acknowledgments

I acknowledge with thanks the invitation from the organizers of this book to make a contribution from my knowledge and experience in this relatively new field of alien plant control.

Appendix

Table 16.3 shows the occurrence and status of alien plants in the natural evergreen forest environment in South Africa as determined from a questionnaire survey via forestry officials and from the literature.

Table 16.3

a. Aliens occurring ONLY along forest margins and in large gaps

Herbs and soft shrubs

Agave spp (Nc T)	*Arundo donax* (T)	*Bidens pilosa* (N T)	*Cenchrus brownie* (Nc)
Cirsium vulgare (C Ec N)	*Coix lachrymal-jobi* (Nc)	*Commelina benghalensis* (Nc)	*Crotalaria agatiflora* (T)
Opuntia aurantiaca (Eu Nc)+	*Rivinia humilis* (Nc)+		

Woody shrubs and lianas

Acacia cyclops (C S Ec)+	*Bambusa balcooa* (T)	*Caesalpinia decapetala* (Eu Nu T)W*	*Carica benghalensis* (Nc)
Cassia bicapsularis (Nc) W	*Cassia coluteoides* (Nc)	*Cassia didymobotrya* (T)	*Cassia floribunda* (Nc T)W
Cestrum laevigatum (Ec Nc)+	*Hakea sericea* (S Eu Nc)	*Lantana camara* (S Ec)+ (N T)*	*Leptospermum laevigatum* (S)
Leucaena leucocephala (Nc)	*Ligustrum japonicum* (T)	*Montanoa bipinnatifida* (Nc)	*Montanoa hibiscifolia* (Nc)
Nicotiana glauca (T)	*Phytolacca* spp (S Nu T)+	*Pyracantha angustifolia* (T)	*Ricinus communis* (Nc T)W*
Rosa sp (T)	*Rubus niveus* (T)+	*Sesbania punicea* (C S) (Nc* T+)W	*Solanum mauritianum* (W Eu) (N T)W*
Tecoma stans (T)	*Vitis* sp (T)		

Trees

Acacia dealbata (T)+	*Acacia decurrens* (T)	*Acacia pycnantha* (W Ec)+	*Acacia saligna* (W S Ec)+
Agathis australis (S)	*Albizia lophantha* (W S)+	*Anacardium occidentale* (Nc)	*Araucaria cunninghamii* (S)
Bauhinia variegate (T)	*Casuarina* spp (Ec Nc T)	*Chamaecyparis lawsoniana* (S)	*Cupressus* sp (T)
Ensete spp (Nc)	*Eucalyptus* spp (C S Eu N T)	*Litsea sebifera* (Nc)	*Mangifera indica* (N)
Melia azedarach (Nc T)*	*Morus alba* (T)+	*Pinus* spp (C S E N T)	*Populus* spp (C S T)+
Psidium cattleianum (Nc)	*Prunus persica* (T)	*Quercus robur* (C S Eu)	*Salix babylonica* (T)+
Schinus terebinthifolius (Nc)+	*Sequoia sempervirens* (C S)	*Syncarpia glomulifera* (C S)	

Table 16.3 *(Continued)*

b. Aliens occurring along forest margins, in small and large gaps and in closed forest

Herbs and soft shrubs

Opuntia ficus-indica
(Eu Nc T)+

Woody shrubs and lianas

Chromolaena odorata (N)W*	*Homalanthus populifolius* (C)	*Passiflora edulis* (S E N T)	*Pereskia aculeata* (Ec Nc)W*
Rubus cuneifolius (C S Eu Nu T)*	*Solanum seaforthianum* (Nc)W*		

Trees

Acacia longifolia (C S Eu N)+	*Acacia mearnsii* (C S E N T)W*	*Acacia melanoxylon* (W S Eu Nu T)W*	*Cedrela toona* (C T)+
Cinnamomum camphora (C S Nc)*	*Citrus* spp (S Eu Nu T)	*Grevillea robusta* (T)+	*Jacaranda mimosifolia* (Nc T*)
Pinus patula (T)*	*Pittosporum undulatum* (C)+	*Populus canescens* (C)+	*Psidium guajava* (Nc T)*

Source: Geldenhuys et al. (1986)

Note: Provinces and Eco-regions: C, Western Cape (Winter rainfall); S, Southern Cape (All-year rainfall); E, Eastern Cape; N, KwaZulu-Natal; T, Mpumalanga and Limpopo (E, N & T, Summer rainfall). Zones within region: u, Upland and inland; c, Coastal. Range and type of invasion: W, widespread; +, passive invader able to grow and reproduce but not dispersed widely; *, active invader able to grow, reproduce, and disperse widely.

References

Berliner, D.D. 2009. *Systematic conservation planning for South Africa's forest biome: An assessment of the conservation status of South Africa's forests and recommendations for their conservation.* PhD thesis, University of Cape Town, Rondebosch, South Africa.

Cawé S.G., and C. J. Geldenhuys. 2007. *Resource status and population dynamics of target species in natural forests of the Port St Johns Forest Estate: A basis for sustainable resource use.* Report for Project 2006-397, 102 pp. Pretoria: Directorate: Forestry Technical Services, Department of Water Affairs and Forestry.

Cooper, K.H. 1985. *The Conservation Status of Indigenous Forests in Transvaal, Natal and Orange Free State, South Africa.* Durban, South Africa: Wildlife Society.

DWAF. 2007. Report on commercial timber resources and primary round wood processing in South Africa: 2005/2006, 140 pp. Pretoria, South Africa: Department of Water Affairs and Forestry.

FAO. 2003. *Workshop on tropical secondary forest management in Africa: Reality and perspectives.* Proceedings of Nairobi Workshop, December 9–13, 2002, 390 pp. Rome, Italy: FAO.

Feely, J.M. 1987. *The early farmers of Transkei, Southern Africa, before A.D. 1870.* Cambridge Monographs in African Archaeology 24, BAR International Series 378. 142 pp. Cambridge, UK.

Geldenhuys, C.J. 1986. Costs and benefits of the Australian Blackwood, *Acacia melanoxylon*, in South African forestry. In *The Ecology and Management of Biological Invasions in Southern Africa*, eds. I.A.W. Macdonald, F.J. Kruger, and A.A. Ferrar, pp. 275–283. Cape Town, South Africa: Oxford University Press.

Geldenhuys, C.J. 1991. Distribution, size and ownership of the southern Cape forests. *South African Forestry Journal* 158:51–66.

Geldenhuys, C.J. 1994. Bergwind fires and the location pattern of forest patches in the southern Cape landscape, South Africa. *Journal of Biogeography* 21:49–62.

Geldenhuys, C.J. 1996. The Blackwood Group System: Its relevance for sustainable forest management in the southern Cape. *South African Forestry Journal* 177:7–21.

Geldenhuys, C.J. 1997. Native forest regeneration in pine and eucalypt plantations in Northern Province, South Africa. *Forest Ecology and Management* 99:101–115.

Geldenhuys, C.J. 1999. Restoration of the biodiversity and productivity of degraded forest by planting useful tree species. In *Gestion Agrobiologique des sols et des systèmes de culture*, eds. F. Rasolo, and M. Raunet, pp. 543–551. *Proceedings of an International Workshop*, Antsirabe, Madagascar, 23–28 Mar 1998, Anae, Cirad, Fafiala, Fifamanor, Fofifa, Tafa, Montpellier, France, Cirad, collection Colloques.

Geldenhuys, C.J. 2002. *Acacia melanoxylon* in South Africa: Commercial and conservation issues in resource management. In *Blackwood Management: Learning from New Zealand*, ed. A.G. Brown, pp. 28–35. Rotorua, New Zealand: International Workshop.

Geldenhuys, C.J. 2008. Can I manipulate alien plant stands to rehabilitate natural forest? *Dendron* 40:38–44.

Geldenhuys, C.J. 2009. Managing forest complexity through application of disturbance-recovery knowledge in development of silvicultural systems and ecological rehabilitation in natural forest systems in Africa. *Journal of Forest Research* 15:3–13.

Geldenhuys, C.J., and L. Bezuidenhout. 2008. *Practical Guidelines for the Rehabilitation of forest-related Streambank vegetation with removal of Invader plant stands along the Berg River, Western Cape.* Unpublished Report FW-02/08, 39 pp. Pretoria, South Africa: Forestwood.

Geldenhuys, C.J., and C. Delvaux. 2007. The *Pinus patula* plantation … A nursery for natural forest seedlings. In *Multiple Use Management of Natural Forests and Woodlands: Policy Refinement and Scientific Progress—Natural Forests and Savanna Woodland Symposium IV*, eds. J.J. Bester, A.H.W. Seydack, T. Vorster, I.J. Van der Merwe, and S. Dzivhani, pp. 94–107. Port Elizabeth, South Africa, May 15–18, 2006. Pretoria, South Africa: Department of Water Affairs and Forestry.

Geldenhuys, C.J., and D.R. MacDevette. 1989. Conservation status of coastal and montane evergreen forest. In *Biotic Diversity in Southern Africa: Concepts and Conservation*, ed. B.J. Huntley, pp. 224–238. Cape Town, South Africa: Oxford University Press. Geldenhuys, C.J., Le Roux, P.J. and Cooper, K.H. 1986. Alien invasions in indigenous evergreen forest. In *The Ecology and Management of Biological Invasions in Southern Africa*, eds. I.A.W. Macdonald, F.J. Kruger, and A.A. Ferrar, pp.119–131. Cape Town: Oxford University Press.

Henderson, L. 2001. *Alien Weeds and Invasive Plants: A Complete Guide to Declared Weeds and Invaders in South Africa*. Plant Protection Research Institute Handbook No. 12, pp. 300. Agricultural Research Council, Pretoria, South Africa: Plant Protection Research Institute.

Kohli, R.K., S. Jose, H.P. Singh, and D.R. Batish. (eds.). 2009. *Invasive Plants and Forest Ecosystems*, pp. 437. Boca Raton, FL: CRC Press.

Lawes, M.J., J.A.F. Obiri, and H.A.C. Eeley. 2004. The use and value of indigenous forest resources in South Africa. In *Indigenous Forests and Woodlands in South Africa: Policy, People and Practice*, eds. M.J. Lawes, H.A.C. Eeley, C.M. Shackleton, and B.G.S. Geach, pp. 227–273. Scottsville, South Africa: University of KwaZulu-Natal Press.

Loumeto, J.J., and C. Huttel. 1997. Understorey vegetation in fast-growing tree plantations on savanna sols in Congo. *Forest Ecology and Management* 99:65–81.

Macdonald, I.A.W., F.J. Kruger, and A.A. Ferrar. (eds.). 1986. *The Ecology and Management of Biological Invasions in Southern Africa*. Proceedings of the National Synthesis Symposium on the Ecology of Biological Invasions, pp. 324. Cape Town, South Africa: Oxford University Press.

Mala, W.A. 2009. *Knowledge systems and adaptive collaborative management of natural resources in southern Cameroon: Decision analysis of agro-biodiversity for forest-agriculture innovations*. PhD thesis, Stellenbosch University, Stellenbosch, South Africa.

Mucina, L., and C.J. Geldenhuys. 2006. Afrotemperate, subtropical and azonal forests. In *The Vegetation of South Africa, Lesotho and Swaziland*, eds. L. Mucina, and M.C. Rutherford, pp. 584–614. Strelitzia 19. Pretoria, South Africa: National Biodiversity Institute.

Parrotta, J.A. 1995. Influence of overstory composition on understory colonization by native species in plantations on a degraded tropical site. *Journal of Vegetation Science* 6:627–636.

Parrotta, J.A., and J.W. Turnbull. 1997. Special Issue: Catalyzing native forest regeneration on degraded tropical lands. *Forest Ecology and Management* 99:1–290.

Parrotta, J.A., J.W. Turnbull, and N. Jones. 1997. Catalyzing native forest regeneration on degraded tropical lands. *Forest Ecology and Management* 99:1–7.

Richardson, D.M., and M.P. Brink. 1985. Notes on *Pittosporum undulatum* in the South-Western Cape. Veld and Flora 71:75–77.

Rutherford, M.C. and R.H. Westfall. 1986. Biomes of southern Africa: An objective categorisation. *Memoirs of the Botanical Survey of South Africa* 54:1–98.

Shackleton, C.M., S.G. Cawe, and C.J. Geldenhuys. 1999. *Review of the definitions and classifications of South African indigenous forests and woodlands*. Report no. ENV-P-C 99007, p. 33. CSIR, Pretoria, South Africa: Division of Water, Environment and Forestry Technology.

Van Daalen, J.C. 1981. The dynamics of the indigenous forest-fynbos ecotone in the southern Cape. *South African Forestry Journal* 119:14–23.

Van Gemerden, B.S., H. Olff, M.P.E. Parren, and F. Bongers. 2003. The pristine rain forest? Remnants of historical human impacts on current tree species composition and diversity. *Journal of Biogeography* 30:1381–1390

Williams, H. 2008. *The Pereskia Flea Beatle (Phenrica guérini)—A Natural Enemy of Pereskia (Pereskia aculeata) in South Africa*. Dossiers on biological control agents available to aid alien plant control No 2. Pretoria, South Africa: Plant Protection Research Institute.

chapter seventeen

Economics of invasive plant management

Damian C. Adams and Donna J. Lee

Contents

17.1 Introduction

Overuse and misuse of natural systems has compromised their resilience and made them more vulnerable to invasions by alien plant and animal species (Lake and Leishman 2004). Invasive alien species (IAS) are now a leading cause of biodiversity loss, which is occurring at a staggering pace with extinction rates far exceeding natural levels (Mack et al. 2000). IAS-induced ecological changes can have profound impacts on economic systems. Natural areas provide critical ecosystem goods and services such as regulating the climate, filtering and storing water, forming and retaining soil, supporting valuable animal and plant species, providing renewable food and fiber resources, and recreation (de Groot et al. 2002). Some of these goods and services are priced through market transactions (e.g., timber) and are explicitly included in gross domestic product and other measures of economic output, but many (e.g., recreation or biodiversity) are not.

Recently, environmental economists have developed and refined nonmarket valuation methods to estimate the worth of ecosystem good and services. Findings based on these methods indicate that natural systems provide tremendous value. For example, Costanza et al. (1997) estimate that the nonmarket value of ecosystem goods and services is worth nearly two times as much as the global gross national product. Despite the inherent dependence of economic systems on healthy and resilient natural systems, the economic value of ecosystem functions is often overlooked or undervalued by policymakers, particularly when economic studies are lacking. This is true for IAS in Florida (Adams 2008).

Economists use benefit-cost analysis to evaluate the net benefits (NB) of policies or projects. This method directly compares costs and benefits, potentially including both market and nonmarket values. In the IAS context, benefit-cost analysis can evaluate prevention and control projects, public awareness campaigns, restoration efforts, and public investment in IAS research. Economic studies of IAS impacts in natural areas are lacking, which can lead to ineffective policy decisions (Adams 2007). This has been demonstrated for invasive plant (IP) maintenance control on public lands in Florida (Lee et al. 2009).

To evaluate the NB of an IP control or management policy, we consider the environmental damages and foregone economic values associated with the invasion as well as management costs. As IAS establish and grow into dense monocultures, they restrict recreation, crowd out native animal and plant species, pose danger to man-made structures (e.g., fire risk from *Melaleuca* or damage to waterway structures from zebra mussels [Adams and Lee 2012]), and cause a variety of deleterious ecological changes. Economic valuation methods put a price on these environmental damages, which can then be weighed against the costs of administering prevention and control programs.

In some cases, there are also important economic benefits associated with IAS that should be included in a benefit-cost analysis (e.g., free trade or aesthetic value). Both accidental and intentional introductions of IAS are linked to economically valuable activities, goods, or services. Accidental introductions happen primarily through international trade (e.g., fire ants, brown tree snakes, and zebra mussels), where species are unintentionally imported along with goods and the movement of people. Roughly four out of five IPs in the United States that arrived during the twentieth century are unintentional by-products of international trade (OTA 1993). Intercountry trade volume is positively correlated with invasions, and trade is expected to continue growing. All else equal, IAS pressure on natural ecosystems is likely to increase in the future (e.g., Levine and D'Antonio 2003). Preventing unintentional introductions will impede trade, so the benefits of IAS prevention efforts must be weighed against the loss in trade. Intentional introductions happen because someone views the species as beneficial (e.g., water hyacinth, *Melaleuca*, and catfish). For example, those in the horticulture and exotic pet trades profit from importing and selling potentially (and in some cases known) invasive species into the United States (e.g., Reichard and White 2001, Lee and Gordon 2006, Greene and Lee 2009). In the recreation context, zebra mussels clear the water column, which improves visibility for anglers (Lee et al. 2007). These factors present an important policy problem and a continued source of IAS pressure on natural systems.

Benefit-cost analysis can inform policy decisions by helping to identify IAS management alternatives that are effective (generate positive NB) and efficient (generate the most positive NB among feasible alternatives) (Stokey and Zeckhauser 1978). This chapter presents approaches that have been used to evaluate the costs and benefits of managing IAS for recreational use in natural areas. The cost and effectiveness of managing invasive aquatic and terrestrial plants has been quantified using dynamic bio-economic models. One model explores control of invasive terrestrial plants in public conservation forestlands (Lee et al. 2009). Another model examines control of aquatic plants (hydrilla, water hyacinth, and water lettuce) in public lakes (Adams and Lee 2007). The benefits of IAS management have been quantified using choice models and statewide surveys of recreational use of managed state parks (Adams et al. 2010, 2011). In these studies, findings showed positive NB from IAS control in natural settings.

17.2   *Bio-economic modeling*

We review two examples of bio-economic models that estimate the economic impact and the value of IAS control in natural areas. In both cases, the policy questions were whether status quo management was economically justified and whether an alternate level of management would increase or decrease social welfare (i.e., NB). Historically, IP management has been underfunded, which has serious ecological, fiscal, and welfare implications that we examine. The first example demonstrates the NB of a coordinated state program to control terrestrial IPs on public conservation forestlands in Florida. The program was established by the Florida Department of Environmental Protection (FDEP) to achieve maintenance control (holding IAS populations at very low levels) on public lands by providing direct resources to land managers, coordinating IAS control efforts of over 500 public entities, and supporting IAS research. When the program began in 1995, roughly 1.5 million acres of public lands were infested with IPs. By 2006, nearly 500,000 acres of public lands had been reclaimed from IP. However, funding for the project was inconsistent during this period, which hindered its success. Using a bio-economic model, we evaluate the NB of the program under status quo and alternative funding levels.

The second bio-economic model examines the impact of aquatic IPs (hydrilla, water hyacinth, and water lettuce) on recreation and state control expenditures on 13 large lakes in Florida. Absent adequate and consistent control efforts, invasive aquatic plants can prevent access to lakes, hinder enjoyment, and negatively impact sport fish populations. In rare cases, swimmers can even get trapped and die in hydrilla-infested waters. In the mid-1990s, lapses in control efforts contributed to lost recreation, larger seed, and tuber banks that increased future plant populations and elevated control costs in the subsequent years to reclaim the lakes. We combined the costs of IP management, biological growth functions for IPs, and demand for recreational freshwater fishing into the bio-economic model and simulated the NB of alternative IP management regimes.

For both cases, we specified bio-economic models to simulate the impacts of IPs on economic value, including the effectiveness and efficiency of the management program. The conceptual models for both cases are similar. Fundamentally, a bio-economic model links the biological characteristics of invasive species (e.g., reproduction, survival, growth, and spread) to the economic costs and benefits associated with their presence and/or management. The models include (1) representations of the IAS population, including plant coverage, plant age, and associated growth/spread rates; (2) management efforts, including type of management (e.g., biocontrol) and relative effectiveness; (3) direct costs of management; (4) a linkage of IP coverage to economic output, recreation, and so on; and (5) a NB equation that is used to net-out the costs and benefits of management alternatives (Figure 17.1). It is important to note that the models are not meant to perfectly represent all aspects of biological growth/spread or economic impacts. Rather, they capture the essential characteristics of both the biological organisms and the economic system, such that the models can evaluate the relative efficiency of management alternatives. For example, Lee et al. (2007) examined the expected economic impacts of freshwater zebra mussels in Lake Okeechobee, Florida, using a bio-economic model. That model uses a zebra mussel population matrix to include fecundity and survival in an economic model with known control strategies and costs, assumed arrival vectors and rates, and expected damages (and benefits) associated with competing management alternatives.

17.2.1   Mathematical model of upland invasive plant control

The terrestrial (or upland) plants bio-economic model as presented in Lee et al. (2009) includes an objective function that maximizes NB to the state and those who engage in nature-based recreation on public lands where IPs are a problem. Linked to the objective function are (1) a dynamic population matrix, which is a stylized representation of IP spread over time; (2) constraints that reflect physical, biological, and fiscal limits; and (3) a benefit function that links IP population levels to recreation. Public efforts to control IP are represented as a total cost function:

$$TC = \Sigma_t TC_t (1+r)^{-t} ; \quad t = 1, \ldots, T \tag{17.1}$$

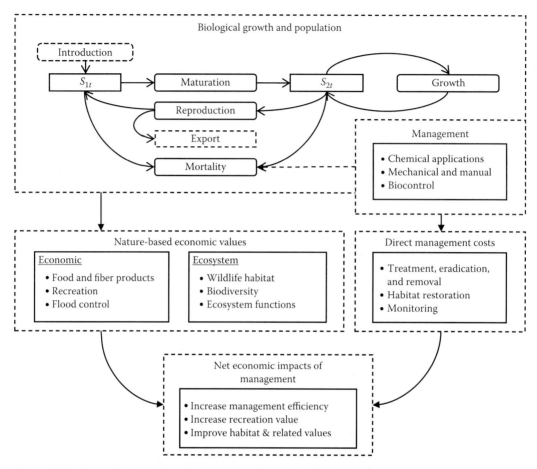

Figure 17.1 Conceptual model of the economic impacts of invasive plants.

where the total cost (TC$_t$) in each time period t is a function of acreage controlled X_t:

$$\text{TC}_t = f(X_t); \quad f' > 0, f'' > 0 \tag{17.2}$$

The population of IPs (S_t) is expressed as a function of the population at time $t - 1$, the growth/spread rate (G), level of control (X_t), and effectiveness of control (Γ). Because the species targeted for removal are well established, vector introductions and exports are of minor significance and are not included in the model. We express a matrix of transition probabilities for IPs as

$$S_t = GS_{t-1} + \Gamma X_t \tag{17.3}$$

where

$$S_t = \begin{bmatrix} s_{1t} \\ s_{2t} \end{bmatrix}, X_t = \begin{bmatrix} x_{1t} \\ x_{2t} \end{bmatrix} \tag{17.4}$$

$$G = \begin{bmatrix} g_{11} & g_{12} \\ g_{21} & g_{22} \end{bmatrix} \tag{17.5}$$

$$\Gamma = \begin{bmatrix} -\gamma_1 & \gamma_2 \varphi^{-1} \\ (1-\gamma_1)\varphi^{-1} & -\gamma_2 \end{bmatrix} \tag{17.6}$$

Equation 17.3 is expressed for different states of the IP population as

$$s_{1t} = g_{11}s_{1t-1} + g_{12}s_{2t-1} + \gamma_2 x_{1t-1} - \gamma_1 x_{2t} \tag{17.7a}$$

$$s_{2t} = g_{21}s_{1t-1} + g_{22}s_{2t-1} + (1-\gamma_1)x_{2t-1} - \gamma_2 x_{1t} \tag{17.7b}$$

where x_1 is acres controlled at first treatment, and x_2 is the level of control of within-year regrowth. The solution to Equation 17.3 is

$$S_t = [S_{t=0} - (I-G)^{-1}\Gamma X_t]G^t + (I-G)^{-1}\Gamma X_t \tag{17.8}$$

where I is the identity matrix, $\lim_{t\to\infty} G^t = 0$, and steady state conditions for S_t are expressed as

$$S_t = (I-G)^{-1}\Gamma X_t \tag{17.9}$$

17.2.2 Empirical model of upland invasive plant control

The cost function for IP management (Equation 17.2) is derived from 1997–2002 statewide acreage and cost data for invasive trees, shrubs and grasses, and vines for FDEP-funded projects from each of the 11 working group regions established under the program. Applying an industry supply and marginal cost function concept, we sort the individual projects by cost in ascending order, compute a cumulative sum (Kim et al. 2007), and specify a second-degree polynomial marginal cost function for IP management:

$$MC_{1t} = c_0 + c_1(x_{1t}) + c_2(x_{1t})^2 \tag{17.10}$$

where c_i for $i = 0, \ldots, 2$ are ordinary least squares regression parameter estimates, and x_{1t} is weed acreage controlled at first instance in year t (Sig. F < 0.01, $R^2 = 0.994$; intercept and first-order coefficients are significant at $p < .01$, but the second-order term is insignificant). Same-year follow-up treatments ($\gamma_2 x_{1t-1}$) are assumed to have static per-acre costs according to a simple average ($9.11/acre):

$$MC_{2t} = c_3 \tag{17.11}$$

Total IP stock at time t is composed of new plant growth (s_{1t}) from IP spread ($g_{12}s_{2t-1}$) and mature stock (s_{2t}), which can be controlled in year t at level x_{1t} (first instance) and x_{2t} (second instance). Integrating Equations 17.10 and 17.11, the total treatment costs are

$$TC_t = f(x_{1t}, x_{2t}) = c_0 x_{1t} + 0.5c_1(x_{1t})^2 + c_3 x_{2t} \tag{17.12}$$

Physical constraints limit the maximum level of control to the stock of IPs:

$$x_{it} \leq s_{it} \quad \text{for } i = 1 \text{ and } 2 \tag{17.13}$$

Public land area free of IPs (Z_t) at time t is given by Equation 17.14 where Z_t is the total public conservation land L_0 minus the public conservation land infested with IPs (above maintenance control levels):

$$Z_t = L_0 - s_{2t} - s_{1t} \tag{17.14}$$

L_0 is taken from FDEP (2006) and s_{it} are recursively calculated for each year t using the empirical model. The trajectory of Z_t varies by management regime.

We assume no significant barriers to introduction between invaded and noninvaded natural areas. Once land has been invaded, the invasion can continue unabated from infested lands to noninfested lands, even if geographically or spatially distant. This assumption is supported by the level of infestation and extent of transmission vectors in Florida. This simplifying assumption allows us to treat public conservation lands as one block, which is necessary given the lack of information on specific locations of IP control, infestation, and geophysical differences that may exist between the natural areas. Recall that the objective of the model is to estimate overall economic impacts, rather than determine the optimal location of IP control efforts.

We add budgetary and target acreage constraints to represent the five policy scenarios. IP management expenditures may not exceed the annual budget:

$$f(x_{1t}, x_{2t}) \leq Y_t \quad \text{where } Y_t = \bar{Y} \tag{17.15}$$

To simulate maintenance control, we constrain noninfested acreage to remain constant:

$$Z_t = Z_0 \tag{17.16}$$

The initial state condition defines the number of acres already infested by IPs at time t_0:

$$s_{20} = \bar{s}_{20} \tag{17.17}$$

Z_0, $\bar{s}_{20}$, and $\bar{Y}$ are taken from FDEP (2006). Annual benefit from public land is expressed as a linear function of noninfested acreage:

$$\text{TB}_t = \left(\frac{Z_t}{Z_0} \right) V_t \tag{17.18}$$

where V_t is the value of public land in year t based on status quo forecasted economic conditions at time $t = 0$. In Equation 17.18, forecasted rather than constant values are used. The rationale is that, ceteris paribus, the willingness-to-pay (WTP) for public land access is assumed increasing over time due to the growing demand for outdoor recreation, rising population numbers, and increasing expenditures by tourists. From Zhang and Lee (2007), a proxy for public land use value is expressed as a linear function of baseline statewide expenditures for hunting (H_0) and wildlife watching (W_0) that increases nonlinearly over time according to inflation factors for hunting and wildlife α_h and α_w:

$$V_t = H_0 \left(1 + \alpha_h \right)^t + W_0 \left(1 + \alpha_w \right)^t \tag{17.19}$$

NB associated with IP control in each time period is defined as

$$NB_t = TB_t - TC_t \qquad (17.20)$$

A typical means of comparing the results under various policies is to assign a discounted dollar value to the resource and sum to compute a present value. The objective function for the empirical model is

$$\max \sum_{t=1}^{T} NB_t (1+r)^{-t} \qquad (17.21)$$

The budget and acreage constraints are imposed or relaxed depending on the policy scenario.

17.2.3 Mathematical and empirical models of aquatic invasive plants

We model the impact of invasive aquatic plants in a similar way (see Adams and Lee 2007). Key differences from the upland plants case include the following: (1) aquatic plant coverage and growth are functions of plant-specific and lake-specific early-, mid-, and late-season growth rates calculated from observed plant coverage levels and (2) the economic linkage includes only benefits related to angler activity. We used econometric analysis to estimate functions for aquatic plant growth/coverage and angler activity based on unpublished state agency data from aquatic plants monitoring and control and creel fishing surveys.

17.2.3.1 Policy scenarios

17.2.3.1.1 Upland plants In the upland plants case, program efficiency and effectiveness are evaluated using five scenarios relating to different levels of fiscal support for the IP management program, which began in 1995 and experienced inconsistent funding. Each of the scenarios represents a distinct funding approach from different periods of the program. The scenarios are as follows: (1) continue annual spending comparable to fiscal years 2001 through 2004 ($6.164 million/year), which can be considered average support (*2001–2004 level*); (2) continue annual spending comparable to the highest level in fiscal year 2005 ($8.697 million/year) (*2005 level*); (3) NB maximization (*Max NPV*); (4) no control (*Do nothing*); and (5) maintenance control only to prevent spread (*Maintenance control*). *2001–2004 level* and *2005 level* scenarios simulate program impacts, given fixed annual expenditures over a 9-year period (2007–2016). The terminal simulation date 2016 is chosen to allow a comparison of outcomes within a 20-year window of the program's inception. With *Max NPV*, the budgetary constraint is relaxed and the model determines a control trajectory that maximizes present value NB. *Do nothing* restricts control efforts to $X_{it} = 0$. *Maintenance control* simulates the level of spending needed to keep IPs at existing levels over the planning horizon. A comparison of the scenarios reveals the value of the program at various expenditure levels and also the expenditures needed to improve the program's efficiency.

The policies are simulated using two hypothetical annual rates of IP growth/spread to noninfested public lands (g_{12}). Limited historical observations of IP spread are available in tropical and subtropical regions (Burnett et al. 2007). Since no estimated growth/spread rates for upland IPs are available, we investigate two assumed rates: (1) if untreated, IP acreage will increase by 50% within 15 years, $g_{12} = 0.028$ and (2) if untreated, IP acreage will double within 15 years, $g_{12} = 0.046$. Equations 17.1 through 17.21 are coded and solved using General Algebraic Modeling System (GAMS) software.

17.2.3.1.2 Aquatic plants   The aquatic plant bio-economic model is limited to three aquatic plants that account for much of the state's management budget: hydrilla, water hyacinth, and water lettuce. In the case of hydrilla, we accounted for tuber banks that help the plants persist even with considerable short-run management efforts, and we assumed that deviations from status quo hydrilla acreage will result in proportional declines in tuber production, with tubers remaining viable for 4 years (Best and Boyd 1996). For example, a regime that allows peak plant coverage on a lake to increase by 32% will cause 32% more tubers to be produced and fully infest the lake much faster in future years. We also assumed that hydrilla populations will be maintenance managed at new peak levels from the following year forward. Thus, early changes in hydrilla control will lead to sustained changes in IP acreage, whereas control efforts may impact only within-year populations of floating plants. Unlike hydrilla, floating IPs do not rely heavily on a seed bank. Similar efforts against floating plants are not expected to achieve long-run population reductions due to the high level of cross-contamination of lakes with water hyacinth and water lettuce by recreational boaters.

We simulate the impacts of three alternative management strategies for 13 large Florida lakes: (1) status quo ("Scenario A"), (2) skipping 1 year of control ("Scenario B"), and (3) adding an additional follow-up application of herbicide in the first year ("Scenario C"). Scenario A allows access for anglers and boaters during much of the year, with recreation access reduced during the summer months. This preserves much of the lake-based recreation value, but may not be the economically efficient choice. Initial steady state conditions were calculated for tuber banks and plant control efforts. Changes in initial control efforts alter the long-run steady state. We report the short-run (1 year) and steady state impacts of Scenarios *A*, *B*, and *C* that were simulated using GAMS software.

17.2.3.2   Results

17.2.3.2.1 Upland plants   Simulated costs vary greatly depending on scenario. From lowest to higher, the order is *Do nothing, Maintenance control, 2001–2004 level, 2005 level,* and *Max NPV. Do nothing* requires no expenditures. *Maintenance control* costs $21.6 million ($g_{12} = 0.028$) to $35.8 million ($g_{12} = 0.046$). *2001–2004 level* and *2005 level* costs are fixed at $49.3 million and $69.6 million respectively. *Max NPV* requires significantly higher expenditures of $210.4 million ($g_{12} = 0.028$, 3.02 times more than *2005 level*) and $214.5 million ($g_{12} = 0.046$, 3.08 times more). This level of spending may be prohibitively high, particularly when considering the front-loaded outlays required by this policy. Expenditures in year 1 would be $89.6–$89.9 million and then taper off below *2001–2004 level* spending by 2015 and below *2005 level* spending by 2016.

Compared to *Do nothing*, the policy alternatives either prevent IPs from spreading or reduce IP coverage significantly, and growth/spread rates significantly impact success. *Maintenance control* keeps IPs from colonizing 101,000 ($g_{12} = 0.028$) to 168,000 additional acres ($g_{12} = 0.046$) by 2016. Higher expenditures show large acreage gains compared with *Do nothing*. *2001–2004 level* keeps IPs from infesting 271,000 ($g_{12} = 0.046$) to 410,000 more acres ($g_{12} = 0.028$). Under *2005 level*, IPs cover 648,000 ($g_{12} = 0.046$) to 752,000 fewer acres ($g_{12} = 0.028$). *Max NPV* shows the largest gains: 1,029,000 ($g_{12} = 0.028$) to 1,093,000 ($g_{12} = 0.046$) fewer acres are infested by 2016. Simulated trajectories for acres infested when the IP spread rate is 2.8% are shown in Figure 17.2 for all five scenarios.

Florida's public forestlands generate large economic values when not infested with IPs. Recreational benefits alone based on annual expenditures for hunting and wildlife watching in Florida is estimated to be $2.6 billion/year (U.S. Fish & Wildlife Service 2002). As IPs are controlled, benefits accrue to the public. Under *Do nothing*, total benefits (TB) are $23.15 billion ($g_{12} = 0.028$) and $23.05 billion ($g_{12} = 0.046$). For all policy alternatives, TB

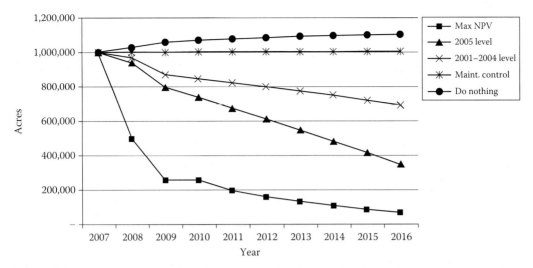

Figure 17.2 IP infestation on public forest lands over time under alternative management at 2.8% plant growth rate.

Table 17.1 Simulation Results from Bio-economic Model of Public Forest IP Management (in millions)

	Spread rate (%)	Do nothing	Maintenance control	2001–2004 level	2005 level	Max NPV
Net	2.8	23,150.4	23,287.5	23,658.0	24,015.5	24,749.0
Benefits ($)	4.6	23,046.4	23,273.3	23,433.7	23,829.3	24,737.0
Δ Net	2.8	—	137.1	507.6	865.1	1,598.6
Benefits ($)	4.6	—	226.9	387.3	782.9	1,690.6
Total	2.8	—	21.6	49.3	69.6	210.4
Costs ($)	4.6	—	35.8	49.3	69.6	214.5
Total	2.8	23,150.4	23,309.0	23,707.3	24,085.1	24,959.3
Benefits ($)	4.6	23,046.4	23,309.0	23,483.0	23,898.9	24,951.5
Δ Total	2.8	—	158.6	556.9	934.7	1,808.9
Benefits ($)	4.6	—	262.6	436.6	852.5	1,905.1
Uninfested	2.8	9.899	10.000	10.309	10.651	10.928
Acres at T	4.6	9.832	10.000	10.103	10.480	10.925
ΔUninfested	2.8	—	0.101	0.410	0.752	1.029
Acres at T	4.6	—	0.168	0.271	0.648	1.093

and NB are positive and improve significantly with expenditure increases, up to the level indicated by *Max NPV* (Table 17.1). Figure 17.3 compares the NB of the policy alternatives to *Do nothing*. *Maintenance control* yields +$158.6 million ($g_{12}$ = 0.028) to +$262.6 million ($g_{12}$ = 0.046) more TB and +$137.1 million ($g_{12}$ = 0.028) to +$226.9 million ($g_{12}$ = 0.046) more NB. Stated another way, *Maintenance control* prevents IPs from reducing the value of service flows from public conservation lands by $262.6 million when IPs double every 15 years.

2001–2004 level expenditures yield additional gains: TB between +$436.6 million ($g_{12}$ = 0.046) and +$556.9 million ($g_{12}$ = 0.028) and NB between +$387.3 million ($g_{12}$ = 0.028)

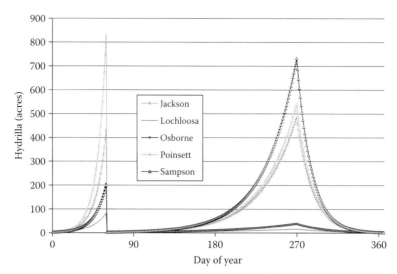

Figure 17.3 IP infestation on five public lakes following a lapse in management control.

and +$507.6 million ($g_{12}$ = 0.046). *2005 level* posts even higher gains: additional TB of +$852.5 million ($g_{12}$ = 0.046) to +$934.7 million ($g_{12}$ = 0.028), and additional NB between +$782.9 million ($g_{12}$ = 0.046) and +$865.1 million ($g_{12}$ = 0.028). *2001–2004 level* and *2005 level* indicate the potential returns to the FDEP program at or near recent spending levels. The program is very effective, allowing public conservation lands to generate additional environmental flows worth nearly a billion dollars and NB worth almost +$783 million to the Florida public by 2016 under simulated conditions. *Max NPV* simulates the most effective policy alternative. Recall the large initial outlays required by this policy (up to $89.9 million in year 1 of the simulation). This front-loaded strategy is far more expensive, but greatly increases the program's gains. Under *Max NPV*, TB are $1.80 billion ($g_{12}$ = 0.028) to $1.90 billion ($g_{12}$ = 0.046) more, and NB are $1.59 billion ($g_{12}$ = 0.028) to $1.69 billion ($g_{12}$ = 0.046) more than *Do nothing*.

All of the policy alternatives are effective—the simulations report positive NB, but positive NB is not sufficient to justify a policy choice. Efficiency must also be assessed. The efficient policy alternative is the one that produces the greatest NB (Stokey and Zeckhauser 1978). The clear efficient policy alternative is *Max NPV*, which boasts significantly greater positive NB—between 1.85 and 2.16 times more than *2005 level* and 3.15–4.37 times more than *2001–2004 level*.

17.2.3.2.2 Aquatic plants Using the aquatic plants bio-economic model, we compared three management alternatives. The short-run (year 1) impacts and long-run steady state impacts are shown in Tables 17.2 and 17.3, respectively. The status quo Scenario *A* establishes a baseline annual value of the 13 lakes with existing management. The 13 lakes are worth $64.78 million annually (or about $4.98 million/lake) to anglers who spend a total of 3.13 million fishing hours/year. The average fishing trip lasts 6 hours, which equates to 521,667 fishing trips/year (Thomas and Stratis 2001). Under Scenario *A*, IP coverage would exceed 80% of surface area approximately 88 days/year/lake, making them largely inaccessible to anglers.

A 1-year lapse in control (Scenario *B*) significantly reduces angler recreation, largely because of reduced access. Under this scenario, steady state peak IP coverage reaches

Table 17.2 Year 1 Cost of Altering Invasive Plant Control on 13 Lakes

Scenario		A	B	C
Total acreage treated	*acres*	13,785	0	16,776
Treatment costs	*$million*	$5.78	0	$4.90
Change in control cost[a]	*$million*	0	−$5.78	+$0.07

[a] Versus status quo Scenario A.

Table 17.3 Steady State Long-Run Impact of Invasive Plant Control on 13 Lakes

Scenario		A	B	C
Fishing hours	*hours*	3,135,966	2,487,857	3,299,093
Acreage treated	*acres*	13,785	23,948	8,193
Peak acreage	*acres*	21,085	43,620	7,163
Fishing benefit	*$million*	$64.78	$51.39	$68.15
Maintenance costs	*$million*	$4.82	$10.15	$1.64
Net fishing benefit	*$million*	$59.95	$41.24	$66.51
Change in net benefit[a]	*$million*	0	−$18.71	$6.55
Days not fished	*Days*	1,153	2,176	802
Mean days not fished/lake	*Days*	88.69	167.38	61.69

[a] Versus status quo Scenario A.

43,620 acres on the 13 lakes (Figure 17.3), fishing activity falls by 20.67%, lost fishing-related recreation worth $13.38 million/year is lost, and the average lake is largely inaccessible about 167 days/year. We note that implied in the steady state angler value calculation is the assumption that fish stocks and catch rates will not drastically change beyond what was reflected by our bio-economic model. The lapse in control initially saves the state $5.78 million, but future annual costs are much higher if the state reverts to maintenance control. With larger seed banks, the state must now treat 23,948 acres/year instead of 13,785 acres, driving control costs to rise by 2.1 times ($10.15 million). When also including the future lost fishing recreation, the steady state annual losses are $18.71 million/year.

A one-time increase in control (Scenario C) has the opposite effect and yields large steady state gains. Annual peak acreage falls to 7,163 from 21,085 on the lakes (Figure 17.4), reducing the number of days that the lakes are inaccessible by 27 days (to 61.69). Angler recreation increases by 5.20% over the status quo, and the value of the lakes jump by $6.55 million/year. This boost in IP control has one-time costs of $5.78 million, but steady state IP control falls from 13,785 to 8,193 acres and associated control costs fall by 66.01% over the status quo. These results suggest that the state could both significantly reduce the budgetary burden of IP control and improve the recreational value of its public lakes by boosting IP control efforts.

The results of both cases suggest that the state is under-investing in IP control. Short-run lapses in IP control lead to higher long-run control costs and drops in recreational activity, perhaps harming local economies. Boosting control efforts (and affiliated spending) pay substantial dividends in both reduced long-run budgetary costs and increased recreational value of terrestrial and aquatic natural areas. The models reveal that, when the rapid growth rates and seed banks of IPs are considered, the efficient alternative to managing IP would be to have large, upfront control efforts. For example, in the upland plants case, this approach generated up to 4.37 times more NB than the status quo.

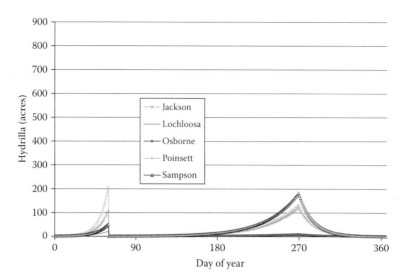

Figure 17.4 IP infestation of five public lakes following an increase in management control.

17.2.4 Choice modeling

For the preceding upland and aquatic plants examples, we relied on existing data and previous work to estimate the recreational use value of natural areas. In the upland plants case, we used statewide expenditures for hunting and wildlife watching as a proxy for the per-acre value of noninfested public lands. In the aquatic plants case, we relied on Florida Fish and Wildlife Conservation Commission creel survey data and earlier estimates of per-day freshwater fishing trip expenditures to approximate lost economic value. Although these approaches provide adequate approximations of lost recreation value, they do not explicitly incorporate the personal preferences of recreational users. They also involved situations that were comparatively data-rich, where established state programs collected necessary data or where adequate assumptions could be made based on the literature.

To establish the benefits of IP control in state parks based on direct observations of preferences of recreational users, we conducted a choice modeling experiment (Adams et al. 2011). This approach allowed us to focus specifically on the values held by those engaged in nature-based recreation. Choice modeling is an attribute-based nonmarket valuation technique that incorporates stated preferences (i.e., what people say they would prefer) and respondent characteristics into a utility model. The approach relies on a series of experimental choice questions that ask survey respondents to choose from pairwise alternatives of an environmental good that vary on the quality or quantity of key attributes. In the state park invasive species example, we asked respondents to indicate their preferences between two state parks that were identical, except for differences in these attributes: (1) park entrance fees, (2) condition and type of facilities, (3) abundance of native animal species, (4) abundance of native plant species, and (5) abundance of IPs.

We model the utility that individual i gets from a state park visit as a function of the park attributes X:

$$U_i = (\beta + \alpha\delta_i)X + \varepsilon_i \tag{17.22}$$

including a systematic component of the attributes and respondent characteristics δ_i and a random error term ε_i. The probability of i choosing park A over park B is

$$P(A) = P[U_i(X^A) > U_i(X^B)] \tag{17.23a}$$

$$P(A) = P[(\beta + \alpha\delta_i)(X^A - X^B) > (\varepsilon_i^A - \varepsilon_i^B)] \tag{17.23b}$$

Over N individuals, the likelihood function becomes

$$L(\beta, \alpha) = \prod_{i=1}^{N}\prod_{j=1}^{C} P_i(j)^{y_{ij}} \tag{17.24}$$

where $y_{ij} = 1$ if the respondent chose alternative j and 0 otherwise. Parameters β and α are chosen to maximize the fit of the observed choices, and when we assume an extreme value distribution on the error term, the likelihood function takes a logit specification (Greene 2000). For C alternatives, the probability that I chooses park A among all possible alternatives is (McFadden 1974)

$$P(A) = \frac{e^{(\beta + \alpha\delta_i)X^A}}{\sum_{j=1}^{C} e^{(\beta + \alpha\delta_i)X^j}} \tag{17.25}$$

For the pairwise park choice, this can be restated as

$$\text{Log}\left(\frac{P(A)}{P(B)}\right) = \sum_{k=1}^{K}\beta_k(X_k^A - X_k^B) + \sum_{k=1}^{K}\sum_{l=1}^{L}\alpha_{kl}\delta_l(X_k^A - X_k^B) \tag{17.26}$$

where β_k are main effects coefficients for the K attributes, l are respondent characteristics, and α_{kl} are attribute-characteristic interaction coefficients (Swallows et al. 1994).

The park entrance fees parameter estimate can be used as a payment vehicle to estimate the marginal WTP (MWTP) for the remaining attributes (Louviere 1988) as a calculated ratio of the changes in marginal utility for attribute k (e.g., more IP coverage) to that of fees:

$$\text{MWTP}_k = -\frac{\partial U/\partial k}{\partial U/\partial \text{Fee}} = -\frac{\beta_k + \alpha_{kl}\delta_l}{\beta_{\text{Fee}} + \alpha_{\text{Fee},l}\delta_l} \tag{17.27}$$

To design the survey, we used preliminary questionnaires, including (1) recreation experts and park managers to determine relevant park attributes ($n = 30$); (2) Florida residents to determine adequate descriptions and background information on IP for the survey ($n = 292$); and (3) Florida residents to establish key attributes and levels for the choice questions ($n = 329$). Most people do not have a good feel for the quantitative side of IP spread, coverage, or density. Instead, they use terms like "numerous and dense" or "few and dispersed" to describe IP where they camp, fish, and sightsee. Our attribute levels reflected this qualitative aspect of how respondents view the quality of state parks. The survey was reviewed by survey experts and pretested ($n = 242$). To significantly reduce respondent fatigue, the conjoint choice questions were split into two surveys that each included park entrance fees, facilities, invasive species, and either native animal or native plant species. This reduced average completion time from over 20 minutes to less than 10 minutes and significantly improved our expected completion rate.

The final instrument included background information (including pictures of IP and pictures and descriptions of typical park amenities), choice questions (e.g., Figure 17.5), and questions about IP-related experiences and knowledge (i.e., whether respondents considered IP beneficial, whether they had specifically avoided certain parks due to IP, or whether they had taken action against IP). In 2007, we implemented the two surveys to 13,330 Florida residents. We screened out those indicating that they had not engaged in nature-based recreation in the last year and were left with 1436 valid responses. Response rates were 10.68% and 10.86% for the native animal species and native plant species surveys respectively. The sampling margin of error for these surveys is ± 4% at the 95% confidence level.

We combined the survey data and parameterized our logit model of park choice. To correct a multicollinearity problem in the demographic variables using factor analysis, we conducted a varimax rotation on the resulting score variables. One full model and one nested model were compared. Both were statistically significant at the 99% confidence level ($p < .0000$), but a likelihood ratio test comparing the models was not statistically significant ($p = .3657$), suggesting that the nested model be used. All parameter estimates in the nested model were statistically significant at the 99% level of confidence, and several attribute-demographic interactions were significant at the 95% or 99% level.

Using fees as a payment vehicle, we calculated MWTP for one-level changes in IP coverage ("None," "Few and dispersed," or "Numerous and dense"), Facilities ("Minimal," "Adequate," or "Excellent"), Native plant species ("Low," "Moderate," or "High"), and Native animal species ("Low," "Moderate," or "High"). We found that the average respondent would pay an additional $3.72 in park entrance fees to improve park facilities, $6.71 to improve native animal species, and $3.73 to improve native plant species. For the IP attribute, the average respondent would pay an additional $5.41 to reduce IP (e.g., from "Numerous and dense" to "Few and dispersed") (Table 17.4). Moreover, the average respondents who feel personally impacted by IP are willing to pay more, $7.88, though if they indicated that IP are beneficial, they would not pay more to reduce IP coverage (and would pay $0.80 to increase IP coverage, Table 17.5). Demographic variables also influenced WTP estimates for IP. Gender and region (South, Central, or North Florida) impacted MWTP by

	Park A	Park B
Facilities condition	Adequate	Excellent
Native plant diversity	Moderate	High
Presence of invasive species	None	Numerous and dense
Fees	$10	Free

Which of the two parks do you prefer?

☐ Park A   ☐ Park B

Figure 17.5 Example of a choice question from recreational use survey.

Table 17.4 Per-Trip Marginal Willingness-to-Pay (MWTP) Estimates for Upland Park Attributes

Main attribute X	Change in attribute	Mean MWTP[a]	95% C.I.	
Facilities	*Improve* condition of facilities from "Minimal" to "Adequate"	$3.72	$0.81	$6.68
Animal species	*Increase* presence of animal species from "Low" to "Moderate"	$6.71	$5.57	$7.87
Plant species	*Increase* native plant species diversity from "Low" to "Moderate"	$3.73	$2.42	$5.07
Invasive plants	*Reduce* invasive plant coverage from "Numerous and dense" to "Few and dispersed"	$5.41	$2.77	$8.08

[a] Interaction variables held at observed means.

Table 17.5 Per-Trip Marginal Willingness to Pay (MWTP) Estimates for Demographic and Experiential Interactions with Invasive Plants

Main attribute X	Interaction δ	Mean MWTP[a]	Δ MWTP from baseline
Invasive plants	Benefit	−$0.80	−$6.21
	Knowledge	$6.24	+$0.83
	Affected	$7.88	+$2.47
	Knowledge and affected	$8.71	+$3.30
	Gender (female)	$5.37	−$0.04
	Education (HS or less)	$4.49	−$0.91
	Education (AA)	$6.38	+$0.97
	Age (18–34 years)	$4.61	−$0.80
	Age (35–54 years)	$6.12	+$0.71
	Income (<$35,000)	$4.27	−$1.13
	Income ($35,000–$99,900)	$5.79	+$0.38
	Region (Central Florida)	$5.42	+$0.01
	Region (South Florida)	$5.44	+$0.03

[a] Interaction variables were held at observed means except where indicated.

less than $0.04/trip, while age, education, and income were more substantial factors that influenced MWTP for IP by ±$0.38–$1.13.

The statewide value of IP control can be inferred by combining our average MWTP estimates for the IP attribute with annual state park attendance data for Florida. Our estimates are for one-level qualitative changes in IP, where the levels were "none," "few and dispersed, and "numerous and dense." Data on the actual condition of IP in state parks were unavailable, but IP are controlled on roughly one-third of state-managed acres in natural areas (DRP 2004). We assume this to mean that IP in the average state park are "few and dispersed" and would become "numerous and dense," absent adequate and consistent public management efforts. From this reference point, our MWTP estimates can be used to evaluate the worth of current and sustained IP control these parks.

Annual attendance data were available for 128 of 160 state parks in Florida (DRP 2004). Using park descriptions, we classified 115 of them as having primarily upland characteristics (i.e., not primarily associated with river, lake, or beach activities). For upland parks, the state recorded 16.5 million visits from October 2003 to September 2004,

Table 17.6 Statewide Willingness-to-Pay to Control Invasive
Plants in 115 Florida State Parks

Level	Annual Attendance	Annual WTP[a]
Per person	11.59	$62.70
Local visitors[b]	5,782,419	$31,283,307
All visitors	16,521,419	$89,380,877

[a] State resident visitors traveling <50 miles from home; assumed to
be 35% of total annual attendance.
[b] Assuming invasive plants are currently "few and dispersed."

with 65% of the visits from "nonlocal" areas more than 50 miles from the park (DRP 2004). An earlier survey by the authors indicated that those recreating in natural areas at least once during the year visited state parks 11.59 times/year on average. Applying our average MWTP estimates to the average trip count and annual attendance data, we infer that the average per-person annual WTP for IP management is $62.70, and statewide aggregate WTP to keep IP from becoming "numerous and dense" is $89.38 million/year ($31.28 million from local visitors, Table 17.6). However, our assessment of the condition of state parks was qualitative. If the current status of IP in state parks was "none" rather than "few and dispersed," then we infer that the aggregate statewide value of controlling IP is $178.76 million/year for all users. The aggregate WTP estimates are plausible. Other studies that evaluated IP impacts report high cost-benefit ratios (e.g., Milon et al. 1986, Colle et al. 1987, Adams 2007). For example, Adams and Lee (2007) evaluated the impact of IP management in just 13 large public Florida lakes and found it to be $30.2 million/year. Our estimates included a broad cross-section of Florida residents, which is expected to have a lower average WTP than a highly specialized segment of park users (e.g., anglers).

17.3   Discussion and conclusions

Economic models can be used to estimate the recreational use value of natural areas, quantify the potential changes in value due to IAS, and project the cost savings from prevention, early eradication, and maintenance management. This chapter reviewed two approaches evaluating the costs and benefits of IAS management: (1) bio-economic modeling and (2) choice modeling. Results of the bio-economic models indicate that IP management in Florida is economically justified based on the NB and that a higher level of control is economically warranted. In the upland plants example, front-loaded IAS efforts could substantially increase present value NB (up to 4.27 times higher). In the case of aquatic IPs, additional IAS management significantly increases steady state NB by improving fishing activity and lowering long-run control costs. Results of the choice modeling study indicate that Florida residents are sensitive to IP where they recreate and would pay considerable amounts (over $31.28 million) for IP control in state parks. Given that the state spends roughly $5 million/year to control IP in all terrestrial natural areas (not just state parks), this implies a very high benefit-cost ratio for IP control.

We note that the benefits estimated by these bio-economic and choice models are lower bounds of the total economic value of IAS control. In these cases, only recreation benefits were considered. The total value of protecting natural areas from invasive species is expected to be much higher, given that other important values (e.g., existence value) commonly associated with natural areas are omitted.

These studies highlight the potential and opportunities for future research. Ideally, decisions should be based on total economic value, which includes an accounting of all types of value associated with environmental changes, including both market and non-market values to all impacted persons. For managing IAS, a full accounting of all economic values may be limited by data availability, knowledge of invading species, modeling and valuation methods, and resources.

First, we note the need for improved integration of economic and ecological research. For example, while there were considerable studies measuring the growth and propagation of IAS that we model earlier, we could not easily incorporate findings from the biology, zoology, and limnology literature. Hydrilla has been studied (mostly in laboratory settings) for decades. Yet, studies of tuber production and growth provide results on scales that were not easily incorporated. Conversations with boaters and anglers indicated that hydrilla problems are perceived in terms of surface coverage of the plant, which was reflected in our bio-economic model. However, biological studies of hydrilla measured plant growth by length, volume, and dry weight and tuber production by size and dry weight, and these were provided over very small time frames (e.g., 45 days) that left us wondering how mature plants would grow/spread/produce.

Also, in many cases, data on lesser-known and potentially IPs are insufficient, and additional work is needed to help establish expected spread and growth rates for potentially IPs. Despite the high level of IAS arrival and invasion in tropical and subtropical regions, there are limited historical observations of IP population growth and spread rates (Burnett et al. 2007). Both the benefits and the costs of invasive species are typically unknown prior to their release into natural areas. Collecting original data is time- and resource-intensive, which may constrain economic studies to well-known IAS. Given that it is often much cheaper to prevent introduction than to manage IAS, this is a significant problem. Researchers have been working to establish appropriate evaluation schemes and models to evaluate and predict invasiveness ex ante, but these have had mixed success (e.g., Goodwin et al. 1999, Fox et al. 2003).

Additional work is also needed to address system complexity, resilience, and nonlinear relationships between IAS and environmental damages. It is known that system resilience can impact the success and impact of invasions by alien plant and animal species (Albers et al. 2006). Yet, economic studies typically assume well-behaved linear relationships for IAS growth, damage, benefits, and costs that do not accommodate complexity or resilience. They have also focused largely on single-service valuation (e.g., only measuring the impact on recreation), ignoring other important services provided by ecosystems, and focused on marginal value rather than total value of environmental functions. These studies may provide results that are irrelevant outside a narrow band of environmental changes.

Finally, more work is needed to address persistent problems with value elicitation. First, impacts of IAS are largely unfamiliar to most people, which creates a difficult challenge for revealed preference methods and stated preference methods like choice modeling. For example, survey respondents may lack adequate information to make consistent and informed choices regarding IAS. This can be addressed by providing information on IAS impacts, costs of management, and so on. However, researchers must be careful not to bias responses. More work is needed to develop and refine valuation methods for unfamiliar environmental goods like IAS and to examine the durability of valuation results in the context of respondents' prior information, knowledge, and experience with them. Second, studies have not adequately addressed how WTP for IAS management changes over spatial distances or time. For example, it is possible that future generations or those living in different regions will value the environment (and the impacts of IAS on it) very differently. Broadly applying preferences from one region and/or applying incorrect social discount rates may result in drastically overestimated or underestimated values. Finally, people find it difficult to monetarily define existence value, intrinsic value, nonuse value, or passive use value of environmental goods and services (Carson et al. 2001) that IAS can erode. As a result, researchers often encounter extreme values in stated preference surveys in the form of protest bids or lexicographic preferences for environmental goods and services (Spash et al. 2000). These have been largely unexamined in the IAS context.

Economic studies provide important guideposts for the value of ecosystem goods and services. Since valuation studies are lacking, there is a risk of many ecosystem functions being severely undervalued or even assuming a default value of zero. Available data, methods, and resources can be used to establish lower bounds of economic value that can inform policy decisions and offer additional context to the discussion over protecting critical ecosystem functions from IAS. However, additional work and resources are needed to address limitations in data and methods, improve the integration of ecological and economic studies, and increase collaboration between scientists, economists, and natural resource managers.

References

Adams, D.C. 2007. *The Economics and Law of Invasive Species Management in Florida*. PhD dissertation. University of Florida, Gainesville, FL.

Adams, D.C. 2008. The economics and law of invasive species management in Florida. *American Journal of Agricultural Economics* 90(5):1353–1354.

Adams, D.C., S. Bucaram, D.J. Lee, and A.W. Hodges. 2010. Public preferences and values for management of invasive aquatic plants in state parks. *Lake and Reservoir Management* 26(3):185–193.

Adams, D.C., F. Bwenge, D.J. Lee, S. Larkin, and J.R.R. Alavalapati. 2011. Public preferences for controlling upland invasive plants in state parks: Application of a choice model. *Forest Policy and Economics* 13(6):465–472.

Adams, D.C., and D.J. Lee. 2007. Estimating the value of invasive aquatic plant control: A bioeconomic analysis of 13 public lakes in Florida. *Journal of Agricultural and Applied Economics* 39:97–109.

Adams, D.C., and D.J. Lee. 2012. Technology adoption and mitigation of invasive species damage and risk: Application to zebra mussels. *Journal of Bioeconomics* 14(1):21–40.

Albers, H., M.J. Goldbach, and D. Kaffine. 2006. Implications of agricultural policy for species invasion in shifting cultivation systems. *Environment and Development Economics* 11:429–452.

Best, E.P.H., and W.A. Boyd. 1996. *A Simulation Model for Growth of the Submersed Aquatic Macrophytehydrilla (Hydrilla verticillata (L.f.) Royle)*. Technical Report A–96–8. Vicksburg, MS: U.S. Army Engineer Waterways Experiment Station.

Burnett, K., B. Kaiser, and J. Roumasset. 2007. Economic lessons from control efforts for an invasive species: *Miconia calvescens* in Hawaii. *Journal of Forest Economics* 13:151–167.

Carson, R.T., N.E. Flores, and N.F. Meade. 2001. Contingent valuation: Controversies and evidence. *Journal of Environmental and Resource Economics* 19(2):173–210.

Colle, D.E., J.V. Shireman, W.T. Haller, J.C. Joyce, and D.E. Canfield, Jr. 1987. Influence of hydrilla on harvestable sport-fish populations, angler use, and angler expectations at Orange Lake, Florida. *North American Journal of Fisheries Management* 7:410–417.

Costanza, R., R. d'Arge, R.S. de Groot, S. Farber, M. Grasso, B. Hannon, K. Limburg et al. 1997. The value of the world's ecosystem services and natural capital. *Nature* 387:253–260.

de Groot, R.S., M.A. Wilson, and R.M.J. Boumans. 2002. A typology for the classification, description and valuation of ecosystem functions, good and services. *Ecological Economics* 41:393–408.

Division of Recreation and Parks (DRP). 2004. *FY 2003/2004 Florida State Park System Economic Impact Assessment*. Tallahassee, FL: Division of Recreation and Parks.

Florida Department of Environmental Protection (FDEP). 2006. *Upland Invasive Exotic Plant Management Program*. Annual report, 2005–2006. Tallahassee, FL: Florida Department of Environmental Protection.

Fox, A.M., D.R. Gordon, and R.K. Stocker. 2003. Challenges of reaching consensus on assessing which non-native plants are invasive in natural areas. *HortScience* 38(1):11–13.

Goodwin, B.J., A.J. McAllister, and L. Fahrig. 1999. Predicting invasiveness of plant species based upon biological attributes. *Conservation Biology* 13(2):422–426.

Greene, G., and D.J. Lee. 2009. Social and economic impacts of the *Loricariid* catfish in Florida. In *Trinational Risk Assessment Guidelines for Aquatic Alien Invasive Species*, pp. 39–49. Montreal Canada: Commission for Environmental Cooperation.

Greene, W.H. 2000. *Econometric analysis.* 4th ed. Upper Saddle River, NJ: Prentice Hall.

Kim, C.S., D.J. Lee, G.S. Schaible, and U. Vasavada. 2007. Multiregional invasive species management: Theory and an application to Florida's exotic plants. *Journal of Agricultural and Applied Economics* 39:111–124.

Lake, J.C., and M.R. Leishman. 2004. Invasion success of exotic plants in natural ecosystems: The role of disturbance, plant attributes and freedom from herbivores. *Biological Conservation* 117:215–226.

Lee, D.J., D.C. Adams, and C.S. Kim. 2009. Managing invasive plants on public conservation forestlands: Application of a bio-economic model. *Forest Policy and Economics* 11:237–243.

Lee, D.J., D.C. Adams, and F. Rossi. 2007. Optimal management of a potential invader: The case of zebra mussels in Florida. *Journal of Agricultural and Applied Economics* 39:69–81.

Lee, D.J., and R.M. Gordon. 2006. Economics of aquaculture and invasive aquatic species—An overview. *Aquaculture Economics and Management* 10:83–96.

Levine, J., and C. D'Antonio. 2003. Forecasting biological invasions with increasing international trade. *Conservation Biology* 17(1):322–326.

Louviere, J.J. 1988. Conjoint analysis modeling of stated preferences: A review of theory, methods, recent developments and external validity. *Journal of Transport Economics and Policy* 10:93–119.

Mack, R.N., D. Simberloff, W.M. Lonsdale, H. Evans, M. Clout, and F.A. Bazzaz, 2000. Biotic invasions: Causes, epidemiology, global consequences, and control. *Ecological Applications* 10(3):689–710.

McFadden, D. 1974. Conditional logit analysis of qualitative choice behavior. In *Frontiers in Econometrics*, ed. P. Zarembka, pp. 105–142. New York, NY: Academic Press.

Milon, J.W., J. Yingling, and J.E. Reynolds. 1986. *An Economic Analysis of the Benefits of Aquatic Weed Control in North-Central Florida.* Economics Report 113. Gainesville, FL: Food and Resource Economics Department, University of Florida.

Office of Technology Assessment (OTA). 1993. *Harmful Non-indigenous Species in the United States.* OTA-F-565. US Government Printing Office: Washington, DC.

Reichard, S.H., and P. White. 2001. Horticulture as a pathway of invasive plant introductions in the United States. *BioScience* 51:103–113.

Spash, C.L., J.D. van der Werff ten Bosch, S. Westmacott, and J. Ruitenbeek. 2000. Lexicographic preferences and the contingent valuation of coral reef biodiversity in Curaçao and Jamaica. In *Integrated Coastal Zone Management of Coral Reefs: Decision Support Modeling*, eds. J. Ruitenbeek, R.M. Huber, and K. Gustavson, pp. 97–118. Washington, DC: The World Bank.

Stokey, E., and R. Zeckhauser. 1978. *A Primer for Policy Analysis.* New York: W.W. Norton & Co.

Swallows, S., T. Weaver, J. Opaluch, and T. Michelman. 1994. Heterogeneous preferences and aggregation in environmental policy analysis: A landfill siting case. *American Journal of Agricultural Economics* 76:431–443.

Thomas, M.H., and N. Stratis. 2001. *Assessing the Economic Impact and Value of Florida's Public Piers and Boat Ramps.* Tallahassee, FL: Florida Fish and Wildlife Conservation Commission.

U.S. Fish & Wildlife Service. 2002. 2001 *National Survey of Fishing, Hunting, and Wildlife-Associated Recreation* (FHWAR). FHW/01-NAT. Shepherdstown, WV: U.S. Fish and Wildlife Service, NCTC Publication Unit.

Zhang, J., and D.J. Lee. 2007. The effect of wildlife recreational activity on Florida's economy. *Tourism Economics* 13(1):87–110.

chapter eighteen

An economic analysis of the invasive plant problem associated with the horticulture industry in North America

Edward B. Barbier, Duncan Knowler,
Johnson Gwatipedza, and Sarah H. Reichard

Contents

18.1 Introduction

The growth in demand for nursery products by consumers has led to the expansion of the global horticultural industry. However, this expansion has also increased the risk of accidental introduction of harmful nonnative species in host ecosystems. Commercial horticultural activities, especially increased imports of exotic plant material and expansion of nursery operations, have become a significant pathway for invasive species to invade natural environments in many regions of the world (Maki and Galatowitsch 2004). Reichard and White (2001) cite research suggesting between 57% and 65% of the naturalized flora in Australia were intentionally introduced via horticulture and argue that the percentage is even higher for woody species introduced to North America.

There is a correlation between the number of plants introduced and the probability of invasion, a phenomenon called "propagule pressure." For instance, an analysis of catalogs from several nurseries in Britain found that the frequency of sale in the nineteenth century predicted naturalization today (Dehnen-Schmutz et al. 2007a). Another study found that the species whose seeds were sold then at lower prices were more likely to be invasive now, suggesting that they may have been purchased more often because they were less expensive (Dehnen-Schmutz et al. 2007b). In the Czech Republic, high levels of propagule pressure, measured as planting intensity, were found to lead to woody species eventually escaping cultivation (Pyšek et al. 2009).

Introduced nonnative invasive plant species cause significant damages to the host environment and are regarded as a major threat to native biological diversity (Mack et al. 2000, Urgenson et al. 2009). The negative consequences of introducing exotic plants include competition for resources with native species, increased nitrogen fixation in natural areas, changes in hydrological cycles, increased sedimentation, alteration of food webs, and increased frequency and intensity of disturbance cycles.

Sinden et al. (2004) estimate the costs from invasive plants in Australia, including damages and control expenditures, to be over A$4 billion per year. Pimentel et al. (2005) estimate total damages to the U.S. economy from nonnative invasive plants in natural areas and agriculture at about $35 billion per year and indicate that more than 5000 alien plant species have escaped and invaded the natural areas of the United States, displacing native species. Alien weeds invade approximately 700,000 hectares of U.S. wildlife habitat each year. Over 1000 introduced exotic plant species have been identified as a threat to the native flora as a result of their aggressive, invasive characteristics (U.S. National Park Service 2007). Bell et al. (2003) indicate that 40% of the endangered native species in North America are at risk from invasive species. They are considered to be second only to habitat loss and degradation in their imperilment of native species (Wilcove et al. 1998).

Thus, while the horticultural industry and its consumers may benefit from selling imported plants, they do not take into account the economic costs from accidental introduction of exotic invasive plant species. Instead, these costs are borne by society as an externality. As a result, the "privately" optimal number of nurseries established by the industry will diverge from the "socially" optimal number, taking into account the additional costs imposed by this externality. Correcting the externality requires government to adopt policy measures that are based on assessing the risk and potential damage costs associated with the introduction of potentially invasive plant species.

Implementing economic instruments to control invasive species has received attention in the recent literature. For example, Horan and Lupi (2005) consider the use of tradable risk permits to control invasive species, while Costello and McAusland (2003) analyze the relationship among volumes of goods traded, import tariffs, and the impact on accidental invasion. In contrast, Barbier and Knowler (2006) address the case where invasive species are not accidentally introduced, while Knowler and Barbier (2005) examine the possible use of "introducer pay" taxes in such cases. In both papers, the authors focus on the nursery sector of the horticulture industry as a significant pathway for the introduction of potentially invasive plants. In Knowler and Barbier (2005), they calculate the socially optimal number of nurseries taking into account both the probability that an ornamental plant can become invasive and the losses to the industry if the invasion occurs. The authors find that the socially optimal number of nurseries is lower than that of the existing nursery market, and they evaluate the use of taxes to restrict the number of nurseries to the social optimum. The optimal level of taxes is shown to be highly dependent on how the probability of invasion changes with even a marginal increase in the number of nurseries.

The purpose of this chapter is to explore ways of countering the problem of propagule pressure arising from the introduction of exotic plants through the horticulture industry. The findings of the study show that high license fees would reduce the number of nurseries significantly and thereby reduce the "propagule pressure" stemming from invasive plant species. More specifically, we examine how a policy intervention in the form of a tax (annual license fee) could induce horticultural nursery firms to internalize the risk of potential invasion and its associated costs and thus bring private incentives in line with socially optimal levels.*

To accomplish our goal, we use the results of a monopolistic competition model of the U.S. and Canadian horticultural industry developed by Barbier et al. (2009) that contrasts the privately optimal decision of the industry to establish nurseries with the socially optimal decision by a government that also considers the risk of an accidental invasion. We then carry out simulations of the potential trade-offs between the commercial profits from the nursery industry and expected social damages from the risk of invasion to determine the appropriate tax to be imposed on the U.S. and Canadian horticultural industry. We assume in our simulations that the North American industry is importing a new exotic species that exhibits potential invasiveness. We base the simulation on the example of a well-known invasive species in North America, purple loosestrife (*Lythrum salicaria*).

The monopolistic competition model of the North American horticultural industry presented in this chapter, and described in more detail in Barbier et al. (2009), allows us to characterize the consumer and producer surplus gained from nursery sales in new locations. However, because each new nursery selling a new exotic plant in a different location increases the risk of a potential plant invasion, social welfare must include not only consumer and producer surplus from sales but also the expected damages from a potentially successful invasion by the species. We assume that the risk of invasion depends on the characteristics of introduced plants and the total number of nurseries selling products based on these plants, which is supported by the ecological literature on past plant invasions (e.g., Rejmánek and Richardson 1996, Reichard and Hamilton 1997, Pheloung et al. 1999). We formulate this risk as a hazard rate and derive the relevant parameters from our earlier analysis of a new ecological dataset on historical introductions of herbaceous species into North America.

In Section 18.2, we sketch out the theoretical model developed by Barbier et al. (2009) for the horticultural industry operating under a monopolistic competitive framework in the presence of an environmental externality. We employ results from our previous empirical work estimating the short-run profit function for a representative nursery firm in the United States and Canada and the invasion hazard rate of a representative herbaceous species. We conclude by conducting policy simulations of the optimal tax (annual license fee) and the optimal number of nursery firms selling a new exotic plant species similar to purple loosestrife.

18.2 Theoretical framework for a horticultural industry model with risk of invasion

The monopolistic competition framework, originally developed by Dixit and Stiglitz (1977) and adopted by Barbier et al. (2009), fits the "stylized facts" of the North American horticultural industry. According to Singh (1999), the horticultural industry in the United States is large, complex, composed of many segments, and produces differentiated multiproducts. Nursery growers are diverse, producing hundreds and even thousands of plant taxa in operations of different sizes. There is the presence of fixed costs (e.g., plant propagation material) and scale economies in the industry (Singh 1999, Brooker et al 2003). The firms

---

* Throughout this chapter, we refer to the "horticultural industry" or "horticulture industry." This terminology is often used in reference to commercial wholesale or retail plant sales, or nurseries, when in fact the industry also includes landscape design and maintenance. In our work, we consider only nurseries.

sell the nursery taxa directly or through retail outlets to consumers who demand a broader selection. In many cases, the nursery firms producing different taxa are vertically integrated, that is, they are linked with specific retail outlets to sell differentiated products to local markets.* Since the production and retail process is vertically integrated and targeted to specific retail markets, we will consider each supplier to a differentiated product market as a single unit, which we will refer to as a "nursery firm." The presence of economies of scale implies there is decreasing average cost in nursery production (Panzar and Willig 1977, Bailey and Friendlaunder 1982). Thus, a small proportionate increase in the level of input factors leads to a more than proportionate increase in the level of output produced (Panzar and Willig 1977). This situation allows us to apply the Dixit–Stiglitz monopolistic competition framework to characterize production and supply by a representative nursery firm.

18.2.1   *Privately optimal conditions*

The key economic assumptions underlying the horticultural industry model by Barbier et al. (2009) are as follows. The model assumes an economy that produces two goods, a nursery product and a composite good representative of all other productions in the economy. Consumers derive utility from the consumption of a homogeneous composite good and a differentiated nursery good, the latter available on the market as many different plant selections or "bundles" that are close substitutes. Production of the composite good is by constant returns to scale. In contrast, each nursery firm produces its own unique selection of plants under increasing returns to scale, targeted to a specific consumer market in a given location. However, by selling the nonnative plant species (or plants based on its material) through establishing a new nursery in a different commercial location, the horticultural industry incurs the risk of a potential plant invasion in the natural environment causing extensive damages. The economic costs of such damages caused by harmful nonnative plant species are not borne by the industry but by society. Thus, the model also incorporates a social objective function that includes the risk and costs of a potential plant invasion to determine the socially optimal number of nursery firms selling a representative exotic species. In turn, this allows us to derive the optimal tax that internalizes the externality associated with the risk of an accidental invasion.

We assume that the horticultural industry consists of n nursery firms, and each firm produces its own selection of plants targeted to a specific consumer market in a given location.[†] Each bundle of plants sold by each nursery, i, is treated as if it were a single commodity, or product, with its quantity denoted by $q(i)$. However, two different nursery firms will not offer the same bundle of plants for sale in their respective markets. This allows

---

[*] The assumption of a vertically integrated nursery firm that imports plant material and sells it in retail markets is supported by evidence that the North American horticultural industry via its nurseries imports nonnative plant species for propagation and delivers plant products directly to the final consumers (Singh 1999, Brooker et al. 2005). A survey conducted by Brooker et al. (2005) estimates the nurseries' total annual sales made between wholesalers and retailers. The results show that for the 44 U.S. states, on average, 75% of the nurseries made some wholesale sales and 58% of all firms made some retail sales. Almost half of nursery firm survey respondents purchased source propagation material from Canada and 31 other countries including Netherlands, Mexico, and Argentina.

[†] Barbier et al. (2009) assume only one nursery firm per bundle, which is standard in the monopolistic competition literature (Baldwin et al. 2003). Suppose there is a firm already producing a bundle of nursery plants i, and another firm contemplating entering the market. If the entrant produces exactly bundle i, then the two firms will split the profit. The profit earned by the entrant will be below those that he/she could earn by producing a unique bundle. Given the assumption that it costs the same to produce each bundle, no entrant will find it profitable to produce an existing bundle.

each firm to sell its own differentiated mix of plants, and thus, it can act in the short run as a monopolist in its own market segment. But in the long run, each nursery firm's profits are driven to zero by free market entry and exit.

The production process of each firm uses labor as the variable input and a fixed cost input. There is a one-time fixed cost (in labor-equivalent units) of importing a new exotic plant species, k, which is part of the overall fixed costs of the nursery firm, that is, $k \in F(i)$. This assumption also implies that the unique bundle of plants, $q(i)$, sold by each nursery i, will contain either imported exotic plants or material propagated from exotic plant material (Avent 2003). The optimal price charged by the nursery firm is a constant markup over marginal cost. The equilibrium output of a firm depends on the presence of substitutes and the ratio of fixed to variable costs. If the nursery firm is faced with high fixed costs and the existence of close substitute, it would have to propagate stock and sell more output in order to break even.

In addition, following a standard assumption in the monopolistic competition literature (Matsuyama 1995, Baldwin et al. 2003), we assume that the equilibrium output of each firm must satisfy a resource constraint, which is the same for all nursery firms operating in a large "political jurisdiction" of markets, such as each state in the United States or province in Canada.

Denoting each jurisdiction as j, the total industry resources or total labor available $L(j)$ in each jurisdiction (i.e., state or province) comprises both the aggregate variable labor input and fixed labor, which include the one-time fixed cost of importing a new exotic plant species expressed in labor equivalent units. Therefore, the total number of firms in each jurisdiction is limited by the scarcity of the factors required for production of the differentiated products. In addition, Barbier et al. (2009) make the simplifying assumption that the fixed cost is likely to be approximately the same for all nursery firms in all jurisdictions, that is, $F(i) = F$.*

Based on these assumptions and the conditions determining the demand for horticultural products by consumers, the long-run number of nursery firms in each jurisdiction, $n^p(j)$, is

$$n^p(j) = \frac{(1-\gamma)L(j)}{F} \tag{18.1}$$

where $0 < \gamma < 1$ measures the substitutability between the different bundles of plants offered by various nursery firms. The privately optimal number of nursery firms established in each jurisdiction j in the long run depends on the degree of substitution between different bundles of plants offered by various nursery firms and fixed costs. The number of nurseries established is also proportional to the total resource supply available in each jurisdiction, $L(j)$. As fixed costs are lowered within the industry, or the amount of labor employed rises, the long-run equilibrium number of nursery firms established in each jurisdiction, $n^p(j)$, is expected to rise.

We can derive an expression for the representative nursery's short-run profits (π^s) as a function of the total number of nurseries $n(j)$ established in each jurisdiction:

$$\pi^s(n(j)) = \frac{1-\gamma}{\gamma}\left(\frac{L(j)}{n(j)} - F\right), \quad \frac{\partial \pi^s}{\partial n(j)} < 0 \tag{18.2}$$

---

* Fixed costs of horticultural nursery firms, as approximated by fixed labor resources per firm, appear to display little variation across these jurisdictions over significant periods of time. For example, see Tables 18.1 and 18.2.

The short-run profits of an individual nursery firm in each jurisdiction are a decreasing function of $n(j)$. Entry of new nursery firms reduces the profits of incumbent firms. The entry process is expected to continue as long as the gross profits exceed the fixed costs, that is, $\pi^s\left(n(j)\right) - F > 0$. Of course, the latter is the long-run profit function of the representative nursery, which can be written as

$$\pi\left(n(j)\right) = \pi\left(n(j)\right)^s - F = \frac{1-\gamma}{\gamma}\left(\frac{L(j)}{n(j)}\right) - \frac{F}{\gamma} \tag{18.3}$$

In the long-run equilibrium, the profits for all nurseries in all jurisdictions will vanish and Equation 18.3 will equal zero. Two outcomes immediately follow. First, setting Equation 18.3 equal to zero and rearranging confirms the long-run equilibrium condition Equation 18.1 for the privately optimal number of nurseries established in each jurisdiction, $n^p(j)$. Second, if this zero profit condition holds for all nurseries across all $1, \ldots, J$ jurisdictions, then from Equation 18.1, we also have an expression for the long-run privately optimal number of nurseries n^p established by the entire horticultural industry:

$$n^p = \sum_{j=1}^{J} n^p(j) = (1-\gamma)\frac{\sum_{j=1}^{J} L(j)}{F}. \tag{18.4}$$

Finally, note that when the number of nurseries in the entire horticultural industry reaches n^p, industry profits are zero, thus satisfying the long-run equilibrium condition.

18.2.2 *Socially optimal conditions*

In the absence of a plant invasion, societal welfare is the sum of the consumers' surplus from consumers' preference for more selections of the nursery good and the horticultural industry's profits or producer surplus. Consequently, we assume the government, or social planner, seeks to choose the number of nurseries established by the industry to maximize social welfare across all markets and jurisdictions. In our model, consumers' surplus is given by: $S(n) = \left(\frac{1-\gamma}{\gamma}\right)nq^\gamma = \left(\frac{1-\gamma}{\gamma}\right)\left(\frac{a}{\gamma}\right)^{\frac{\gamma}{\gamma-1}}n = Dn$. In the short run, price in the horticultural market is a fixed markup of variable costs so that output q is determined by parameters, that is, $q = \left(\frac{a}{\gamma}\right)^{\frac{1}{\gamma-1}}$, and D is essentially a parameter. As a result, aggregate consumer surplus is proportionate to the total number of nurseries operating in the industry. Denoting $\Pi(n)$ as aggregate industry profit, and expressing this as a function of the number of nurseries in the industry, social welfare is given by the sum of consumer and producer surplus as

$$W(n) = Dn + \Pi(n), \quad \Pi(n^p) = 0, \quad \Pi' > 0, \quad \Pi'' < 0 \tag{18.5}$$

It follows from our discussion of the long-run equilibrium condition of the horticultural industry that aggregate industry profits must equal zero when the number of nurseries reach n^p.* In the absence of any consideration of the risks and potential damages of a plant invasion, the social planner would choose the socially optimal number of nurseries, n^s, that would maximize consumer and producer surplus, which from Equation 18.5 satisfies the condition $D + \Pi'(n^s) = 0$. The n^s chosen by the social planner would be different from the privately optimal number of nurseries, n^p, established by the industry under the zero-profit condition (Equation 18.4).

However, if the exotic plant species may escape at some future time and become established successfully in the host environment, the social planner needs to consider this possibility in determining the socially optimal number of nurseries. As noted earlier, there is a need for policy measures that internalize this externality and limit the private number of nursery firms to the socially optimal number. This social regulatory framework must balance the consumer and producer surplus benefits from establishing new nursery firms with the expected social costs from the risk of a potentially harmful plant invasion. Adapting an approach from Reed and Heras (1992), we incorporate the risk of invasion in the model as a hazard rate (Knowler and Barbier 2005, Barbier et al. 2009).

A hazard rate is characterized as the probability that the plant invasion occurs at any time given that it has not yet invaded at that time. Barbier et al. (2009) assume that the hazard rate, h, depends on the number of nurseries established by the horticultural industry, n, the attributes of the exotic plant species, a_k, as well as time t:

$$h(t) = \varphi(n(t), a_k), \quad \frac{\partial \varphi}{\partial n} = \varphi_n > 0 \tag{18.6}$$

As each nursery firm is located in a unique jurisdiction serving the local market, it is reasonable to assume that these sites serve as dispersal points for the exotic plant species. This implies that the probability of invasion depends positively on the number of nursery firms, n, established by the industry. The characteristics of the exotic species, such as its leaf type, length of flowering period, germination, and reproduction attributes, may also influence the success rate at which the exotic plant invades the environment (Rejmánek and Richardson 1996, Reichard and Hamilton 1997, Pheloung et al. 1999).

If invasion occurs at some future time, τ, then damages of magnitude $G(A(\tau))$ are realized. These damages can be expressed as the product of the total area invaded, $A(t)$, and the average losses per hectare (ha) invaded, c. When discounted to the time of invasion, τ, using the instantaneous discount rate, δ, the present value of these losses is

$$G(\tau) = \int_\tau^\infty e^{-\delta(t-\tau)} CA(t) \, dt \tag{18.7}$$

The social planner maximizes the expected present value of welfare by choosing the number of nursery firms, $n(t)$, taking into account the sum of consumer and producer surplus in Equation 18.5, the hazard rate in Equation 18.6, and the expected present value of damages in Equation 18.7, subject to the growth in area invaded by the exotic plant species and $n \geq 0$. Solving the social welfare maximizing problem, Barbier et al. (2009) derive the following long-run equilibrium condition:

$$S'(n) + \Pi'(n) - \varphi_n \frac{\left[\Pi(n) + S(n) + \delta G\right]}{\delta + \varphi} = 0 \tag{18.8}$$

---

* In addition, the sufficient condition for social welfare to have a maximum is that the profit function must be concave.

Equation 18.8 is the long-run condition for establishing another nursery in the industry at the social optimum. In equilibrium, the contribution of a new nursery to consumer surplus and industry profits is $S'(n) + \Pi'(n) = D + \Pi'(n)$, and this term must equal the expected marginal social costs of this extra nursery, $\varphi_n \dfrac{[\Pi(n) + S(n) + \delta G]}{\delta + \varphi}$. These social costs consist of the increased likelihood of invasion due to an additional nursery, φ_n, multiplied by the penalty if the invasion occurs. The latter penalty includes the loss in profits and consumer surplus due to suspension of sales of the exotic plant species, $\Pi(n) + S(n)$, plus the annualized value of damages from the invasion, δG. These social costs are converted into a present value using an effective discount rate, $\delta + \varphi$, which comprises the social discount rate plus a risk premium represented by the hazard rate function.

If the social planner ignores the expected damages from invasion, that is, treats the second expression on the left-hand side of Equation 18.8 as zero, then we obtain the original long-run equilibrium condition from maximizing $W(n)$ in Equation 18.5, that is, $D + \Pi'(n^s) = 0$. In contrast, the privately optimal zero-profit condition was simply $\Pi(n^p) = 0$.* Denoting n^* as the socially optimal number of long-run nurseries that satisfies the social planner's condition in Equation 18.8, it follows that it is unlikely that n^* will be equal to either n^s or n^p.

By rearranging Equation 18.8, it is possible to get an expression for the long-run socially optimal number of nursery firms, n^*:

$$n^* = \left(\frac{\Pi'(n) + D}{D} \right) \frac{\delta + \varphi}{\varphi_n} - \frac{\Pi(n) + \delta G}{D} \tag{18.9}$$

Equation 18.9 suggests that the socially optimal number of nurseries is likely to be lower if the increased likelihood of invasion due to an additional nursery, φ_n, and the annualized value of damages from the invasion, δG, are high. In contrast, if the effective discount rate, $\delta + \varphi$, is large, then the socially optimal number of nurseries is likely to be greater. Due to these offsetting effects, it is difficult to determine a priori whether the long-run socially optimal number of nurseries, n^*, will be less than the privately optimal number of nurseries, n^p. But in general, if the risk of an invasion from establishing additional nurseries is high and the potential damages of the invasion are large, then one would expect $n^* < n^p$ and $n^* < n^s$.

Equations 18.8 and 18.9 imply that there is scope for employing a tax on the nursery industry to internalize the expected social cost of accidental invasion associated with establishing new nursery firms. Denoting this tax as χ, then it can be derived from Equation 18.8 as

$$\chi = (\Pi'(n) + D) \frac{\delta + \varphi}{\varphi_n} - \delta G - Dn \tag{18.10}$$

Note that this tax is a "net tax." It includes the optimal tax to internalize the increased hazard associated with the risk of invasion as the industry expands its nurseries, $(\Pi'(n) + D) \dfrac{\delta + \varphi}{\varphi_n} - \delta G$. However, the tax must be adjusted for consumer surplus, Dn, generated by the industry. Once the optimal tax is imposed on the industry, its long-run equilibrium condition becomes $\Pi(n) - \chi = 0$. Rearranging this expression yields both equilibrium condition (Equation 18.8) and the socially optimal nurseries, n^*, as governed by Equation 18.9. Note as well that in the absence of the tax, the horticultural industry

---

* The private industry ignores both the consumer surplus and the social costs of a potential invasion and thus will establish n^p nurseries, ensuring in the long run that the last nursery drives industry and each nursery's profits to zero.

will simply establish its long-run privately optimal number of nurseries, n^p, that is, the long-run profit condition reverts to $\Pi(n^p) = 0$.

Given the structure of the horticultural industry model, the optimal tax or fee imposed on the industry, χ, is not an output tax per se. Instead, it is a license fee paid by the industry, or equivalently, an optimal fee $x^* = \chi/n^*$ paid by each firm. There are two reasons why such a license fee rather than an output tax is optimal in the model. First, the hazard rate increases with the establishment of an additional nursery by the industry, as can be seen from Equation 18.6. Thus, the risk of an invasion is related more to the presence of another nursery than to the size of output from that nursery. Second, the privately optimal number of firms, n^p, is determined by the industry's zero-profit condition $\Pi(n^p) = 0$, which results from the standard outcome for a monopolistically competitive industry. That is, the equilibrium output of an individual firm is constant so that the equilibrium output for the industry expands only if additional nursery firms are established in a new location.*

18.3   Data sources and parameter estimation

Simulating empirical results for the model presented earlier requires selecting, or in some cases, estimating a number of parameters. In the latter case, econometric procedures were used to determine a key parameter associated with the horticultural industry model and to establish the risk of invasion, as captured by an individual species' hazard rate. We also required several other parameters for our simulations, such as a social discount rate and time horizon.

18.3.1   Estimation of the short-run profit function for the North American horticultural industry

We used the monopolistic competition model of the horticultural industry as a basis for estimating the profits of a representative nursery firm as a function of the number of nurseries established by the industry. Following an approach suggested by Panzar and Ross (1987), we estimated the short-run profit function to derive the key parameter, γ, which represents the substitutability between plant bundles.† We estimated separate short-run profit functions for the horticultural industry in the United States and Canada at jurisdiction levels for which data were available. For the United States, the estimation was conducted using panel data for all 50 states covering the years 1979, 1988, and 1998, based on the *Census of Horticultural Specialties* conducted for these years (USDC 1982, 1991; USDA 2001). For Canada, the estimation was conducted using panel data for five regional groupings of provinces, the Atlantic Provinces, Quebec, Ontario, the Prairie Provinces, and British Columbia, and time series data for 1997–2006. Tables 18.1 and 18.2 provide summary statistics for the relevant variables for the estimation applied to the U.S. and Canadian horticultural industries.

A random effects model worked best with our data. For the United States, this procedure yielded a value for the substitutability parameter, γ, of 0.7757, which is relatively close to 1.

---

* Of course, it follows that if the monopolistically competitive industry expands its equilibrium number of nurseries, total industry output will increase. But to accommodate additional nurseries, they have to be established in new locations (markets). Again, this supports the assumption of the model that the likelihood of an invasion occurring is associated with an additional nursery firm establishing at a new location rather than increased output per firm at existing locations.

† We estimate the relationship between profits and the number of nurseries, fixed labor resources and total labor resources, as given by (see text for variable definitions and Barbier et al., 2009, for more details):

$$\pi_{jt} = b_o + b_1\left(\frac{L_{jt}}{n_{jt}} - F_{jt}\right) + \varepsilon_{jt}, \quad j = 1,...J, \quad b_1 = \frac{(1-\gamma)}{\gamma}s$$

Table 18.1 Summary Statistics for the Horticultural Industry in the United States

	1978	1987	1998
Average number of firms per state	166	140	176
Total labor resources (L) per state	31,612	13,965	10,658
Fixed labor resources per firm (F) per state	8	4	5
Average profit per firm ($000) per state	30.31	15.06	20

Sources: U.S. Department of Commerce, *1978 Census of Agriculture vol. 5. Special Report.* Part 7.
   1979 Census of Horticultural Specialties, U.S. Department of Commerce, Bureau of
   the Census, Washington, DC, 1982; U.S. Department of Commerce (USDC), *1987
   Census of Agriculture. Vol. 4. 1988 Census of Horticultural Specialties*, U.S. Department of
   Commerce, Bureau of the Census, Washington, DC, 1991; and U.S. Department of
   Agriculture (USDA), *1998 Census of Horticultural Specialties*, USDA, National
   Agricultural Statistic Services (NASS), Washington, DC, 2001.

Table 18.2 Summary Statistics for the Horticultural Industry of Canada

Year	Average number of firms/region	Total Labor resources (L)/region	Fixed labor resources (F)/firm	Average profit (C$000)/firm
1997	312	2599	2	56.019
1998	274	2565	3	68.604
1999	252	2182	3	78.424
2000	252	2676	3	111.836
2001	229	2635	3	127.841
2002	320	3094	4	100.286
2003	327	3212	5	945.552
2004	249	3019	6	123.607
2005	237	2967	7	111.557
2006	226	2812	7	147.213

Source: Statistics Canada, Agricultural Division, *Greenhouse, Sod and Nursery Industry*, Department of
   Industry, Canada, 1997–2006.

Note: The five regional groupings of provinces are the Atlantic provinces, Quebec, Ontario, the Prairie
   provinces, and British Columbia.

This result implies that the bundles of nursery products offered by different nursery firms
are close substitutes, suggesting relatively little differentiation between nursery products in the U.S. market. In comparison, the estimate of γ for Canada was 0.1154, which is
relatively close to zero. This result indicates that the various selections of products offered
by different nurseries are not close substitutes. Therefore, we can conclude that there is
more differentiation in nursery products for the Canada market compared to the U.S. market. The elasticity of substitution in our model is $\sigma = 1/(1 - \gamma)$, and for Canada this is 1.134,
which is lower than the U.S. value of 4.458. Such a finding implies that economies of scale
in the Canadian nursery industry are more pronounced than in the U.S. nursery industry.

From the U.S. horticultural industry summary data (see Table 18.1), estimates of total
resources and the fixed resources for the 3 years of data are $L = 937,250$ and $F = 6$. For Canada,
across all 10 years, $L = 13,880$ and $F = 4$ (Table 18.2). Since profits of the industry are denoted in
$1000 (or C$1000) units, our estimate of $G(\tau)$ is also in the same units and apply to both United
States and Canada.*

* We used an exchange rate of $1 = C$1.06 to convert U.S. damage estimates into Canadian prices.

18.3.2   Estimation of the hazard rate and damage costs for an exotic plant based on purple loosestrife

For our simulations of the introducer pay tax, we assume the case of a representative exotic species with characteristics similar to purple loosestrife. First, we describe the procedure used to estimate the general hazard function relationship and then we use this relationship to determine specific values for the hazard rate and damage costs for our representative species. Following Knowler and Barbier (2005), we assume that the hazard function in Equation 18.6, capturing the risk of invasion associated with imports of exotic plants by the nursery industry, can be represented as a standard duration model. In such models, the "spell" refers to the number of periods after some initial period before "failure" takes place. In our model, the spell represents the number of periods subsequent to introduction of the species without invasion taking place.

Given the preceding considerations, and in light of limited data, we represent the hazard function as $h(t) = \varphi(n(t), a_k) = \varphi(a_k) f(n(t))$, so that the function $\varphi(n(t), a_k)$ is assumed to be a product of two functions, $\varphi(a_k)$ and $f(n(t))$. This approach implies that the underlying "hazard rate" associated with the inherent invasiveness of an exotic plant is distinct from the influence of the number of nursery firms on the overall likelihood that a new exotic plant species will become invasive. As modeled here, the presence of nursery firms selling the exotic plant serves to "scale" the probability that an invasion will occur and this increases as the industry establishes nurseries in new locations.

The baseline hazard rate, $\varphi(a_k)$, is determined by plant attributes.* We estimated this hazard rate assuming that it has two possible functional forms, an unrestricted standard Weibull distribution, or an exponential relationship, which is a restricted version of the Weibull hazard model.[†] Modeling of the probability that a herbaceous plant will become invasive required a sample of both invasive and noninvasive species that have been introduced to North America. Since species may not begin to invade for some time after introduction, we selected species that have been established for some time without invading as the "noninvaders." In effect, the duration data series for these species are truncated, since we do not know if or when the "noninvaders" might become invasive. We selected taxonomic, eco-geographic, and biological traits describing the species in our samples based on numerous studies of the traits associated with invasive and noninvasive plants (e.g., Rejmánek and Richardson 1996, Reichard and Hamilton 1997, Pheloung et al. 1999). The final herbaceous species dataset contained 106 observations, after several were removed because of missing variable values. Of these 106 herbaceous species observations, 77 (72.6%) were invasive species and 29 (27.4%) were noninvasive species.[‡] Barbier et al. (2009) provide a description of the plant characteristic variables and the estimation procedures used in the herbaceous species hazard analysis.

We found that the exponential hazard model provides a good approximation of the hazard rate for our data, which greatly improves the tractability of our empirical model.[§]

---

* Recall that this represents the likelihood that the new exotic species will invade during the current time interval given that it has not invaded previously.
† The two models are expressed as (1) the Weibull hazard function, which is $h = \varphi p (\varphi t)^{p-1}$ and (2) the exponential hazard function, which is $h = \varphi$. Note that if $p = 1$, then the Weibull function converges to the exponential hazard rate. See Kiefer (1988) for further discussion of different duration models and their relationships.
‡ Interestingly, the invasive herbaceous species seem to have been more studied than noninvasives, so that there is more readily available information on the former. Thus, our dataset does not contain as many noninvasive observations as we might have liked but this is unavoidable.
§ Our estimate of the p parameter using the standard unrestricted Weibull model was 1.145, which falls sufficiently close to 1.00 to allow us to use the exponential model as an approximation.

Based on the mean values of the covariates, we computed an estimate of 0.005 for the baseline hazard rate, φ. This suggests that for a newly imported exotic herbaceous species, assuming that "on average" it shares the typical plant characteristics of herbaceous species contained in our dataset, the probability that the new exotic species will invade during the current time interval given that it has not invaded before is 0.005.* Next, we apply our hazard function to a representative exotic plant similar to purple loosestrife.

Purple loosestrife is a Eurasian wetland perennial introduced to North America in the early 1800s as an ornamental plant. It has now become a prolific invader of wetlands throughout U.S. temperate zones and the southern portions of Canada, from Newfoundland to British Columbia (Brown 2005).† While it is illegal to sell purple loosestrife in some states and provinces, in others, the plant is still being sold as an ornamental plant and is widely available throughout North America via Internet sales (Kay and Hoyle 2001). If the values of the covariates in our hazard function relationship corresponding to purple loosestrife are employed, the estimated baseline hazard rate is calculated as 0.00904. As expected, a newly imported plant with the characteristics of purple loosestrife appears substantially more invasive than the average herbaceous species imported into North America contained in our dataset.

Finally, we use estimates of the damage costs inflicted by purple loosestrife to estimate the potential damages that would be associated with a successful invasion by our new herbaceous species. Direct economic costs from purple loosestrife include reduced palatability of hay containing purple loosestrife and the reduction in water flow in irrigation systems. Indirect losses include reduction in waterfowl viewing and hunting opportunities (Blossey 2002). The Government of Canada estimated that, in the Great Lakes and St Lawrence River Basins, $500 million is spent each year on efforts to eradicate invasive species such as purple loosestrife (Lindgren 2003). In the United States, purple loosestrife is estimated to cost $45 million per year in control costs and forage losses (Malecki et al. 1993). Based on this damage estimate for the United States and a total area invaded of 131,152 ha, the annual damage cost from purple loosestrife is about $340 per ha. Given the long gestation period from initial introduction to establishment as an invader, and assuming a carrying capacity of 500,000 ha, we can solve to get an intrinsic growth rate for purples loosestrife of $r^A = 0.11$. Using a discount rate of 5% and an infinite time horizon, the present value of damages from purple loosestrife in the United States would be approximately $13.8 million, measured from the date of its establishment as an invader (τ). We use this estimate as a proxy for damages from a newly introduced herbaceous plant species.‡

---

* As noted earlier, the second component in the full hazard function relationship represents the "scaling" influence of the number of nurseries established by the horticultural industry, $f(n(t))$. We assume that this scaling effect is a simple proportional relationship $f(n(t)) = fn(t)$ and that $f = 0.02$ for North America. Thus, $fn = 1$ when 50 nurseries are present and more than 50 firms augment the scaling effect, while less than 50 firms diminish it.

† The plant rapidly forms nonspecific stands, displacing native plant species that provide food, cover, and breeding areas for a number of wild species. Purple loosestrife is estimated to occur in nine provinces in Canada (Blossey 2002). In Manitoba, purple loosestrife is estimated to cover 5575 ha of habitat (Lindgren 2003, Henne et al 2005). In the United States, Malecki et al (1993) indicate that it is found in 47 states and has been spreading at the rate of 115,000 ha/year. Alaska, Hawaii and Florida are the exceptions. Brown (2005) estimates that about 1.2 million acres (499,000 ha) of wetlands in the Atlantic and Mississippi flyway are considered at risk of invasion, and that as of 2003, the area infested in the United States was about 324,000 acres (131,152 ha).

‡ The relatively low figure for the present value of damages from purple loosestrife stems from the long gestation period of many decades over which the discounting calculation is applied. Of course, taking even a single current year, the undiscounted damages are much higher than this figure.

Table 18.3 Parameters Used in the Analysis of a
Newly Introduced Herbaceous Plant Species
Similar to Purple Loosestrife (*Lythrum salicaria*)

	Parameters	
Variable	United States	Canada
F	6	4
L	937,250	13,880
δ	0.05	0.05
γ	0.7757	0.1154
A	0.3	0.2
$G(\tau)$	13,800	14,634
$\varphi(a_k)$	0.009041	0.009041
f	0.02	0.02
r	0.0005	0.0003

Table 18.3 lists the complete set of ecological and economic parameters for the United States and Canada, which is used in the simulations described in Section 18.4.

18.4   Simulation of socially optimal nurseries and taxes

In this section, we simulate the socially optimal number of nursery firms and the appropriate introducers pay tax for a representative exotic species similar to purple loosestrife. For simulation purposes, it would be easy to assume that sales of the newly imported exotic species contribute 100% to the profits of the horticultural firms. In practice, nursery firms sell a mixture of both native and exotic nursery products as a single bundle. Therefore, it is not likely that all firm profits would come from a single plant species. As a result, we simulate the socially optimal number of nurseries and the annual firm tax, defined as $x = \chi/n^*$, for profit shares from the exotic species ranging from 1% to 100%. As this share of profit rises, the socially optimal number of nurseries would be expected to fall and the optimal tax (license fee) rise. Taking this approach, our simulation results for the United States and Canada are depicted in Table 18.4.

For the United States, depending on the profit share of the new exotic species, the socially optimal number of nurseries lies between 4,000 and 35,000 firms. The firm tax is as low as $45 per year, in the case of a 1% profit share, and reaches as high as $4500 per year if the profit share of the exotic species is 100%. For Canada, the socially optimal number of nurseries ranges between 900 and just over 3000 firms. The firm tax is C$49 per year in the 1% profit share case and is nearly C$5,000 per year in the 100% profit share case.

For both the United States and Canada, we also simulate the social-welfare maximizing level of nurseries in the absence of an invasion risk, n^s, the privately optimal level of nurseries, n^p, as well the actual number of nursery firms, n, as reported in the current industry survey data (Table 18.4). Again making use of the parameter values in Table 18.3, we calculate $n^s = 44,244$ nursery firms for the United States and 3,086 for Canada. For the United States, the long-run privately optimal number of nursery firms is obtained by substituting the relevant parameter values from Table 18.3 in Equation 18.1 to yield $n^p = 35,035$ firms. Similarly, for Canada, the long-run private optimal number of nursery firms is obtained from Equation 18.1 to yield $n^p = 3070$ firms. In contrast, the survey data indicates that for the United States, there are currently 8050 nurseries, and for Canada, there are 1345 nurseries

Table 18.4 Simulations of the Socially Optimal Tax and Number of Nurseries
Selling a New Exotic Herbaceous Plant Species in North America

Share of exotic species in profits	1%	10%	25%	50%	75%	100%
United States						
Nurseries, n^*	34,727	31,949	27,319	19,603	11,886	4,170
Firm tax ($/year)	45	446	1,116	2,231	3,347	4,462
n^*/n^s	78%	72%	62%	44%	27%	9%
n^*/n^p	99%	91%	78%	56%	34%	12%
n^*/n	431%	397%	339%	244%	148%	52%
Canada						
Nurseries, n^*	3,048	2,856	2,536	2,002	1,468	934
Firm tax (C$/year)	49	491	1,228	2,456	3,684	4,913
n^*/n^s	99%	93%	82%	65%	48%	30%
n^*/n^p	99%	93%	83%	65%	48%	30%
n^*/n	227%	212%	189%	149%	109%	69%

Notes: n^* is the social optimal number of nurseries taking into account the risk and damages of a potential invasion, n^s is the number of nurseries that maximizes social welfare (producer and consumer surplus) but ignores the risk and damages of a potential invasion, n^p is the long-run private optimum number of nursery firms, and n is the actual number of nursery firms reported in industry data. For the United States, $n^s = 44,244$, $n^p = 35,035$, and $n = 8,050$; for Canada, $n^s = 3,086$, $np = 3,070$, and $n = 1,345$.

(Tables 18.1 and 18.2). Thus, the long-run privately optimal number of nursery firms in the United States is just over four times the current number, whereas for Canada, it is less than 2.5 times the current number of nurseries.*

As expected, n^* converges to the privately optimal number of nurseries, n^p, as the contribution to profits from the newly imported species declines toward zero. However, as long as there is still a risk of invasion, n^* is always less than n^s, although in the case of Canada, n^* and n^s appear to converge as the profit share of the exotic species declines to 1%. In all simulations for the United States and Canada, except for the case of a 100% profit share, the socially optimal number of nurseries, n^*, exceeds the current number of nurseries, n.

Our simulations of the introduction of an exotic herbaceous species to North America via the horticultural industry indicate that the results are highly sensitive to the share of the exotic plant sales in final profits. Careful consideration must be given to this profit share before designing any tax policy to discourage the introduction of a potentially invasive exotic plant species by the North American horticultural nursery industry. The design of an appropriate tax (annual license fee) policy must take into account the implications of the tax on consumer surplus, producer surplus, and the risks and potential damages associated with any likely invasion. Although some of these factors appear to be difficult to measure, our simulations show that such an analysis can be conducted.

---

* This outcome is relatively common for a monopolistically competitive industry. In such an industry, if firms currently make substantial profits, then more firms will enter the industry until in the long run industry profits are zero. Tables 18.1 and 18.2 indicate that on average current nurseries in Canada and the United States do make sizable profits, which suggests that the long-run number of nurseries, n^p, is likely to be significantly larger than the current number, n. Using a different model of the horticultural nursery industry than the one developed here, Knowler and Barbier (2005) estimate the privately optimal number of nurseries in the United States in the long run to be 36,226.

In addition, there are two additional advantages from such a tax policy. First, as long as the newly introduced plant's contribution to profits is known, it is possible to construct a differentiated and, therefore, more optimal tax structure. For example, if a U.S. nursery receives only 10% of its profits from the newly introduced exotic species, then the optimal annual fee is $446 per year, whereas a firm with sales of the new exotic representing 75% of profits should pay about $3350 per year. For two equivalent Canadian firms, the tax would be similarly differentiated. Second, an annual license fee along the lines suggested here could provide funds for combating the various damages associated with an unwanted invasion. For example, we calculated the present value of the damages inflicted by purple loosestrife to be $13.8 million in the United States, which results in an annualized value of $690,000 for a new herbaceous plant species invasion using a social discount rate of 5%. Based on our simulations, an annual license fee in the United States would raise anywhere from $1.6 to $44 million, depending on the share of the exotic species sales in industry profits. Such annual revenues would not only cover the costs of annualized damages but also fund screening programs for all newly introduced species, education and scientific research on plant invasives, and eradication of past exotic plant invasions. Similarly, the annual revenues from the license fee would amount to C$150,000–C$5.4 million in Canada.

18.5 Conclusion

In this chapter, we have modeled the privately and socially optimal conditions derived from the industry's decision to import a new exotic plant species that might become a damaging invader. We used parameter estimates from our modeling of the North American horticultural industry and the hazard rate associated with exotic species imported by that industry, basing the potential invasion damages on purple loosestrife, a known invasive herbaceous plant species. Then, we carried out simulations of the trade-offs between the commercial profits and the expected social damages from importation of a new exotic plant species similar to purple loosestrife. We used these simulations to estimate the optimal tax (i.e., annual license fee) per nursery selling the exotic species in the form of an annual license fee and solved for the posttax number of socially optimal nurseries established by the industry. The simulations were performed for both the U.S. and Canadian horticultural nursery industries.

The simulations produced similar outcomes for the United States and Canada with respect to the optimal tax and the socially optimal number of nurseries. All outcomes from the simulations show that as the exotic plant's contribution to profits increases, the size of the optimal tax rises and the socially optimal number of nurseries falls. Thus, our simulations suggest that the optimal tax and socially optimal number of nurseries are highly sensitive to the share of the exotic plant sales in profits. Given that the magnitude of the tax is highly sensitive to the profit share, careful consideration must be given to determining this share before imposing such a fee on the North American horticultural industry. This would appear to be particularly important in the United States, as our simulations show that a high license fee would reduce the number of nurseries significantly.

However, there are important advantages with an annual license fee compared to other policy options. First, the license fee can be adjusted optimally to the share of profits earned by each firm from sales of the new exotic plant, and second, the fee could be employed not only to cover the costs of damages but also to fund screening programs for all newly introduced species, education and scientific research on plant invasives, and eradication of past plant invasions For example, we estimate that the annual revenues could range from $1.6 to $44 million in the United States (C$150,000–C$5.4 million in Canada), depending on the share of the exotic species sales in industry profits.

Acknowledgments

Research for this chapter was funded by financial support from the Program of Research on the Economics of Invasive Species Management (PREISM) and the Economic Research Service (ERS) at the U.S. Department of Agriculture (under Cooperative Agreement 43-3AEM-5-80066). The chapter has benefited from the research assistance of Katie Barndt, Joanne Burgess, Kristin Dust, Arianne Ransom-Hodges, and Lizbeth Seebacher and from the project assistance of Margie Reis. We thank the assistance at ERS provided by our project liaison, Daniel Pick, and also by Linda Calvin, Barney Caton, Robert Dismukes, Andy Jerardo, Craig Osteen, Craig Ramsey, and Donna Roberts.

References

Avent, T. 2003. *So You Want to Start a Nursery*. Beverly, MA: Timber Press.

Bailey, E.E., and A.F. Friendlaender. 1982. Market structure and multiproduct industries. *Journal of Economic Literature* 20(3):1024–1048.

Baldwin, R., R. Forslid, P. Martin, G. Ottaviano, and F. Robert-Nicoud. 2003. *Economic Geography and Public Policy*. Princeton, NJ: Princeton University Press.

Barbier, E.B., J. Gwatipedza, D. Knowler, and S. Reichard. 2011. The North American horticultural industry and the risk of plant invasion. *Agricultural Economics* 45:743–759.

Barbier, E.B., and D. Knowler. 2006. Commercialization decisions and the economics of introduction. *Euphytica* 148(1/2):151–164.

Bell, C.E., C.A. Wilen, and A.E. Stanton. 2003. Invasive plants of horticultural origin. *HortScience* 38:14–16.

Blossey, B. 2002. Purple loosestrife. In *Biological Control of Invasive Plants in the Eastern United States*, eds. R. Van Driesche, S. Lyon, B. Blossey, M. Hoddle, and R. Reardon, USDA Forest Service Publication FHTET-2002-04 Morgantown, West Virginia. USDA Forest Service. Available at http://wiki.bugwood.org/Archive:BCIPEUS/Purple_Loosestrife.

Brooker, J., D. Eastwood, C. Hall, K. Morris, A. Hodges, and J. Haydu. 2003. *Trade Flows and Marketing Practises within the United States Nursery Industry: 2003*. Southern Cooperative Series Bulletin 404, 2005. Available at http://economics.ag.utk.edu/publications/marketing/SCB404.pdf.

Brown, M.L. 2005. Purple loosestrife - *Lythrum salicaria* L. In *Invasive Plants of range and Wildlands and Their Environmental, Economic, and Societal Impacts*, eds. C.L. Duncan, and J.K. Clark, pp. 128–146. WSSA Special Publication. Lawrence, KS: Weed Science Society of America.

Costello, C., and C. McAusland. 2003. Protectionism, trade, and measures of damage from exotic species introductions. *American Journal of Agricultural Economics* 85:964–975.

Dehnen-Schmutz, K., J. Touza, C. Perrings, and M. Williamson. 2007a. The horticultural trade and ornamental plant invasions in Britain. *Conservation Biology* 21(1):224–231.

Dehnen-Schmutz, K., J. Touza, C. Perrings, and M. Williamson. 2007b. A century of the ornamental plant trade and its impact on invasion success. *Diversity and Distribution* 13:527–534.

Dixit, A.K., and J.E. Stiglitz. 1977. Monopolistic competition and optimum product diversity. *American Economic Review* 67(3):297–308.

Henne, D.C., C.J. Lindgren, T.S. Gabor, H.R. Murkin, and R.E. Roughley. 2005. Evaluation of an integrated management strategy for the control of purple loosestrife (*Lythrum salicaria* L.) in the Netley-Libau Marsh, Southern Manitoba. *Biological Control* 32:319–325.

Horan, R.D., and F. Lupi. 2005. Tradeable risk permits to prevent future introductions of invasive alien species in the Great Lakes. *Ecological Economics* 52(3):289–304.

Kay, S.H., and S.T. Hoyle. 2001. Mail order, the internet, and invasive aquatic weeds. *Journal of Aquatic Plant Management* 39:88–91.

Kiefer, N.M. 1988. Economic duration data and hazard functions. *Journal of Economic Literature* XXVI (June):646–679.

Knowler, D., and E.B. Barbier. 2005. Importing exotic plants and the risk of invasion: Are market-based instruments adequate? *Ecological Economics* 52:341–354.

Lindgren, C.J. 2003. A brief history of Purple Loosestrife, *Lythrum salicaria*, in Manitoba and its status in 2001. *Canadian Field-Naturalist* 117:100–109.

Mack, R.N., D. Simberloff, W.M. Lonsdale, H. Evans, M. Clout, and F.A. Bazzaz. 2000. Biotic invasions: Causes, epidemiology, global consequences and control. *Ecological Applications* 10:689–710.

Maki, K., and S. Galatowitsch. 2004. Movement of invasive aquatic plants into minnesota (USA) through horticultural trade. *Biological Conservation* 118:389–396.

Malecki, R.A., B. Blossey, S.D. Hight, D. Schroeder, D.T. Kok, and J. R. Coulson. 1993. Biological control of purple loosestrife. *BioScience* 43:680–686.

Matsuyama, K. 1995. Complementarities and cumulative processes in models of monopolistic competition. *Journal of Economic Literature* 33:701–729.

Panzar, J.C., and J.N. Rosse. 1987. Testing for monopoly equilibrium. *Journal of Industrial Economics* 35(4):446–456.

Panzar, J.C., and R.D. Willig. 1977. Economies of scale in multi-output production. *The Quarterly Journal of Economics* 91(3):481–493.

Pheloung, P.C., P.A. Williams, and S.R. Halloy. 1999. A weed risk assessment model for use as a biosecurity tool evaluating plant introductions. *Journal of Environmental Management* 57:239–251.

Pimentel, D., R. Zuniga, and D. Morrison. 2005. Update on the environmental and economic costs associated with alien-invasive species in the United States. *Ecological Economics* 52:273–288.

Pyšek, P., M. Křivánek, and V. Jarošík. 2009. Planting intensity, residence time, and species traits determine invasion success of alien woody species. *Ecology* 90:2734–2744.

Reed, W.J., and H.E. Heras. 1992. The conservation and exploitation of vulnerable resources. *Bulletin of Mathematical Biology* 54(2/3):185–207.

Reichard, S., and C.W. Hamilton. 1997. Predicting invasions of woody plants introduced into North America. *Conservation Biology* 11:193–203.

Reichard, S.H., and P. White. 2001. Horticulture as a pathway of invasive plant introductions in the United States. *BioScience* 51:103–113.

Rejmánek, M., and D.M. Richardson. 1996. What attributes make some plant species more invasive? *Ecology* 77:1655–1661.

Sinden, J., R. Jones, S. Hester, D. Odom, C. Kalisch, R. James, and O. Cacho. 2004. *The Economic Impact of Weeds in Australia*. CRC for Australian Weed Management, Technical Series Report No. 8. Glen Osmond, South Australia: University of Adelaide.

Singh, S.P. 1999. The changing structure of the U. S. green industry: Factors influencing opportunities and issues. *Journal of International Food 7 Agribusiness Marketing* 10:65–83.

Statistics Canada, Agricultural Division. 1997–2006. *Greenhouse, Sod and Nursery Industry*. Canada: Department of Industry.

Urgenson, L., S. Reichard, and C. Halpern. 2009. Community and ecosystem consequences of giant knotweed (*Polygonum sachalinense*) invasion into riparian forests of western Washington, USA. *Biological Conservation* 142:1536–1541.

US Department of Agriculture (USDA). 2001. *1998 Census of Horticultural Specialties*. Washington, DC: USDA, National Agricultural Statistic Services (NASS).

US Department of Commerce (USDC). 1982. *1978 Census of Agriculture vol. 5. Special Report. Part 7. 1979 Census of Horticultural Specialties*. Washington, DC: US Department of Commerce, Bureau of the Census.

US Department of Commerce (USDC). 1991. *1987 Census of Agriculture. Vol. 4. 1988 Census of Horticultural Specialties*. Washington, DC: US Department of Commerce, Bureau of the Census.

US National Park Service. 2007. Weeds gone wild. Available at http://www.nps.gov/plants/outreach/brochures/pdf/pca-invasiveplants.pdf.

Wilcove, D.S., D. Rothstein, J. Dubrow, A. Phillips, and E. Losos. 1998. Quantifying threats to imperiled species in the United States. *BioScience* 48:607–615.

Index